Microbiology Monographs
Volume 17

Series Editor: Alexander Steinbüchel
Münster, Germany

AF739774

Microbiology Monographs

Volumes published in the series

Inclusions in Prokaryotes
Volume Editor: Jessup M. Shively
Vol. 1, 2006

Complex Intracellular Structures in Prokaryotes
Volume Editor: Jessup M. Shively
Vol. 2, 2006

Magnetoreception and Magnetosomes in Bacteria
Volume Editor: Dirk Schüler
Vol. 3, 2007

Predatory Prokaryotes – Biology, Ecology and Evolution
Volume Editor: Edouard Jurkevitch
Vol. 4, 2007

Amino Acid Biosynthesis – Pathways, Regulation and Metabolic Engineering
Volume Editor: Volker F. Wendisch
Vol. 5, 2007

Molecular Microbiology of Heavy Metals
Volume Editors: Dietrich H. Nies and Simon Silver
Vol. 6, 2007

Microbial Linear Plasmids
Volume Editors: Friedhelm Meinhardt and Roland Klassen
Vol. 7, 2007

Prokaryotic Symbionts in Plants
Volume Editor: Katharina Pawlowski
Vol. 8, 2009

Hydrogenosomes and Mitosomes: Mitochondria of Anaerobic Eukaryotes
Volume Editor: Jan Tachezy
Vol. 9, 2008

Uncultivated Microorganisms
Volume Editor: Slava S. Epstein
Vol. 10, 2009

Microbial Megaplasmids
Volume Editor: Edward Schwartz
Vol. 11, 2009

Endosymbionts in Paramecium
Volume Editor: Masahiro Fujishima
Vol. 12, 2009

Alginates: Biology and Applications
Volume Editor: Bernd H. A. Rehm
Vol. 13, 2009

Plastics from Bacteria: Natural Functions and Applications
Volume Editor: Guo-Qiang Chen
Vol. 14, 2010

Amino-Acid Homopolymers Occurring in Nature
Volume Editor: Yoshimitsu Hamano
Vol. 15, 2010

Biology of Rhodococcus
Volume Editor: Héctor M. Alvarez
Vol. 16, 2010

Structures and Organelles in Pathogenic Protists
Volume Editor: Wanderley de Souza
Vol. 17, 2010

Wanderley de Souza
Editor

Structures and Organelles in Pathogenic Protists

Springer

Editor
Prof. Dr. Wanderley de Souza
Instituto de Biofísica Carlos Chagas Filho
Universidade Federal do Rio de Janeiro
CCS-Bloco G, Ilha do Fundão
21941-900, Rio de Janeiro-RJ
Brasil
wsouza@biof.ufrj.br

Series Editor
Professor Dr. Alexander Steinbüchel
Institut für Molekulare Mikrobiologie und Biotechnology
Westfälische Wilhelms-Universität
Corrensstr. 3
48149 Münster
Germany
steinbu@uni-muenster.de

ISSN 1862-5576 e-ISSN 1862-5584
ISBN 978-3-642-26496-2 ISBN 978-3-642-12863-9 (eBook)
DOI 10.1007/978-3-642-12863-9
Springer Heidelberg Dordrecht London New York

© Springer-Verlag Berlin Heidelberg 2010
Softcover reprint of the hardcover 1st edition 2010
This work is subject to copyright. All rights are reserved, whether the whole or part of the material is concerned, specifically the rights of translation, reprinting, reuse of illustrations, recitation, broadcasting, reproduction on microfilm or in any other way, and storage in data banks. Duplication of this publication or parts thereof is permitted only under the provisions of the German Copyright Law of September 9, 1965, in its current version, and permission for use must always be obtained from Springer. Violations are liable to prosecution under the German Copyright Law.
The use of general descriptive names, registered names, trademarks, etc. in this publication does not imply, even in the absence of a specific statement, that such names are exempt from the relevant protective laws and regulations and therefore free for general use.

Cover design: SPi Publisher Services

Printed on acid-free paper

Springer is part of Springer Science+Business Media (www.springer.com)

Preface

Parasitic protozoa comprise a large number of species, including some which are agents of human and veterinary diseases such as malaria, leishmaniasis, Chagas disease, African trypanosomiasis, amebiasis, trichomoniasis, giardiasis, toxoplasmosis, coccidiosis, theileriosis, and babesiosis, to mention only the more important ones. Some of these protozoa, as is the case with *Trichomonas*, present a simple life cycle. For others, however, as occurs with Apicomplexa (which includes *Plasmodium, Toxoplasma, Eimeria*, etc.) and some trypanosomatids, the life cycle is relatively complex, displaying several developmental stages in the vertebrate host and, in some cases, in invertebrate hosts. These protozoa are also of interest from a cell biology point of view, as they present special cytoplasmic structures and organelles that have been studied in some detail during the last several years, providing new information of general biological interest. These studies have discovered new metabolic pathways that take place in these organelles and open up alternate possibilities for the identification of different drug targets and innovative drugs to be used for the treatment of patients and animals with diseases caused by protozoa.

There are currently only a few publications that review the available data on the cell biology of pathogenic protozoa. This Microbiology Monographs volume covers the current and most recent advances made on relevant cytoskeletal structures and organelles found in parasitic protists. Renowned scientists, some of whom were directly involved in the discovery and characterization of these organelles, have contributed reviews that incorporate recent results obtained using modern cell biology and molecular approaches, including genomics and proteomics. Some important organelles such as the hydrogenosome and mitosomes were not reviewed here as they were examined in detail in a previous volume of this series (Tachezy 2008).

The first group of reviews deals with cytoskeletal structures such as the mastigont system found in trichomonads (written by Marlene Benchimol), the subpellicular microtubules, better characterized in trypanosomatids and in some Apicomplexa (written by Wanderley de Souza and Marcia Attias), and the paraflagellar rod, a characteristic feature of the flagellum of some protists (written by Johana Buisson and Philippe Bastin).

The second group deals with structures and organelles involved in the synthesis and secretion of macromolecules, as well as in the uptake of molecules through an endocytic process. These include the flagellar pocket of trypanosomatids (written by Paul McKean and Keith Gull), the reservosome of *Trypanosoma cruzi* (written by Narcisa Leal Cunha-e-Silva, Celso Sant'Anna, Miria Pereira, and Wanderley de Souza), the megasome found in *Leishmania* (written by Diane McMahon-Pratt, Tania Ueda-Nakmura, and Yara Traub-Cseko), the various organelles and the traffic of vesicles in *Entamoeba histolytica* (written by Sherri Smith and Nancy Guillen), the secretory organelles found in members of Apicomplexa (written by Jean François Dubremetz), and the secretory events that take place during the process of encystation of *Giardia lamblia* (written by Fernando Rivero, Dana Muller, and Hugo Lujan).

The final group of reviews deals with various organelles, which are characteristic features of protozoa. These include the kinetoplast–mitochondrion complex of trypanosomes and related flagellates (written by Julius Lukes, Hassan Hashimi, Zdenek Verner, and Zdenka Cicov), the apicoplast, an ancient organelle found in Apicomplexa (written by Swaiti Agrawal, Sethu Nair, Lilach Sheiner, and Boris Striepen), the glycosomes found in Kinetoplastida (written by Fred Opperdoes), and the acidocalcisome found in several protozoa (written by Paul Ulrich, Rozana Cintrón-Moret, and Roberto Docampo).

Rio de Janeiro, Brazil Wanderley de Souza

Reference

Tachezy J (ed) (2008) Hydrogenosomes and mitosomes: mitochondria of anaerobic eukaryotes. In: Microbiology monographs, vol. 9. Springer, Berlin Heidelberg, New York

Contents

Contributors

Swati Agrawal Center for Tropical and Emerging Global Diseases & Department for Cellular Biology, University of Georgia, 500 D.W. Brooks Drive, Athens, GA 30602, USA

Marcia Attias Laboratório de Ultraestrutura Celular Hertha Meyer, Instituto de Biofísica Carlos Chagas Filho, Universidade and Diretoria de Programas, Inmetro, Federal do Rio de Janeiro, CCS-Bloco G, 21941-900 Ilha do Fundão, Rio de Janeiro, Brasil

Philippe Bastin Trypanosome Cell Biology Unit, Institut Pasteur & CNRS, 25 rue du Docteur Roux, 75015 Paris, France, philippe.bastin@pasteur.fr

Marlene Benchimol Universidade Santa Úrsula, Rua Jornalista Orlando Dantas 59, CEP 222-31-010, Rio de Janeiro, Brazil, marlenebenchimol@gmail.com

Johanna Buisson Trypanosome Cell Biology Unit, Institut Pasteur & CNRS, 25 rue du Docteur Roux, 75015 Paris, France

Zdeňka Čičová Biology Centre, Institute of Parasitology, Czech Academy of Science and Faculty of Science, University of South Bohemia, České Budějovice (Budweis), Czech Republic

Roxana Cintrón-Moret Center for Tropical and Emerging Global Diseases and Department of Cellular Biology, University of Georgia, Athens, GA 30602, USA

Narcisa Leal Cunha-e-Silva Laboratório de Ultraestrutura Celular Hertha Meyer, Instituto de Biofísica Carlos Chagas Filho, Universidade Federal do Rio de Janeiro, Av. Carlos Chagas Filho, 373, bloco G subsolo, Cidade Universitria, Ilha do Fundão, Rio de Janeiro 21941-902, Brazil, narcisa@biof.ufrj.br

Wanderley de Souza Laboratório de Ultraestrutura Celular Hertha Meyer, Instituto de Biofísica Carlos Chagas Filho, Universidade Federal do Rio de Janeiro, CCS-Bloco G, Ilha do Fundão, Rio de Janeiro 21941-900, Brazil, wsouza@biof.ufrj.br; Diretoria de Programas, Instituto Nacional de Metrologia, Normalização e Qualidade Industrial-INMETRO, Av. Nossa Senhora das Graças, 50, Xerém, Duque de Caxias 25250-020, Rio de Janeiro, Brazil, wsouza@biof.ufrj.br;

Roberto Docampo Center for Tropical and Emerging Global Diseases and Department of Cellular Biology, University of Georgia, Athens, GA 30602, USA, rdocampo@uga.edu

Jean François Dubremetz UMR CNRS 5235, Bt 24, CC 107, Université de Montpellier 2, Place Eugène Bataillon, Montpellier cedex 05 34095, France, dubremet@univ-montp2.fr

Nancy Guillen Institut Pasteur, Cell Biology of Parasitism Unit, 25–28 rue du Docteur Roux, 75015 Paris, France, nguillen@pasteur.fr; Inserm U786, 25–28 rue du Docteur Roux, 75015 Paris, France

Keith Gull Sir William Dunn School of Pathology, University of Oxford, Oxford OX1 3RE, UK, keith.gull@path.ox.ac.uk

Hassan Hashimi Biology Centre, Institute of Parasitology, Czech Academy of Science and Faculty of Science, University of South Bohemia, České Budějovice (Budweis), Czech Republic

Hugo D. Lujan Laboratory of Biochemistry and Molecular Biology, School of Medicine, Catholic University of Cordoba, Jacinto Rios 571, CP X5004ASK Cordoba, Argentina, hlujan@ucc.edu.ar

Julius Lukeš Biology Centre, Institute of Parasitology, Czech Academy of Science and Faculty of Science, University of South Bohemia, České Budějovice (Budweis), Czech Republic, jula@paru.cas.cz

Paul G. McKean Division of Biomedical and Life Sciences, School of Health and Medicine, Lancaster University, Lancaster LA1 4YQ, UK, p.mckean@lancaster.ac.uk

Diane McMahon-Pratt Yale University School of Public Health, New Haven CT, USA

Dana Müller Laboratory of Biochemistry and Molecular Biology, School of Medicine, Catholic University of Cordoba, Jacinto Rios 571, CP X5004ASK Cordoba, Argentina

Sethu Nair Center for Tropical and Emerging Global Diseases & Department for Cellular Biology, University of Georgia, 500 D.W. Brooks Drive, Athens, GA 30602, USA

Fred R. Opperdoes Research Unit for Tropical Diseases, de Duve Institute and Biochemistry Unit, Université catholique de Louvain, Avenue Hipprocrate 75, 1200 Brussels, Belgium, fred.opperdoes@uclouvain.be

Miria G. Pereira Laboratório de Ultraestrutura Celular Hertha Meyer, Instituto de Biofísica Carlos Chagas Filho, Universidade Federal do Rio de Janeiro, Av. Carlos Chagas Filho, 373, bloco G subsolo, Cidade Universitria, Ilha do Fundão Rio de Janeiro 21941-902, Brazil

Fernando D. Rivero Laboratory of Biochemistry and Molecular Biology, School of Medicine, Catholic University of Cordoba, Jacinto Rios 571, CP X5004ASK Cordoba, Argentina

Celso Sant'Anna Laboratório de Ultraestrutura Celular Hertha Meyer, Instituto de Biofísica Carlos Chagas Filho, Universidade Federal do Rio de Janeiro, Av. Carlos Chagas Filho, 373, bloco G subsolo, Cidade Universitria, Ilha do Fundão Rio de Janeiro 21941-902, Brazil; Diretoria de Programas, Instituto Nacional de Metrologia, Normalização e Qualidade Industrial-INMETRO, Av. Nossa Senhora das Graças, 50, Xerém, Duque de Caxias 25250-020, Rio de Janeiro, Brazil

Lilach Sheiner Center for Tropical and Emerging Global Diseases & Department for Cellular Biology, University of Georgia, 500 D.W. Brooks Drive, Athens, GA 30602, USA

Sherri S. Smith Institut Pasteur, Cell Biology of Parasitism Unit, 25–28 rue du Docteur Roux, 75015 Paris, France; Inserm U786, 25–28 rue du Docteur Roux, 75015 Paris, France

Boris Striepen Center for Tropical and Emerging Global Diseases & Department for Cellular Biology, University of Georgia, 500 D.W. Brooks Drive, Athens, GA 30602, USA, striepen@cb.uga.edu

Yara M. Traub-Csekö Laboratório de Parasitas e Vetores, Instituto Oswaldo Cruz Fiocruz, Manguinhos, Rio de Janeiro, Brazil, ytraub@ioc.fiocruz.br

Tania Ueda-Nakamura Departamento de Análises Clínicas, Universidade Estadual de Maringá, Centro de Ciências da Saúde, Maringá Paraná, Brazil

Paul Ulrich Center for Tropical and Emerging Global Diseases and Department of Cellular Biology, University of Georgia, Athens, GA 30602, USA

Zdeněk Verner Biology Centre, Institute of Parasitology, Czech Academy of Science and Faculty of Science, University of South Bohemia, České Budějovice (Budweis), Czech Republic

The Mastigont System in Trichomonads

Marlene Benchimol

Contents

Abstract Trichomonads are protists that are found in several environments. Some of them are parasites, whereas others live in the gut of different animals without provoking any apparent infection. The trichomonad mastigont system is formed by the basal bodies and several structures composed of rootlets and proteinaceous fibrils; the function of many of these structures is not yet determined. The main structures in the mastigont system consist of the following: the pelta–axostyle system, made of microtubules; the costa, a periodic rootlet; the parabasal and sigmoid filaments; and several other filaments. The axostyle supports the axis of the cell and participates in karyokinesis, the costa supports the flagellar movements of the recurrent flagellum, and several fibrils provide an anchoring system for the Golgi and the nucleus. The pelta, a microtubular structure, supports the flagellar

---

M. Benchimol
Universidade Santa Úrsula, Rua Jornalista Orlando Dantas 59, CEP 222-31-010, Rio de Janeiro, Brazil
e-mail: marlenebenchimol@gmail.com

W. de Souza (ed.), *Structures and Organelles in Pathogenic Protists*,
Microbiology Monographs 17, DOI 10.1007/978-3-642-12863-9_1,
© Springer-Verlag Berlin Heidelberg 2010

canal. The trichomonad mastigont system has been studied with different techniques such as plasma membrane extraction, high-voltage electron microscopy, field emission scanning electron microscopy, the cell-sandwich technique, freeze-etching, and immunocytochemistry.

1 Introduction

Trichomonads are protists found in several environments. Some are parasites, whereas others live in the gut of different animals, such as mammals, birds, snakes, and insects, without provoking any apparent infection. Among the most important trichomonads are *Tritrichomonas foetus* (Figs. 1 and 2) and *Trichomonas vaginalis* (Fig. 3), which are flagellated parasites of the urogenital tracts of cattle and humans, respectively. Trichomoniasis, caused by *T. vaginalis*, is the most common non-viral sexually transmitted infection associated with adverse consequences for women's health (World Health Organization 2001). Molecular phylogenetic studies using large and small subunit ribosomal RNAs indicate that these organisms are among the most early diverging eukaryotes (Cavalier-Smith and Chao 1996), belonging to the Phylum Parabasalia, although it is possible that phylogenetic repositioning took place (Embley and Hirt 1998). It was demonstrated that *T. foetus* and *Tritrichomonas suis* belong to the same species (Mattos et al. 1997; Tachezy et al. 2002).

T. foetus (Figs. 1a, b and 2) and *T. vaginalis* (Figs. 3a, b and 4) are extracellular parasites, and they must overcome the mucus barrier and parasitise the vaginal epithelium for infection to occur. When grown in axenic medium (Figs. 1a, b, 8a, and 9a–d), *T. foetus* is characterised by a pear-shaped body that is 16.5-μm long and 5.0–6.7-μm wide (Mattos et al. 1997), with one anterior nucleus (Figs. 1, 2, and 9c, d).

2 The Mastigont System

Trichomonads present unusual organelles, such as the hydrogenosomes, mitochondria-like organelles that produce ATP and molecular hydrogen, the pelta, the axostyle, the costa, and several rootlet fibres concentrated in the mastigont system, as shown in the illustration in Fig. 2. This system is localised to the anterior region of the protist's body, as we will present in detail below.

The complex mastigont system (Figs. 1b, 2a, b, 5a, b, 6a, b, 9a–d, and 10–14) is comprised of several skeletal structures, including the following (1) the flagellar axonemes and their respective basal bodies (kinetosomes), located at the most anterior region of the cell body (Figs. 1b, 5a, b, 6a, b, and 12–14); (2) several appendages and rootlet filaments around the basal bodies (Figs. 1b, 2, 5a, b, 6a, b,

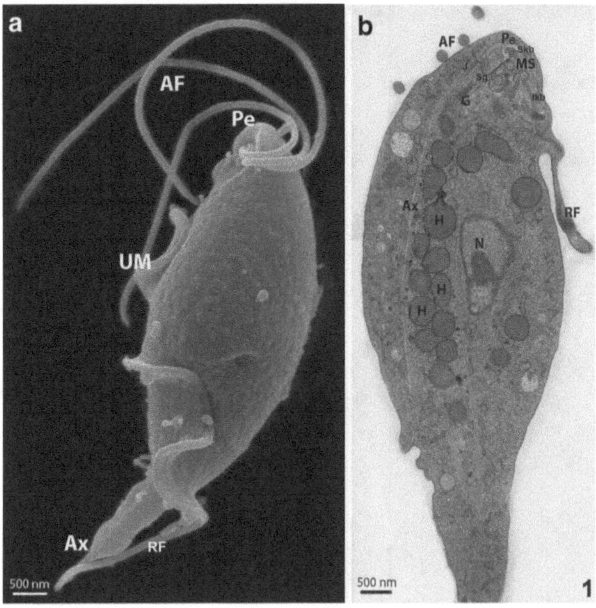

Fig. 1 Routine preparation for scanning (**a**) and transmission electron microscopy (**b**) of *Tritrichomonas foetus*. Figure (**a**) shows a general view with the anterior flagella (AF), the recurrent flagellum (RF) forming the undulating membrane (UM), the flagellar canal supported by the pelta (Pe), and the region of origin of the anterior flagella. The axostyle (Ax) forms the posterior tip of the cell. Figure (**b**) shows the cell interior and it is possible to note the following structures: an anterior single nucleus (N), the axostyle (Ax), the pelta (Pe), and the mastigont system (MS), which includes kinetosomes (basal bodies) and associated structures such as the supra-kinetosomal body (Skb) and the infra-kinetosomal body (Ikb). Note that the sigmoid filaments end in the pelta–axostylar junction (J). The anterior flagella (AF), the recurrent flagellum (RF), the Golgi complex (G), and the hydrogenosomes (H) aligned near the axostyle (Ax) can be seen clearly. *Bars*, 500 nm (Pereira-Neves and Benchimol unpublished)

9d, 13, and 14); (3) the costa (Figs. 2a, b, 7a–c, 9a–d, and 14), a periodic proteinaceous structure that underlies the recurrent flagellum along the undulating membrane (UM); (4) the parabasal filaments (PBF) (Figs. 2a, b, 3b, 5a, b, and 9b–d), which support a single Golgi complex, forming a Parabasal apparatus (Figs. 9b, c, and 14); (5) a pelta (Figs. 1a, b, 2a, b, 5a, b, 6a, b, 7a, 9b–d, 13, and 14) supporting the flagellar canal (Figs. 1a and 9b); and (6) a single ribbon of microtubules forming the axostyle (Figs. 1–5, 9a–d, 10, and 14), which runs from the basal bodies to the cell tip, forming, together with the pelta, the pelta–axostyle complex. Trichomonads also present other structures not directly related to the mastigont system, such as a nucleus, hydrogenosomes, glycogen granules, a Golgi complex, lysosomes, an endoplasmic reticulum, and a cell matrix that contains an extensive and elaborate three-dimensional skeletal network (Figs. 1b, 2a, b, 3, and 5a, b).

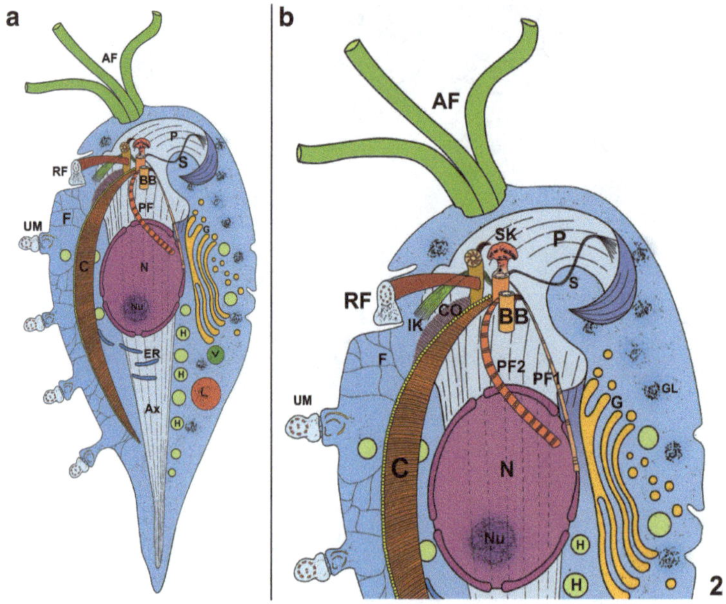

Fig. 2 Schematic diagrams of a whole view of *Tritrichomonas foetus* (**a**) and detail of the anterior region (**b**) showing the main cell structures and a close view of the mastigont system. *AF* anterior flagella; *Ax* axostyle; *BB* basal body; *C* costa; *Co* comb; *ER* endoplasmic reticulum; *F* filaments connecting the costa to the plasma membrane, next to the recurrent flagellum; *G* Golgi; *GL* glycogen granules; *H* hydrogenosomes; *IK* infra-kinetosomal filament; *L* lysosome; *N* nucleus; *Nu* nucleolus; *P* pelta; *PF* parabasal filament; PF_1 parabasal filament number 1; PF_2 parabasal filament number 2; *RF* recurrent flagellum; *S* sigmoid filaments; *SK* supra-kinetosomal body; *UM* undulating membrane; *V* vacuole (Benchimol unpublished)

Several procedures that provide three-dimensional images have been employed to better analyse the mastigont system in trichomonads; these include routine preparation for either conventional (Figs. 1b, 3, 5, and 6) or high-voltage transmission electron microscopy (HVTEM) (Fig. 9a), detergent-extraction procedures and observation with scanning electron microscopy (SEM) (Fig. 9b–d), physical fixation by freezing methods (Figs. 7c, 10, and 11b, c), and techniques such as immunocytochemistry (Fig. 6a) and tomography (Benchimol 2004, 2005, in press; Benchimol et al. 1992, 1993, 2000; Lee et al. 2009).

2.1 Basal Bodies and Associated Filaments

2.1.1 Basal Body (kinetosomes)

The kinetosomes or basal bodies are microtubular structures located in the most anterior region of the cell, where the flagella, the spindle, and several other rootlet

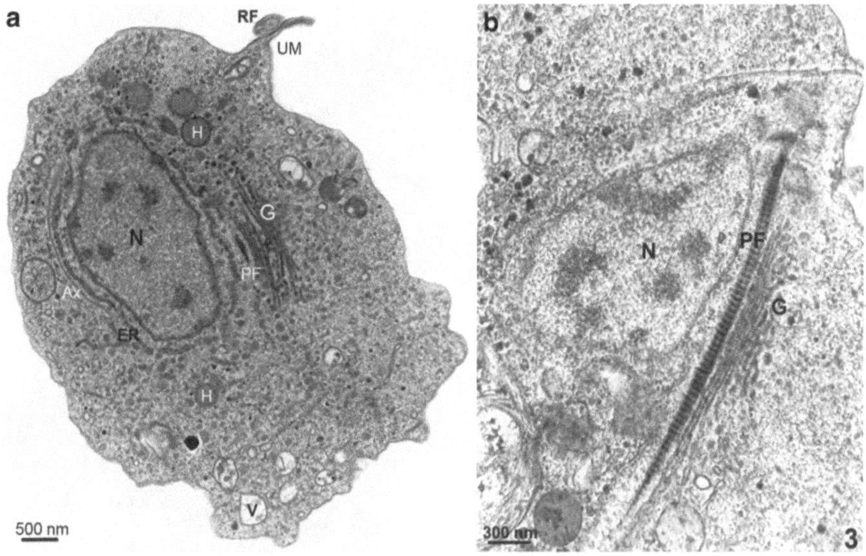

Fig. 3 Routine preparation for transmission electron microscopy of *Trichomonas vaginalis* showing a general view of the cell (**a**) and detail of the Golgi complex (G) and its parabasal filament (PF), a periodic structure that follows the Golgi in trichomonads in (**b**). *N* nucleus, recurrent flagellum (RF), forming the undulating membrane (UM), the endoplasmic reticulum (ER), Golgi and its parabasal filament (PF), vacuoles (V), hydrogenosomes (H), and the axostyle (Ax). *Bar*, 500 nm (**a**), 300 nm (**b**) (Benchimol unpublished)

structures originate (Figs. 1b, 2a, b, 5a, b, 6a, b, 9c, d, 13a, b, and 14). The number of kinetosomes varies according to the *Trichomonas* species; for instance, there are five kinetosomes in *T. vaginalis* (Fig. 4) and four in *T. foetus* (Figs. 1b and 2a, b). All these kinetosomes are embedded in a peripheral material of greater density compared to the surrounding cytoplasm.

Several proteinaceous accessory structures, such as striated roots and filaments in the form of hook-shaped lamellae, are associated with the basal bodies (Figs. 1b, 2a, b, 5a, b, and 6b); they include the sigmoid filaments (Figs. 1b, 2a, b, 5, 6a, 9d, 13a, b, and 14), the supra-kinetosomal (Figs. 1b, 2a, b, 5a, b, and 6b) and infra-kinetosomal bodies (Figs. 1b, 5a, b, and 6b), the PBF (Figs. 3a, b, 5a, and 14), the costa (Figs. 2a, b, 7a–c, 9a–d, and 14), the comb (Figs. 2a, b and 5b), and the pelta–axostyle complex (Figs. 1b, 2a, b, 5a, b, 9a–d, 14, and Table 1). None of these fibres are considered contractile (Honigberg and Brugerolle 1990; Bricheux et al. 2000; Viscogliosi and Brugerolle 1994), with the few exceptions outlined below.

The basal bodies present a structure common to the basal bodies of many organisms, including higher eukaryotic cells, with nine triplets of microtubules (Figs. 1b, 5a, and 6b). In the transitional zone, the basal bodies present a wheel-like aspect comprising a central hub and spokes (Benchimol 2004). Each flagellar axoneme emerges, with its typical structure consisting of 9+2 microtubules.

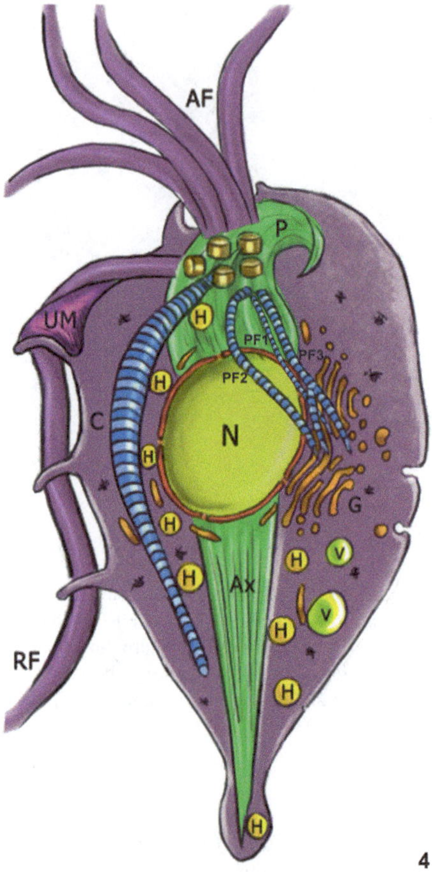

4

Fig. 4 Schematic diagram of *T. vaginalis* showing the undulating membrane (UM) which is formed by a fold of the plasma membrane in contact with the recurrent flagellum (RF). Note the difference with the undulating membrane of the *T. foetus* in Figs. 1 and 2. *AF* anterior flagella; *Ax* axostyle; *BB* basal body; *C* costa; *H* hydrogenosome; *N* nucleus; *PF1* parabasal filament 1; *PF2* parabasal filament 2; *PF3* parabasal filament 3

In trichomonads, the basal bodies/kinetosomes are assembled during the S phase (Ribeiro et al. 2000).

2.1.2 Kinetosome R

The recurrent flagellum (Figs. 1a, 2a, b, 3a, 4, 9a–c, and 11a) originates from the R kinetosome (Figs. 4, 6a, and 7a), which is situated at a right angle to the kinetosomes of the anterior flagella.

2.1.3 Kinetosomes #1, #2, #3, #4

Depending on the species (Figs. 1b, 2a, b, 5a, b, 6a, b, and 13a, b), there are several other kinetosomes that are located in the anterior region and connect to

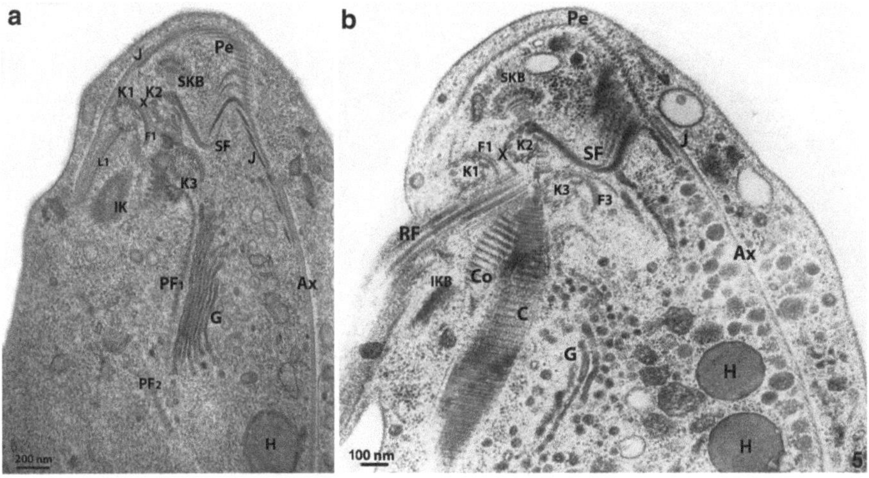

Fig. 5 (**a**, **b**) Routine thin sections of *T. foetus* through the anterior region. These two sequential sections display the complex structures that form the mastigont system. The basal bodies or kinetosomes, K_1, K_2, and K_3, are seen. Originating from kinetosome 1 (K_1), lamella 1 (L_1) is located towards the posterior region. The sigmoidal filaments (SF) emerge from kinetosome 2 towards the pelta (Pe). The costa (C) originates in the region of kinetosomes R (not seen here) and K_2–K_3, close to the comb (Co). The pelta (Pe) and the axostyle (Ax) overlap in the pelta–axostylar junction (J). The basal body of the recurrent flagellum (RF) is in an orthogonal position in relation to the other three basal bodies of the anterior flagella (K_1–K_3). Note the presence of rootlet filaments (F_1–F_3) emerging from basal bodies 1 and 3. Another fibrillar structure (**X**) courses between and connects kinetosome #2 and F_1. The sigmoid filaments (SF) emerge from the region of basal body 2 and the supra-basal body (SB). The supra-basal (or supra-kinetosomal) body (SKB) appears as a stalk connected to basal body 2, whereas the infra-basal (or infra-kinetosomal) body (IKB) lies below the whole basal body complex. The parabasal filaments (PF_1–PF_2) follow the Golgi complex (G). *H* hydrogenosome. (**a**, **b**) Benchimol (unpublished). *Bars* (**a**) 200 nm, (**b**) 100 nm

several other mastigont structures All kinetosomes are formed by microtubule triplets (Fig. 6b) in a configuration similar to that seen in the centrioles of higher eukaryotes.

2.1.4 Clockwise Filaments

There are several types of filaments and lamellae originating from the kinetosome region (Figs. 1b, 2a, b, 5a, b, 6a, b, and 9d). These filaments have a clockwise orientation and emerge from each kinetosome. Kinetosome #1 is the origin of the clockwise F_1 (filament 1) as well as other filaments known as the marginal lamellae (ML) of the UM (Figs. 1a, 2, and 5a, b), which emerge from the left side of kinetosome #1. There is also IK1, a filament that emerges from kinetosome #1 to the right of the ML (Figs. 2a, b and 5a).

Kinetosome #2 is the origin of the F_2 sigmoid filaments and the X filament (Figs. 1b, 2a, b, 5a, b, 6a, b, 9d, and 14). The sigmoid filaments course ventrally and

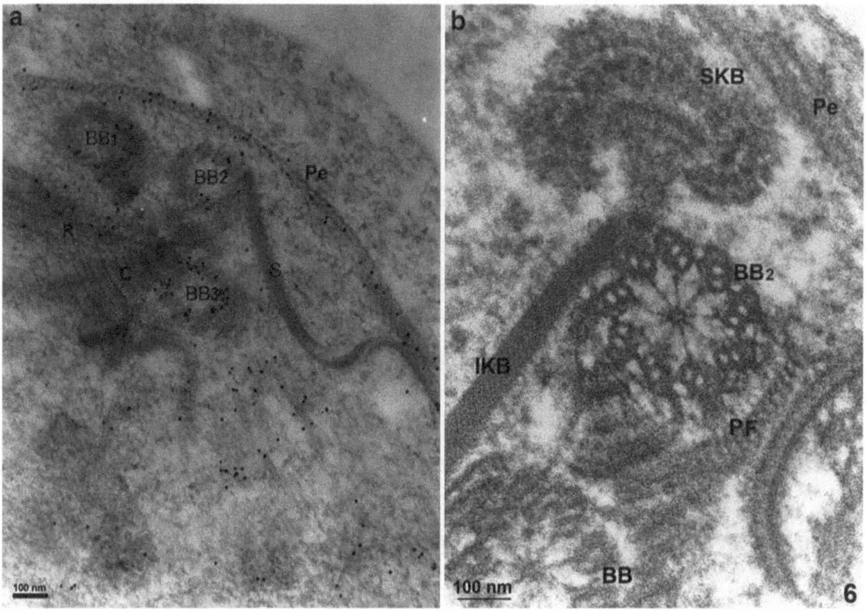

Fig. 6 Immunolabelling (**a**) and routine thin section (**b**) through the anterior region of *T. foetus*, showing some structures that form the mastigont system. The basal bodies, BB_1, BB_2, and BB_3, are seen labelled by gold-conjugated anti-acetylated tubulin monoclonal antibody. The sigmoidal filaments (SF) emerge from the basal bodies towards the pelta (Pe). The costa (C) emerges in the region of the R kinetosome (*not seen here*) and BB_2–BB_3. The pelta (Pe) is formed by microtubules and is also gold-labelled. The recurrent flagellum (R) is situated in an orthogonal position in relation to the other three basal bodies of the anterior flagella. The supra-basal body (SB) and the sigmoid filaments (SF) emerge from the region of basal body 2. Neither the costa nor the sigmoid filaments present labelling for anti-tubulin antibody; therefore, they are not made of microtubules. The supra-basal (or supra-kinetosomal) body (SKB) appears as a stalk connected to basal body 2, whereas the infra-basal (or infra-kinetosomal) body (IKB) lies below the whole basal body complex. The parabasal filaments (PF) are also seen. *Bars*, 100 nm (Benchimol unpublished)

towards the pelta–axostylar complex until they reach the pelta–axostylar junction (Figs. 1b, 5a, b, and 14). The X filament courses between and connects kinetosome #2 and F_1 (Fig. 5a, b). Kinetosome #3 is the origin of the clockwise periodic filament F_3 (Fig. 5a, b).

The costa originates between kinetosomes R and #2 (Figs. 2a, b, 5b, 6a, and 7a), and its broad base (Fig. 5a) is better seen in that region. In cells that present a comb (Fig. 5b), such as those of *T. foetus*, this structure originates from the R kinetosome and the costal base.

2.1.5 Sigmoidal Filaments

Several filamentous lamellae anchor the basal bodies and connect them to other cell structures such as the pelta and the cytoplasm. The largest of the lamellae is a

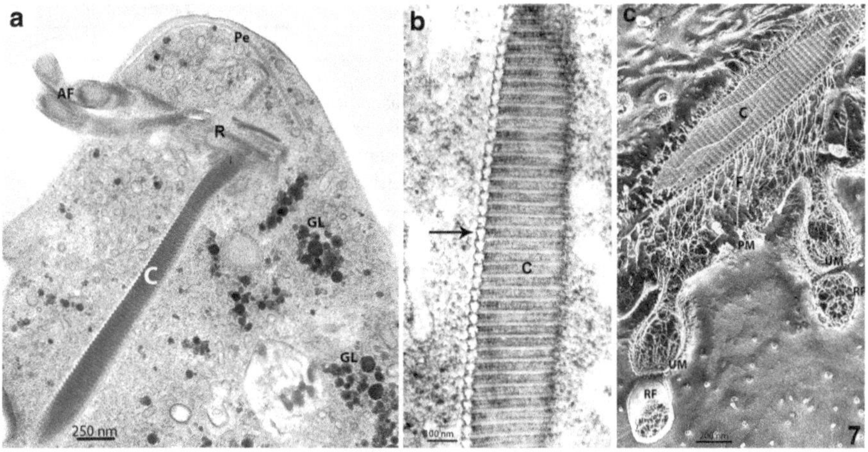

Fig. 7 Three different views (**a–c**) of the costa, a proteinaceous periodic structure present in trichomonads. This structure originates from the basal body region, between kinetosomes #2, R, and #3; it courses near the cell surface, in the region of the undulating membrane. Its periodic structure can be seen clearly (**a–c**). (**b**) Longitudinal thin section of the *T. foetus* costa (C) under high magnification, showing its periodic structure with dense bands alternating with lucent bands. Pits (*arrow*) are seen on the peripheral border of the costa. (**c**) Costa (C) of *T. foetus* cell treated by fast-freezing and freeze-fracture followed by freeze-etching. This preparation allowed the visualisation of 15-nm-thick filaments (F) emanating from the costa (C) and connecting it with the plasma membrane (PM). *GL* glycogen granules; *Pe* pelta; *AF* anterior flagella. *Bars*, (**a**) 250 nm; (**b**) 100 nm, (**c**) 200 nm [(**a**, **b**) Benchimol (unpublished); (**c**) from Benchimol et al. (1993)]

sigmoid sheet (Figs. 1b, 2a, b, 5a, b, 6a, 9d, 13a, b, and 14) composed of parallel filaments in a curved distribution that connect kinetosome #2 to the pelta and fan out at the first bend. These filaments end just in the region of the pelta–axostylar junction (Figs. 1b, 5a, b, 6a, and 14). A part of the sigmoid sheet of filaments originates from kinetosome #2, extends towards the pelta, and continues along the pelta–axostylar junction (Figs. 1b, 5a, b, and 14), which presented a positive reaction for centrin, and an unknown function. Tomographic images obtained by Lee et al. (2009) showed that the sigmoid filaments are composed of several individual filaments. The curve of each sigmoid filament gradually becomes smooth (Lee et al. 2009).

2.1.6 Supra-Kinetosomal Body or Supra-Basal Body

The Supra-kinetosomal bodies (SKB) (Figs. 1b, 2a, b, 5a, b, and 6b) is a stalked structure restricted to *Tritrichomonadinae*. It is connected to basal body #2 in the region of attachment of the sigmoid filaments (Figs. 1b, 2a, b, and 5a, b). In many views, it is observed to be crescent-shaped. Its function and composition are unknown but anti-tubulin antibodies do not stain this structure.

Table 1 Fibrillar structures in *Trichomonas*

Name	Characteristics	Function
Axostyle	Microtubular ribbon	Participates in karyokinesis
Pelta	Microtubular sheet; overlaps with the axostyle; anterior region of the cell	Reinforces the wall of the periflagellar canal
Costa	Periodic proteinaceous structure; courses near the cell surface; connects basal bodies #2 and #3	Supports the movement of the recurrent flagellum; serves as an axis for trichomonad cells
Comb	Periodic structure located between the costal base and basal body R; present only in *Tritrichomonadinae*	Unknown
Basal bodies (kinetosomes)	Nine microtubule triplets	Origin of the flagellar axonemes
MTOC (atractophore)	Electron-dense material found in the costal base (*T. foetus*) or at the origin of the parabasal filaments (*T. vaginalis)*	Origin of the spindle microtubules
Axonemes	Motile microtubule structures of flagella in typical 9+2 arrangement; four in *T. foetus* and five in *T. vaginalis*	Participates in cell movement and division
Parabasal filaments PF_1 –PF_3	Periodic structures; follow the Golgi	Probable function in supporting the Golgi
Parabasal filament PF_1	Connects to the kinetosome #2; courses to the right of lamella $F_{3;}$ extends towards the end of the cel; PF_1 is a long filament that splits into two strands	Probable function in supporting the Golgi
Parabasal filament PF_2	Presents a common base with the costa; origin in kinetosomes #2 and #R	Unknown
Parabasal filament PF_3	Presents a common base with the costa; origin in kinetosomes #2 and #R ; runs closely to PF_1	Probable function in supporting the Golgi
Sigmoidal filaments	Filaments in curved distribution; connect basal body #2 to the pelta; course ventrally and towards the pelta–axostylar complex	Unknown; positive for centrin
Infra-basal body (Infra-kinetosomal body)	Large and dense structure, below the basal body complex	Contributes to the proximal marginal lamella; restricted to *Tritrichomonadinae*
Supra-kinetosomal body (Supra-basal body)	Stalked structure; connects to the basal body #2 in the region of the attachment of the sigmoid filaments; crescent-shaped in many views	Unknown; restricted to *Tritrichomonadinae*
J	Junction between the pelta and the axostyle	Unknown
Clockwise F_1 (filament 1)	Origin from the kinetosome #1; comma-shaped	Unknown
Marginal lamellae (ML)	Origin from the kinetosome #1	Filaments following the undulating membrane
IK1	Filament that emerges from kinetosome #1 on the right of the ML	Unknown

(*continued*)

Table 1 (continued)

Name	Characteristics	Function
X filament	Origin from the kinetosome #2; courses between and connects the kinetosome #2 and F_1	Unknown
Filament F_3	Origin from the kinetosome #3; a clockwise turning periodic filament; comma-shaped	Unknown
Undulating membrane	Contains the proximal and distal marginal lamella	Supports the beating of the recurrent flagellum
Marginal lamella	Origin from the kinetosome #1; raises the plasma membrane; forms a thin, finlike dorsal fold	Adheres to the recurrent flagellum

2.1.7 Infra-Kinetosomal Body or Infra-Basal Body

The infra-kinetosomal body (IKB) (Figs. 1b, 2b, 5a, b, and 6b) is a large and dense structure located below the basal body complex and restricted to the *Tritrichomonadinae*. The IKB appears to contribute to the proximal ML (Fig. 5a) of the UM.

2.1.8 Marginal Lamella

Kinetosome #1 presents a left rootlet filament or lamella that runs ventrally to the recurrent flagellum (Fig. 5a). This structure is hardly seen in conventional preparation and exhibits a stack of fine filaments when observed properly in cross-sections. Its function and composition are unknown, but anti-tubulin antibodies do not stain this structure.

2.2 Costa

The costa (Figs. 2a, b, 5b, 6a, 7a–c, 9a–d, and 14) is a rod-shaped skeletal birefringent structure found only in trichomonads that possess an UM. It is assumed that its function is to provide mechanical support to the UM. This large striated root fibril emerges from the basal body region, between the kinetosome of the recurrent flagellum (R) (Figs. 5b, 6a, and 7a) and kinetosomes #2 and #3 (Honigberg et al. 1971). The costa is broader at the point at which it emerges. This proteinaceous structure courses under the UM and extends towards the posterior region of the cell under the dorsal cell surface. There are connections between the costa and the plasma membrane in the region of the UM (Fig. 7c).

2.2.1 Structure

The costa is formed by a complex array of filaments and globular structures (Fig. 7a–c). Through the use of fast-freezing methods, it was observed that the costa is connected to the recurrent flagellum through a complex network formed by 15- and 10-nm wide filaments that emerge from the peripheral region of the costa and penetrate into the surface projections of the protozoan body to which the recurrent flagellum is attached (Fig. 7c) (Benchimol et al. 1993). As demonstrated by several methods, such as thin-sections, freeze-fracture, deep-etching, and negative staining, the costa is a complex structure in which 60-nm-wide bands alternate with thinner, 26-nm bands (Figs. 7a–c and 14) (Benchimol et al. 1993). The complex array of filaments that form the costa is more clearly seen in the thinner bands. The costa also contains incomplete secondary transverse bands. This proteinaceous structure presents basic proteins, as demonstrated by cytochemical methods such as the use of ethanolic phosphotunstic acid (Benchimol et al. 1982a).

Freeze-etching experiments (Fig. 7c) demonstrated that, in those regions facing the areas where the recurrent flagellum establishes a specialised junction with a surface projection of the cell body, a large number of filaments, about 15 nm thick, emerge from pits in the costa and make contact with a network of other filaments present in the surface projections. These latter filaments are shorter and thinner and about 10 nm wide (Benchimol et al. 1993; Monteiro Leal et al. 1993). In addition, in some species such as *T. foetus*, a filamentous branch can be seen originating from the costa; it is known as C_1F_1. Pits (Fig. 7b) are seen in the peripheral border of the costa. Moreover, the costa was observed to be connected to the cytoskeleton via filaments (Fig. 7c).

2.2.2 Types of Costa

Costae are classified into two types depending on the striation pattern, size, and shape at the structure's origin: type A (also known as type C) and type B (or C_1) (Honigberg 1978). *T. foetus* has an A-type costa (Figs. 7a–c, 9a–d, and 14) whereas *T. vaginalis* as well as all *Trichomonas* and *Pentatrichomonas* have B-type costae.

The periodicity of B-type costae is estimated at about 42–60 nm (Honigberg and Brugerolle 1990), depending on the preservation method used. In *Trichomonas* (B-type costa), the longitudinal filaments are arranged in a "herringbone" pattern, a finding based on observed alteration in horizontal rows (Honigberg and Brugerolle 1990).

2.2.3 Association with Other Structures

The close association of the costa with the UM and the recurrent flagellum (Figs. 7c and 9a–c) led to the hypothesis that its function is to support the UM. Such an association is especially seen in electron micrographs of mildly detergent-extracted cells (Benchimol et al. 1993).

The hydrogenosomes (Figs. 1b, 2a, b, 3a, 5a, b, 13a, b, and 14), which are important ATP-producing organelles in trichomonads, are observed to be associated with the costa and the axostyle (Fig. 9c), which explains the old nomenclature for these organelles (paracostal and paraxostylar granules).

2.2.4 Function and Motility

It is assumed that the costa is a non-motile structure but provides mechanical support to the membrane of the recurrent flagellum. However, in some large species such as *Trichomonas gigantea* and *T. termopsidis*, the costa is a motile structure (Amos et al. 1979). In parasitic protists such as *T. vaginalis* and *T. foetus*, the costa seems to be an immotile structure, although in pseudocysts the costa is curved (Fig. 14) (Benchimol unpublished).

2.2.5 Chemical Composition

Studies by Viscogliosi and Brugerolle as well as by our group, in which monoclonal antibodies were produced (Viscogliosi and Brugerolle 1994; Monteiro Leal et al. 1993; Benchimol unpublished) and biochemical techniques such as cell fractionation were used to obtain pure costa fractions, revealed that B-type costa proteins in trichomonads are composed of several major polypeptides with molecular weights between 100 and 135 kDa, similar to those found in A-type costae. SDS-PAGE analysis showed that the costa contains several proteins with major bands corresponding to apparent molecular masses of 122, 115, 112, 93, 87, 82, 59, 44, 41, 32, and 26 kDa (Monteiro Leal et al. 1993; Viscogliosi and Brugerolle 1994). In addition, these techniques allowed the separation of several major protein components with molecular weights between 100 and 150 kDa (Viscogliosi and Brugerolle 1994). Monoclonal antibodies recognised five polypeptides with molecular weights of 135, 125, 114, 88, and 47 kDa by immunoblotting and the costa was labelled by immunofluorescence using these antibodies (Viscogliosi nd Brugerolle 1994). The authors did not find immunological cross-reactivity with other trichomonad genera; this indicates that the costae are not identical in their biochemical composition, which is in agreement with the differences in their respective fine structures. However, some proteins were similar, indicating the existence of shared epitopes (Viscogliosi and Brugerolle 1994).

2.3 Comb

In *T. foetus* and other related members of its subfamily, the simpler basal costa is replaced by a comb-like structure (Fig. 5b). This structure is located between the region of origin of the costa on the right and the IKB, close to the R kinetosome, on the left (Fig. 5b). The comb is also a cross-striated structure formed by periodic fibrils or lamellae that are thicker when seen towards the posterior area. The comb also interacts with the IKB, on its posterior and ventral side. The function of the comb is unknown, and until now there have been no studies focused on this structure.

2.4 Axostyle and Pelta

2.4.1 Structure

The axostyle is an axial ribbon of longitudinally oriented microtubules, and in trichomonads (Benchimol et al. 2000) it runs from the anterior region to the other end of the cell (Figs. 1b, 2, 5a, b, 8, 9a–d, 10, and 14), The anterior part of the axostyle is wider and forms the capitulum. Posterior to the nucleus, the axostyle turns upon itself, forming a tube known as the axostylar trunk (Fig. 9a–d). The axostyle appears to narrow progressively until its terminal segment, which protrudes from the posterior cell as a thin tip covered by the cell membrane (Figs. 1 and 9a, b) (Honigberg et al. 1971; Ribeiro et al. 2000). The *T. foetus* axostyle contains a ribbon of about 150 microtubules with a diameter of 24 nm and is spaced by 40 nm (Fig. 10) (Benchimol 2004). These microtubules are connected by thin filamentous bridges that are 30–40-nm long and 10-nm thick, maintaining a uniform distance of 25 nm between them (Benchimol et al. 1993; Benchimol 2004). The chemical composition of these proteinaceous bridges has not been elucidated yet. Just posterior to the nucleus, the microtubular sheet of microtubules of the axostyle turns upon itself to form the axostylar trunk, which runs near the antero-posterior axis of the flagellum and extends into its caudal tip projection (Fig. 9a–d).

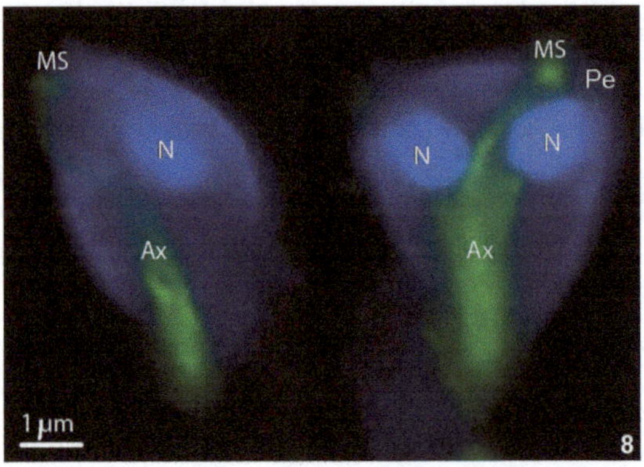

Fig. 8 Immunofluorescence image of an interphasic (*left*) and a dividing *T. foetus* cell using an anti-glutamylated tubulin monoclonal antibody. The nucleus (N) and microtubular structures such as the axostyle (Ax) are seen duplicated. The nuclei (N) were stained with DAPI. The axostyle (Ax) is seen running along the main axis of the cell, and is seen duplicated when the cell is dividing. The mastigont system (MS) is located on the upper anterior side of the cell. *Bar*, 1 μm (Benchimol unpublished)

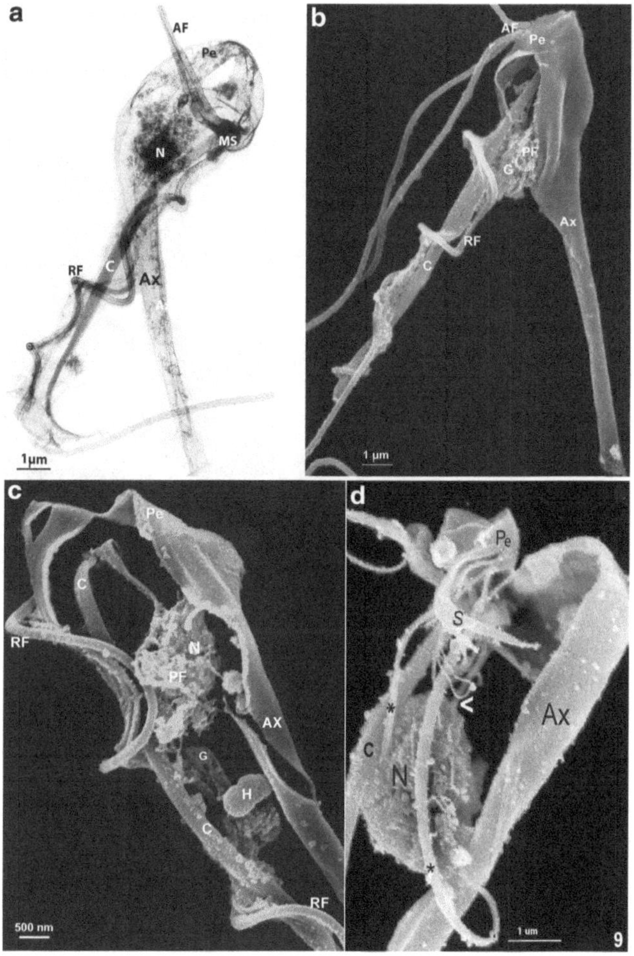

Fig. 9 Internal view of *T. foetus* using high-voltage transmission electron microscopy (HVTEM) (**a**) or detergent treatment to remove the plasma membrane and exhibit the cytoskeleton (**b–d**). (**a**) General view of Triton X-100-extracted trichomonads as observed with HVTEM. The microtubule ribbons that form the pelta (Pe) and the axostyle (Ax) are seen; the anterior part of the axostyle is wider and forms the capitulum. Posteriorly, the axostyle turns upon itself forming a tube, the axostylar trunk. Note that the costa (C) runs alongside the recurrent flagellum (RF). The anterior flagella (AF) emerge from the same point on the costa as the mastigont system (MS) does. *N* nucleus. Bar, 1 μm. (Benchimol and DeSouza unpublished). (**b**, **c**) Mastigont system of *T. foetus* after plasma membrane extraction, which allows better visualisation of the pelta (Pe), the anterior flagella (AF), and the axostyle (Ax) trunk. Note that the costa (C) follows the recurrent flagellum (RF), forming the undulating membrane. The costa bands can be seen in *panel* (**c**). Note the Golgi complex (G) associated with the parabasal filaments (PF) and the still-preserved nucleus (N). The hydrogenosomes (H) are depicted in panel (**c**); one of them (*asterisk*) is seen in the process of division. *Bars*, (**b**) 1 μm; (**c**) 500 nm (Benchimol unpublished). (**d**) High-resolution field emission scanning electron microscopy (FESEM) of *T. foetus* after detergent extraction. When the plasma membrane is removed and the cell is observed by FESEM, several fibrillar structures are seen forming the mastigont system. The axostyle (Ax) can be observed as a ribbon of microtubules

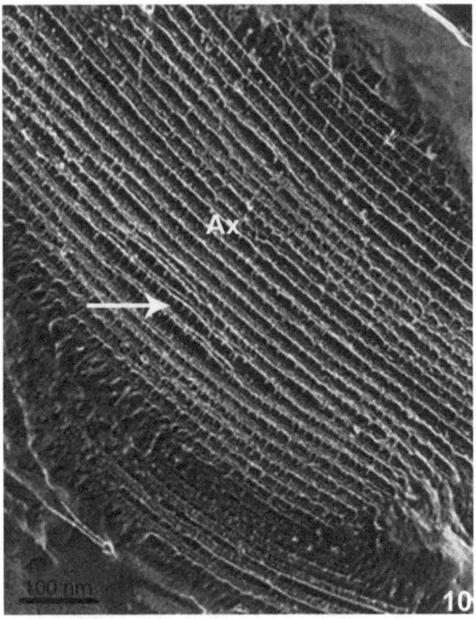

Fig. 10 *T. foetus* axostyle (Ax) observed by freeze-etching after quick-freezing by slam-freezing. The axostyle microtubules are seen connected by thin filaments (*arrow*). *Bar*, 100 nm (Benchimol unpublished)

A previous report (Benchimol et al. 2000) demonstrated that the trichomonad axostyle microtubules have a lateral projection formed by two protofilaments in addition to the 13 protofilaments normally found in microtubules. In addition, when different types of anti-tubulin antibodies were used to label microtubules in trichomonads (Fig. 6a), it was observed that the axostyle–pelta junction is a structure with high affinity for antibodies that recognise glutamylated tubulins in *T. foetus* (Lopes et al. 2001).

2.4.2 Pelta

The pelta (Figs. 1a, b, 2, 5a, b, 6a, b, 8, 9a–d, and 14) is a crescent-shaped sheet also formed by microtubules, and it overlaps with the axostyle in the anterior region of the cell. The pelta supports the wall of the anterior region of the cell and the flagellar canal from which the flagella emerges (Honigberg et al. 1971; Benchimol 2004) (Figs. 1a and 9b). At its broadest part, the pelta is composed of 35–40 microtubules (Figs. 5a, b and 9c), each one 25 nm in diameter. The pelta microtubules are internal to those of the axostyle (Figs. 5a, b and 9b–d), and when they encounter the axostyle microtubules they form the pelta–axostylar junction

---

Fig. 9 (continued) running from the anterior to the posterior region of the cell. The nucleus (N) is seen connected to the basal body region through an anchoring fibrillar structure (*arrowhead*), and it is also associated with the costa (C), an axial proteinaceous structure. Two parabasal filaments (*asterisks*), the pelta (Pe), the rootlet structures, and sigmoidal filaments (S) are also seen. *F* flagella. *Bar*, 1 μm (Benchimol unpublished)

(Fig. 14), which may also include sigmoidal filaments (Figs. 5a, b, 6a, and 14). As the pelta courses to the posterior region of the cell, the number of microtubules progressively decreases.

2.4.3 Functions

The main function of the axostyle in trichomonads is to support the axis of the cells. In addition, the axostyle also participates in karyokinesis. It was demonstrated in a previous study of cell division in *T. foetus* (Fig. 8b) (Ribeiro et al. 2000) that the axostyle in *T. foetus* presses the dividing nucleus, aiding in its division (karyokinesis). These processes were not observed by other authors in *T. vaginalis* (Viscogliosi and Brugerolle 1994; Delgado-Viscogliosi et al. 1996).

2.4.4 Stability

Some authors indicated that the axostyle is formed by labile microtubules since they depolymerised when the protozoa were incubated in the presence of colchicine early in the cell cycle (Brugerolle 1975; Juliano et al. 1986; Viscogliosi and Brugerolle 1994; Delgado-Viscogliosi et al. 1996). However, our group demonstrated that the microtubules of the axostyle are stable since they exhibited acetylated tubulins, did not depolymerise when treated with drugs affecting the cytoskeleton (e.g. vinblastine, colchicine, nocodazole, and taxol) (Silva-Filho and De Souza 1986; Batista et al. 1988; Ribeiro et al. 2000, 2002; Madeiro da Costa and Benchimol 2004), and were stable during mitosis (Ribeiro et al. 2000).

2.4.5 Movement

Unlike axostyles found elsewhere, the axostyle of trichomonads does not contract (Monteiro-Leal et al. 1996).

2.4.6 Composition

The axostyle is labelled after incubation with several antibodies recognising α- and β-tubulin (Figs. 6a and 8a, b) as well as glutamylated and acetylated α-tubulin (Delgado-Viscogliosi et al. 1996; Lopes et al. 2001; Boggild et al. 2002). It has been demonstrated that the most anterior region of the *T. foetus* pelta–axostyle exhibits glutamylated tubulin, whereas acetylated tubulin predominates in the posterior region (Lopes et al. 2001). Tyrosinated and polyglycylated microtubules were not found in trichomonads (Delgado-Viscogliosi et al. 1996; Schneider et al. 1999; Lopes et al. 2001; Boggild et al. 2002), and Brugerolle et al. (2000) have suggested that the sigmoid filaments could constitute an MTOC (microtubule-organising centre) of the pelta–axostylar system.

2.4.7 Association with Other Cell Structures

The axostyle is also associated with other cell structures such as the hydrogenosomes, the endoplasmic reticulum, the sigmoid filaments, and glycogen particles (Figs. 1b, 5a, b, and 9b, c) (Benchimol et al. 2000). The association of the hydrogenosomes with the axostyle explains the old nomenclature for these organelles (paraxostylar granules). These associations are aided by thin filaments that were only observed after the use of special techniques such as fast-freezing, followed by freeze-etching (Benchimol et al. 2000), and high-voltage electron microscopy (Benchimol and De Souza 1987; Benchimol 2004).

2.5 *The Parabasal Filaments*

Trichomonas vaginalis present three PBF: PF_1, PF_2, and PF_3 (Figs. 2a, b, 3b, 5a, 9b–d, and 14) (Honigberg and Brugerolle 1990; Lee et al. 2009). All PFs originate in the kinetosome region, near the costa's base (Fig. 5a). PF_1 arises between kinetosomes #2 and #3, whereas PF_2 has a common origin with the costa's base and originates from kinetosomes #2 and R (from the recurrent flagellum) (Honigberg and Brugerolle 1990). Lee et al. (2009) demonstrated by tomography that PF_1 is a long filament that extends towards the posterior region of the cell, where it splits into two strands, one of which curves towards the interior of the cell from the split point. They have shown that PF_1 and PF_3 appeared to be very close to each other. However, in 1971, Honigberg stated that PF_1 could be seen as two or three separate structures (Fig. 2a, b).

These two filaments present, at first glance, identical periodicity to that of the costa; however, their band pattern is quite different (Honigberg and Brugerolle 1990). The PFs have a periodic structure (Figs. 3b and 5a) that involves alternating transverse electron-dense and electron-lucent regions, and the dense area consists of four thin, dense lines. The PFs are located above the nucleus and below the well-developed Golgi complex in trichomonads (Figs. 3b, 9b–d, and 14). In the early literature, the term parabasal apparatus was applied to both PF_1 and PF_2, both of which are associated with the Golgi complex. It is believed that PFs accompany the Golgi, and thus that their function is to support the Golgi complex. However, until now, there is no clear evidence for this assumption.

2.6 *Flagella*

The flagella of trichomonads vary in number and size in each individual species. *T. foetus* has three anterior flagella (Figs. 1a and 2a, b), whereas *T. vaginalis* has four (Fig. 4); both have a recurrent flagellum (Figs. 1a, b, 3a, 4, 7a, 9a–c, and 11a–c) that runs towards the posterior region of the cell and adheres to the cell body, forming an UM (Figs. 1a, 4, 9a–c, and 11a) (Honigberg and Brugerolle 1990).

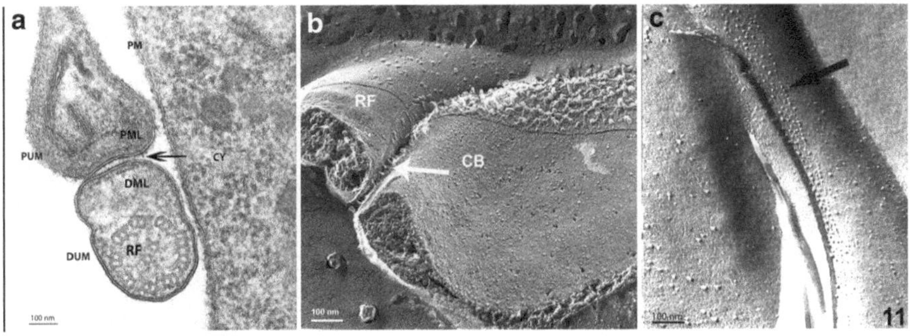

Fig. 11 *The undulating membrane (UM) observed by TEM.* (**a**) Thin section of the undulating membrane in the region of contact with the recurrent flagellum (RF); it is possible to see that it is formed of two parts: the proximal (PUM) and the distal (DUM) undulating membranes. The PUM is a dorsal fold of the plasma membrane (PM) and contains membranous profiles of the endoplasmic reticulum, known as proximal marginal lamellae (PML). The distal part of the undulating membrane presents the flagellum axoneme and the distal marginal lamella (DML). An *arrow* points to fibrils that connect the PUM to the DUM. *Cy* cytoplasm. *Bar*, 100 nm (Benchimol unpublished). (**b**) Recurrent flagellum (RF) of *T. foetus* observed by freeze-etching after quick-freezing by slam-freezing. Note the filamentous structures (*arrow*) connecting the flagellar membrane to the cell body (CB) membrane. (from Benchimol et al. 1992). *Bar*, 100 nm. (**c**) Freeze-fracture of the recurrent flagellum of *T. foetus* observed by MET. A linear array of intramembranous particles is seen in the region where the flagellum attaches to the plasma membrane, forming the undulating membrane (from Benchimol et al. 1982b). *Bar*, 100 nm

2.6.1 The Undulating Membrane

The UM is found in all species of trichomonads but it differs in Trichomonadinae such as *T. vaginalis* (Figs. 3a–4) and *T. gallinae,* compared to Tritrichomonadinae-like *T. foetus* (Figs. 1a, b, 2a, b, 9a–c, and 11a–c) and *T. augusta*. The UM is a cytoplasmic fold that consists of two parts: a proximal and a distal part (Fig. 11a). The proximal part (PUM) belongs to the cell surface and is a fold-like differentiation of the dorsal region of the cell surface that contains the proximal ML (Fig. 11a). The distal part (DUM) is formed by the axonemes of the recurrent flagellum and the distal ML. This part of the UM connects to the fold of the cell body through filamentous structures (Fig. 11a, b) of unknown composition. A special array of intra-membranous particles has been observed by freeze-fracture (Benchimol et al. 1982b) (Fig. 11c) in the region of contact of the cell body plasma membrane and the recurrent flagellum. The ML originates from kinetosome #1 and raises the plasma membrane, forming a thin, fin-like dorsal fold.

The proximal lamella contains membranous profiles of the endoplasmic reticulum (Fig. 11a), that react positively for glucose-6-phosphate (Benchimol unpublished).

2.6.2 Flagellar Structure

The axonemes of these flagella present characteristics typical of eukaryotic flagella, with a 9+2 arrangement of microtubules (Fig. 11a), and they originate from basal bodies located in the most anterior region of the cell (Figs. 1b, 5a, b, 6a, and 7a). They emerge from the cell through the flagellar canal, which is supported by the pelta (Figs. 1a, 9b, and 11). The recurrent flagellum also emerges from this region, but through a different opening. It bends and projects towards the posterior region, making contact with the cell body and thus forming the UM (Figs. 1a, 3a, 4, 9a–c, and 11a). Conventional freeze-fracture or freeze-etching techniques have shown that the anterior flagella, but not the recurrent flagellum of *T. foetus* and *T. vaginalis,* present a special array of nine to twelve intra-membranous particles forming rosettes (Fig. 12) (Benchimol et al. 1981). An array of particles that form a flagellar necklace (Fig. 12) in the region from which the flagella emerge has also been found. The function of this specialised structure is unknown, although there are several hypotheses (Benchimol et al. 1992).

2.6.3 Flagellar Movement

Filamentous bridges were seen connecting the cell body surface to the membrane of the recurrent flagellum (Fig. 11a, b). It has been shown by videomicroscopy (Benchimol et al. 1992; Monteiro-Leal et al. 1996) that all flagella participate in the cell's movement. The anterior flagella beat in a ciliary pattern, displaying

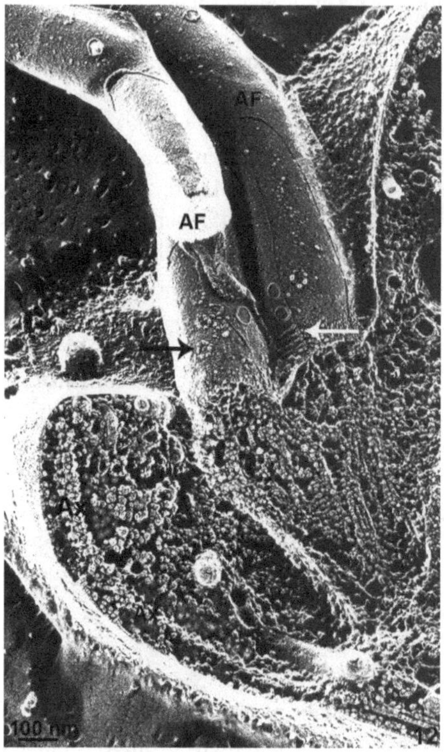

Fig. 12 Anterior region of *T. foetus* observed by freeze-etching after quick-freezing by slam-freezing. The anterior flagella (AF) present a special array of intra-membranous particles forming rosettes (*black arrow*) and an arrangement of linear particles in the flagellar base (*white arrow*) forming a flagellar necklace. The axostyle microtubules (Ax) are also seen connected. *Bar*, 100 nm (from Benchimol et al. 1992)

effective and recovery strokes, whereas the recurrent flagellum beats in a typical flagellar wave pattern (Monteiro-Leal et al. 1995). On the other hand, these studies also showed that one of the three anterior flagella beats straight in the forward direction, opposite from the other two flagella (Fig. 9b). Frequency measurements obtained from cells swimming in a viscous medium show that the beating frequency of the recurrent flagellum is approximately twice the frequency of the three anterior flagella (Monteiro-Leal et al. 1995). The authors of this study also believed that the costa and the axostyle form a cytoskeletal base for the anchoring and orientation of the flagella. In another study (Monteiro-Leal et al. 1996), it was shown that, during 1 s of recorded movement, *T. foetus* performs four complete anterior flagellar beating cycles (with active-like and recovery-like beatings). In each cycle, the cell swims ±6.5 μm forward, after the recovery of ±1.5 μm of receding movement. These observations led the authors to conclude that the estimated average speed of *T. foetus* is 25 μm/s and that all flagella participate in the cell's movement. The cell also performs rotational movements and the recurrent flagellum contributes continuously to the forward movement of the protozoan (Monteiro-Leal et al. 1995, 1996).

2.6.4 Flagellum Internalisation

The flagella in *T. foetus* and *T. muris* are promptly internalised (Fig. 14) when the cells are in an unfavourable or stress situation (Granger et al. 2000). In this form, the trichomonads are referred to as pseudocysts since they do not present a true cell wall. The flagella are able to externalise when the environment becomes favourable.

3 The Spindle

All trichomonads present a distinct type of mitosis, closed mitosis, in which the nuclear envelope remains intact throughout mitosis and the spindle is extranuclear. This spindle (Figs. 13 and 14) is formed during the migration of the duplicated basal bodies to opposite poles of the cell (Brugerolle 1975; Ribeiro et al. 2000). During the migration of basal bodies, two rod-shaped structures known as atractophores, located at the base of kinetosome #2, emerge from the spindle (Figs. 13 and 14), and the microtubules that form the spindle extend between the cell poles and elongate, pressing the nuclear envelope (Brugerolle 1975; Ribeiro et al. 2000).

3.1 The Atractophore

The MTOCs of trichomonads are named atractophores and are located at the base of the costa, underneath kinetosome #2 (Figs. 13 and 14). Together with the basal

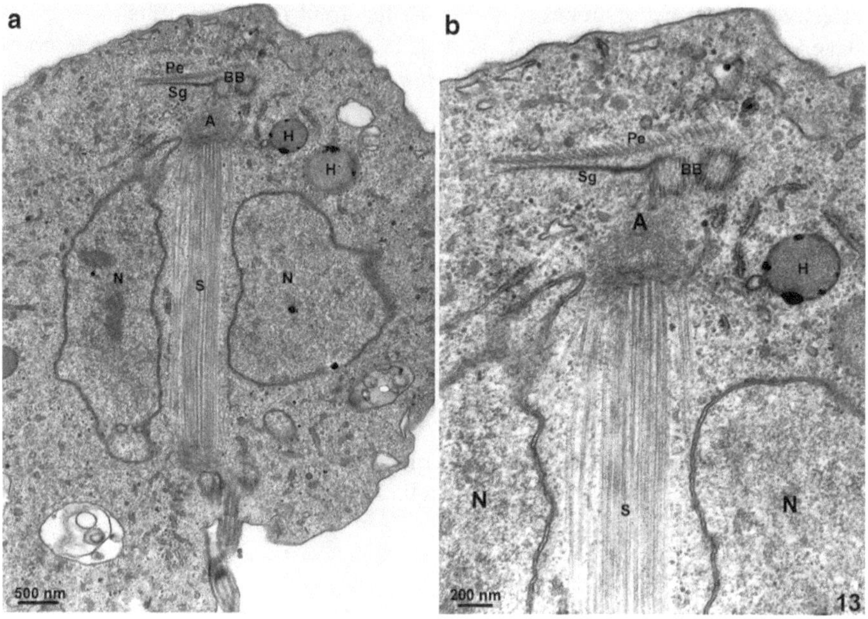

Fig. 13 Mitosis in trichomonads. (**a**) General view of a mitotic cell and (**b**) close view of the anterior region of *T. foetus,* showing the extranuclear spindle (S) originating from the atractophore (A). Note that the atractophore is an electron-dense fibrillar structure that is adjacent to the basal body (BB), and that it is a microtubule-organising centre. Two nuclei (N) are seen. Mitosis is closed in trichomonads, since the nuclear envelope does not fragment. The sigmoidal filament (Sg) and the pelta (Pe) are also seen. *H* hydrogenosome. *Bar*, (**a**), 500 nm; (**b**) 200 nm (Benchimol unpublished)

body, the atractophores form centrosome-like structures. In *Trichomonas,* the spindle presents an unusual arc shape during the initial phases of mitosis in motile trophozoites, but the pseudocyst form displays some differences in its mitotic process (Pereira-Neves and Benchimol 2009).

In a study performed by Bricheux et al. (2007), a monoclonal antibody recognising the atractophores of *Trichomonas* was raised and used to screen an expression library. A protein, named P477, was characterised as a novel protein associated with the atractophore of *T. vaginalis.* The sequence obtained was completed using the TIGR *T. vaginalis* genome sequence database (Carlton et al. 2007). Sequence analysis showed that P477 belongs to the family of large coiled-coil proteins, which share a highly versatile protein folding motif adaptable to many biological functions. Specific domain searches did not provide evidence of catalytic or binding activities. The Bricheux group suggested that P477 could play a role providing support or acting as a scaffold in the centrosome-like atractophore and interact with other proteins using specific coiled-coil binding. Thus, P477 could act as an anchor to transport cellular activities and components to the Golgi centrosomal region; it could represent a new class of structural proteins since similar proteins were found in many protozoa.

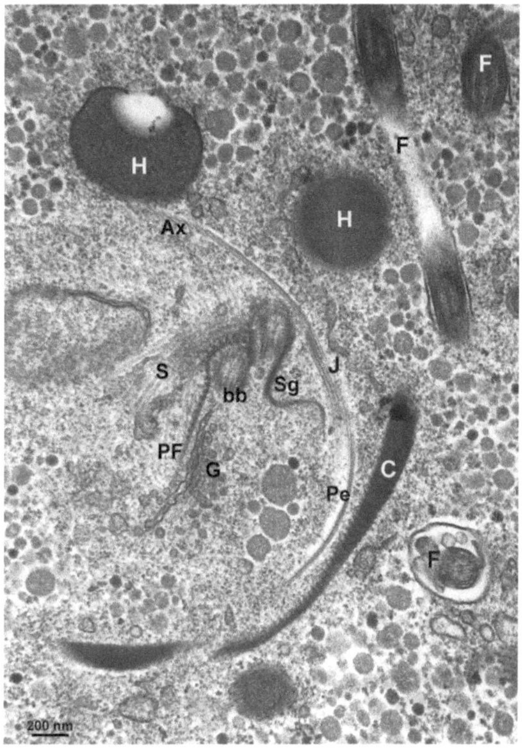

Fig. 14 *The pseudocyst in T. foetus*. A cell presenting internalised flagella (F) under stress conditions. This pseudocyst is dividing, as can be seen by the presence of the mitotic spindle microtubules (S). The axostyle (Ax) and pelta (Pe) overlap in *J*, where the sigmoid filaments (Sg) emanate. One parabasal filament (PF) originates from the basal body (bb) region and follows the Golgi complex. *C* costa; *H* hydrogenosome. *Bar*, 200 nm (Pereira-Neves and Benchimol unpublished)

4 Centrin

Centrin (also known as caltractin) is a 20-kDa acidic cytoskeletal protein that has been localised to centrosomes and the basal bodies of all flagellated or ciliated cell types (Brugerolle et al. 2000), including mammalian cells; it has also been localised to fibrous structures associated with basal bodies in many flagellated or ciliated unicellular organisms. This protein is an important member of the EF-hand family of calcium-binding proteins and it is known to show calcium-sensitive contractile behaviour. Centrin was described in trichomonads (Brugerolle et al. 2000) as a structural component of basal bodies or simply associated with them, as in all other eukaryotes. In trichomonads, this protein was found apparently associated with several other structures attached to basal bodies, such as the hooked fibres F1, F3, and X-fibre, rather than to a component of these structures.

This same group reported that the region of the pelta–axostyle junction could function as a MTOC because they found centrin immunolabelling at this site. These authors also reported labelling with anti-centrin antibody in the region of the UM. Centrin was also found close to the sigmoid fibres and at the pelta–axostyle junction and appeared to be a component of the nine basal body arms that anchor basal bodies/flagella to the plasma membrane. However, centrin was not found in the costa, parabasal fibres, or pelta–axostyle complex (Brugerolle et al. 2000). In the draft genome sequence of *T. Vaginalis*, genes for centrin were identified (Carlton et al. 2007).

5 Genome Sequence

In the draft genome sequence of the sexually transmitted pathogen *T. vaginalis*, genes for the following flagellar proteins were identified (Carlton et al. 2007): PF16 homologue, PF20 homologue, LF4, Intra-flagellar Transport Particle Protein IFT20 homologue, Intra-flagellar Transport Particle Protein IFT57 homologue, Intra-flagellar Transport Particle Protein IFT172 homologue, RIB43, and RIB72. The analysis also found motor proteins such as dyneins and kinesins, including genes for the Dynein light chain (axonemal, cytoplasmic, and Roadblock domain), Dynein intermediate chain, Dynein heavy chain, and Dynein beta chain. Regarding tubulins, this group also found α-tubulin, β-tubulin, γ-tubulin, δ-tubulin, and ε-tubulin. Among γ-tubulin Complex proteins were γ-tubulin complex component 2 (GCP-2) and γ-tubulin complex component (Spc97/Spc98 homologues). Finally, the study also found Microtubule Decoration proteins (CLASP homologues, MAP1 light chain and MAP-65 homologue Radial Spoke Protein homologue, Outer Dynein Arm-Docking Complex), Microtubule Capping proteins (EB family and CLIP-170), and Microtubule-Severing proteins (Katanin p60 subunit and Katanin p80 subunit).

Acknowledgements This work was supported by the Conselho Nacional de Desenvolvimento Científico e Tecnológico (CNPq), Fundação Carlos Chagas Filho de Amparo a Pesquisa do Estado do Rio de Janeiro (FAPERJ), Programa de Núcleos de Excelência (PRONEX), Coordenação de Aperfeiçoamento de Pessoal de Ensino Superior (CAPES), and Associação Universitária Santa Úrsula (AUSU).

References

Amos WB, Grimstone AV, Tothschild LJ (1979) Structure, protein composition and birefringence of the costa: a motile flagellar root fibre in the flagellate *Trichomonas*. J Cell Sci 35:139–164

Batista CM, Benchimol M, Cunha E, Silva NL, De Souza W (1988) Localization of acetylated alpha-tubulin in *Tritrichomonas foetus* and *Trichomonas vaginalis*. Cell Struct Funct 13:445–453

Benchimol M (2004) Trichomonads under microscopy. Microsc Microanal 10:1–23

Benchimol M (2005) New ultrastructural observations on the skeletal matrix of *Tritrichomonas foetus*. Parasitol Res 97:408–416

Benchimol M (in press) Cytoskeleton in Trichomonads. Trends Cell Mol Biol

Benchimol M, de Souza W (1987) Structural analysis of the cytoskeleton of *Tritrichomonas foetus*. J Submicrosc Cytol 19:139–147

Benchimol M, Elias CA, De Souza W (1981) Specializations in the flagellar membrane to *Tritrichomonas foetus*. J Parasitol 67:174–178

Benchimol M, Elias CA, De Souza W (1982a) *Tritrichomonas foetus*: ultrastructural localization of basic proteins and carbohydrates. Exp Parasitol 54:135–144

Benchimol M, Elias CA, De Souza W (1982b) *Tritrichomonas foetus*: fine structure of freeze-fractured membranes. J Protozool 29:348–353

Benchimol M, Kachar B, De Souza W (1992) Surface domains in the pathogenic protozoan *Tritrichomonas foetus*. J Protozool 34:480–484

Benchimol M, Kachar B, De Souza W (1993) The structural organization of the pathogenic protozoan *Tritrichomonas foetus* as seen in replicas of quick-frozen, freeze-fractured and deep etched cells. Biol Cell 77:289–295

Benchimol M, Picanço Diniz JA, Ribeiro KC (2000) Fine structure of the axostyle and its associations with other cell organelles in *Trichomonads*. Tissue Cell 32:178–187

Boggild AK, Sundermann CA, Estridge BH (2002) Post-translational glutamylation and tyrosination in tubulin of *Tritrichomonas* and the diplomonad *Giardia intestinalis*. Parasitol Res 88:58–62

Bricheux G, Coffe G, Bayle D, Brugerolle G (2000) Characterization, cloning and immunolocalization of a coronin homologue in *Trichomonas vaginalis*. Eur J Cell Biol 79:413–422

Bricheux G, Coffe G, Brugerolle G (2007) Identification of a new protein in the centrosome-like "atractophore" of *Trichomonas vaginalis*. Mol Biochem Parasitol 153:133–140

Brugerolle G (1975) Étude de la cryptopleuromitose et de la morphogenése de division chez plusiers Genres de la trichomonadines primitives. Protistologica 4:457–468

Brugerolle G, Bricheux G, Coffe G (2000) Centrin protein and genes in *Trichomonas vaginalis* and close relatives. J Eukaryot Microbiol 47:129–138

Carlton JM, Hirt RP, Silva JC, Delcher AL, Schatz M, Zhao Q, Wortman JR, Bidwell SL, Alsmark UC, Besteiro S, Sicheritz-Ponten T, Noel CJ, Dacks JB, Foster PG, Simillion C, Van de Peer Y, Miranda-Saavedra D, Barton GJ, Westrop GD, Muller S, Dessi D, Fiori PL, Ren Q, Paulsen I, Zhang H, Bastida-Corcuera FD, Simoes-Barbosa A, Brown MT, Hayes RD, Mukherjee M, Okumura Y, Schneider R, Smith AJ, Vanacova S, Villalvazo M, Haas BJ, Pertea M, Feldblyum TV, Utterback TR, Shu CL, Osoegawa K, de Jong PJ, Hrdy I, Horvathova L, Zubacova Z, Dolezal P, Malik SB, Logsdon JM Jr, Henze K, Gupta A, Wang CC, Dunne RL, Upcroft JA, Upcroft P, White O, Salzberg SL, Tang P, Chiu CH, Lee YS, Embley TM, Coombs GH, Mottram JC, Tachezy J, Fraser-Liggett CM, Johnson PJ (2007) Draft genome sequence of the sexually transmitted pathogen *Trichomonas vaginalis*. Science 315:207–212

Cavalier-Smith T, Chao EE (1996) Molecular phylogeny of the free-living archeozoan *Trepomonas agilis* and the nature of the first eukaryote. J Mol Evol 43:551–562

Delgado-Viscogliosi P, Brugerolle G, Viscogliosi E (1996) Tubulin post-translational modifications in the primitive protist *Trichomonas vaginalis*. Cell Motil Cytoskeleton 33:288–297

Embley TM, Hirt RP (1998) Early branching eukaryotes? Curr Opin Genet Dev 8:624–629

Granger BL, Warwood SJ, Benchimol M, De Souza W (2000) Transient invagination of flagella by *Tritrichomonas foetus*. Parasitol Res 86:699–709

Honigberg MB (1978) Introduction. In: Honigberg BM (ed) Trichomonads of veterinary importance in parasitic Protozoa. Academic, New York, pp 164–273

Honigberg MB, Brugerolle G (1990) Structure. In: Honigberg BM (ed) Trichomonads parasitic in human. Springer, New York, pp 5–35

Honigberg MB, Mattern CF, Daniel WA (1971) Fine structure of the mastigont system in *Tritrichomonas foetus*. J Protozool 18:183–198

Juliano C, Rubino S, Zicconi D, Cappuccinelli P (1986) An immunofluorescent study of the microtubule organization in *Trichomonas vaginalis* using anti-tubulin antibodies. J Protozool 33:56–59

Lee KE, Kim JH, Jung MK, Arii T, Ryu J, Han SS (2009) Three-dimensional structure of the cytoskeleton in *Trichomonas vaginalis* revealed new features. J Electron Microsc 58:305–313. doi:10.1093/jmicro/dfp019

Lopes LC, Ribeiro KC, Benchimol M (2001) Immunolocalization of tubulin isoforms in the protists *Tritrichomonas foetus* and *Trichomonas vaginalis*. Histochem Cell Biol 116:17–29

Madeiro da Costa RF, Benchimol M (2004) The effect of drugs in *Tritrichomonas foetus*. Parasitol Res 92:159–170

Mattos A, Solé-Cava AM, DeCarli G, Benchimol M (1997) Fine structure and isozymic characterization of trichomonadid protozoa. Parasitol Res 83:290–295

Monteiro Leal LH, Cunha-e-Silva NL, Benchimol M, De Souza W (1993) Isolation and biochemical characterization of the Costa of *Tritrichomonas foetus*. Eur J Cell Biol 60:235–242

Monteiro-Leal LH, Farina M, Benchimol M, Kachar B, De Souza W (1995) Coordinated flagellar and ciliary beating in the protozoon *Tritrichomonas foetus*. J Eukaryot Microbiol 42:709–714

Monteiro-Leal LH, Farina M, De Souza W (1996) Free movement of *Tritrichomonas foetus* in a liquid medium: a video-microscopy study. Cell Motil Cytoskeleton 34:206–214

Pereira-Neves A, Benchimol M (2009) *Tritrichomonas foetus*: budding from multinucleated pseudocysts. Protist 160:536–551

Ribeiro KC, Monteiro-Leal LH, Benchimol M (2000) Contributions of the axostyle and flagella to closed mitosis in the protists *Tritrichomonas foetus* and *Trichomonas vaginalis*. J Euk Microbiol 47:481–492

Ribeiro KC, Pereira-Neves A, Benchimol M (2002) The mitotic spindle and associated membranes in the closed mitosis of trichomonads. Biol Cell 94:157–172

Schneider A, Plessmann U, Felleisen R, Weber A (1999) Alpha-tubulins of *Tritrichomonas mobilensis* are encoded by multiple genes and are not post-translationally tyrosinated. Parasitol Res 85:246–248

Silva-Filho FC, De Souza W (1986) Effect of colchicine, vinblastine, and cytochalasin B on cell surface anionic sites of *Tritrichomonas foetus*. J Protozool 33:6–10

Tachezy J, Tachezy R, Hampl V, Sedinová M, Vanacová S, Vrlík M, Van Ranst M, Flegr J, Kulda J (2002) Cattle pathogen *Tritrichomonas foetus* (Riedmüller, 1928) and pig commensal *Tritrichomonas suis* (Gruby and Delafond, 1843) belong to the same species. J Eukaryot Microbiol 49:154–163

Viscogliosi E, Brugerolle G (1994) Cytoskeleton in trichomonads. III Study of the morphogenesis during division by using monoclonal antibodies against cytoskeletal structures. Eur J Protistol 30:129–138

World Health Organization (2001) Trichomoniaisis. Global prevalence and incidence of selected curable sexually transmitted infections. WHO, Geneva, Switzerland

Subpellicular Microtubules in Apicomplexa and Trypanosomatids

Wanderley de Souza and Marcia Attias

Contents

W. de Souza (✉) and M. Attias
Laboratório de Ultraestrutura Celular Hertha Meyer, Instituto de Biofísica Carlos Chagas Filho, Universidade Federal do Rio de Janeiro, CCS-Bloco G, Ilha do Fundão, Rio de Janeiro 21941-900 Brazil
e-mail: wsouza@biof.ufrj.br

W. de Souza (ed.), *Structures and Organelles in Pathogenic Protists*,
Microbiology Monographs 17, DOI 10.1007/978-3-642-12863-9_2,

© Springer-Verlag Berlin Heidelberg 2010

Abstract The cytoskeleton plays a fundamental role in various processes such as the establishment of cell shape, cell locomotion, and the intracellular motility of various structures found in eukaryotic cells. Microtubules are among the most conspicuous structures in the cytoskeleton. They can be found free in the cytoplasm, forming the mitotic spindle or assembled in various structures. A special type of microtubule arrangement is found in some protozoa, whereby they are organized as a single layer located immediately below the plasma membrane, constituting what is generally referred to as subpellicular microtubules. This special array of microtubules is found in members of the Kinetoplastida family and in the Apicomplexa phylum. Here, we review basic aspects of subpellicular microtubules, emphasizing their visualization as a whole network, their structural organization, their heterogeneity as analyzed using an immunocytochemical approach, some of their biochemical properties, and their sensitivity to drugs, as well as their functional role.

1 Introduction

Microtubules can be defined as a macromolecular complex in which the major proteins (tubulins) associate with each other, building a polymeric structure that interacts with minor proteins; these, in turn, control the stability and the dynamic assembly–disassembly process of microtubules, modulating their functional properties (Fig. 1) (Reviewed in Li and Gundersen 2008). The length of the polymer, which in some cases is essential for its function, is regulated by the addition and/or loss of tubulin subunits at the plus and minus ends of the microtubule, respectively, and breakage of the polymers by severing enzymes such as katanin, spastin, and fidgetin (Casanova et al. 2009).

In most eukaryotic cells, the microtubules appear as individualized structures. However, they may associate with each other and to other cell structures to form more complex assemblies. Cilia and flagella, found in many cell types in organisms ranging from protists to mammalians, are two very representative and well-known examples of such structures. However, protists' microtubules can assemble in a wider variety of patterns and form unique and complex structures. Examples include the pelta–axostylar system found in the members of the Trichomonadidae family (see chapter "The Mastigont System in Trichomonads" by Benchimol), the adhesive disk of *Giardia*, and the subpellicular microtubules found in Apicomplexa and Kinetoplastida.

This chapter will deal with subpellicular microtubules. These are found in two groups of protists of high relevance, as they include the agents of some very important human and veterinary diseases: Kinetoplastida and Apicomplexa.

The Apicomplexa phylum comprises a large and heterogeneous group of species that are distributed worldwide and in a variety of environments. Many of them are agents of important diseases affecting humans, as is the case for the agents of malaria (*Plasmodium* genus) and toxoplasmosis (*Toxoplasma gondii*). Others cause disease in animals of economic interest (*Eimeria* in chickens and cattle, *Theileria* in

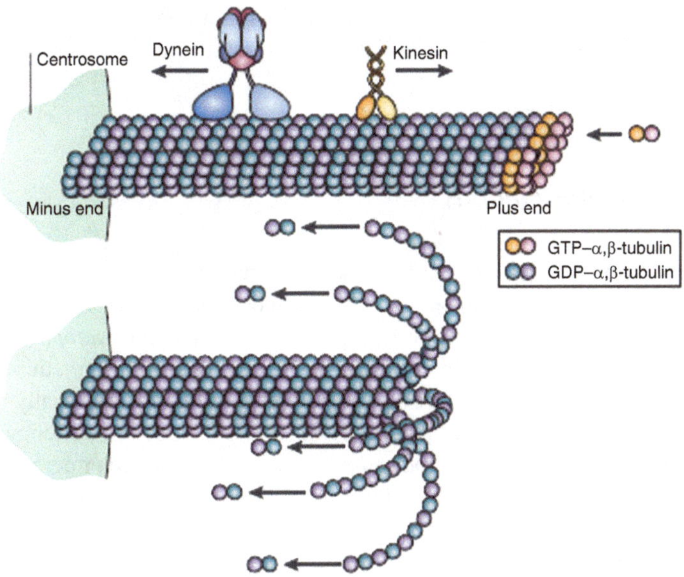

Fig. 1 Schematic representation of the main features of microtubules: heterodimers of αβ-tubulin are assembled from the centrosome, or MTOC, with the minus ends toward it and the plus ends to the cell periphery. Motor proteins may use microtubules as tracks to transport molecules or organelles in specific directions: toward the centrosome, in the case of dyneins, or to the cell periphery, for kinesins. The microtubule grows as cytoplasmic dimmers of αβ-tubulin associated with GTP are added to the plus ends. After assembly, GTP is hydrolyzed to GPD weakening the microtubule and favoring its quick depolymerization. This alternation between growth (*upper panel*) and shrinkage (*bottom*) is known as dynamic instability (Adapted from Li and Gundersen 2008 with permission)

cattle). The Apicomplexa phylum also includes *Cryptosporidium*, a parasite that infects animals and is also an opportunistic parasite of humans, and nonpathogenic protozoa such as *Gregarina*. Among the pathogenic species, some present a life cycle that involves a single animal species. Others, as is the case of *T. gondii*, have a more complex life cycle involving certain animal species in which only the asexual cycle takes place and other species in which the life cycle is complete. Still others, as is the case of *Plasmodium*, have an even more complex life cycle involving vertebrate and invertebrate (insect) hosts.

The Kinetoplastidae order comprises a large number of species, some of which are the causative agents of highly prevalent diseases in humans, such as leishmaniasis, Chagas' disease and sleeping sickness. Species of the genus *Phytomonas* also affect plants of economic interest, such as coconut, oil palm, and cassava.

2 Subpellicular Microtubules

The basic scaffold that determines the cell shape in several protists, including Trypanosomatids and Apicomplexans, is composed of parallel rows of microtubules that run under the pellicle – the subpellicular microtubules. This and the fact that in both cases, tubulin dimmers are the main constituents are two of the few characteristics shared by subpelicular microtubules of Trypanosomatids and Apicomplexa.

2.1 In Trypanosomatids

Trypanosomatids belong to the order Kinetoplastida – that comprises the suborders Bodonina and Trypanosomatina. The first includes both parasitic and free-living genera, such as *Cryptobia* and *Bodo*, respectively. Although *Cryptobia* species are parasites of salmonid fishes with considerable impact on fish-farming, this group has not received much attention so far. However, Trypanosomatids, especially the trypanosomes, were already notorious in the early history of electron microscopy. In 1950, Hertha Meyer started to study *Trypanosoma cruzi* with an electron microscope in the laboratory of Keith Porter at the Rockefeller Institute in New York (Meyer and Porter 1954; Review in De Souza 2008). At that time, it was not possible to obtain thin sections, although many groups were trying to adapt a conventional microtome to cut thinner sections. The first attempts used tissue cultures infected with *T. cruzi* or culture forms of the protozoan. The samples were fixed with osmium tetroxide and washed several times, and then one drop of the cell suspension was placed on a copper grid previously coated with Parlodion and allowed to dry in dust-protected dishes. The first results obtained were rather disappointing. In spite of the small size of the flagellates, much smaller than a tissue cell, nothing could be seen of their inner structure. Only at the periphery was the cell thin enough to be traversed by the electron beam, revealing fine striations in a parallel array in the cytoplasm (Meyer and Porter 1954). Trypsinization or prolonged fixation with osmium tetroxide destroyed this fine striation in the periphery of the cell as well as the fiber bundle of the flagellum. The above-mentioned striations correspond to what are now known as subpellicular microtubules, the major component of the cytoskeleton of trypanosomatids. Subsequently, with the methodological improvements in the fixation of biological specimens for electron microscopy, these microtubules were better visualized, especially after the introduction of glutaraldehyde as a fixative.

Transmission electron microscopy of thin sections showed that, in all species, the microtubules have a diameter of around 24 nm and they keep a fixed distance of about 44 nm between each other and about 15 nm with the plasma membrane (Fig. 2). A detailed study of *T. cruzi* showed that the number of microtubules varied according to the regions of the protozoan's body. In trypomastigote forms, the maximum number of microtubules found was 120 in the region where the Golgi complex is located (Meyer and De Souza 1976). At the cell extremities, as few as 40 microtubules were counted. In the proliferative amastigote forms, the highest number of microtubules found was 222, observed in dividing cells. This observation indicates that there is an increase in the number of microtubules during the cell division process, and suggests that the assemblage of microtubules is organized and controlled in such a way that, as the volume of the cell increases, new microtubules

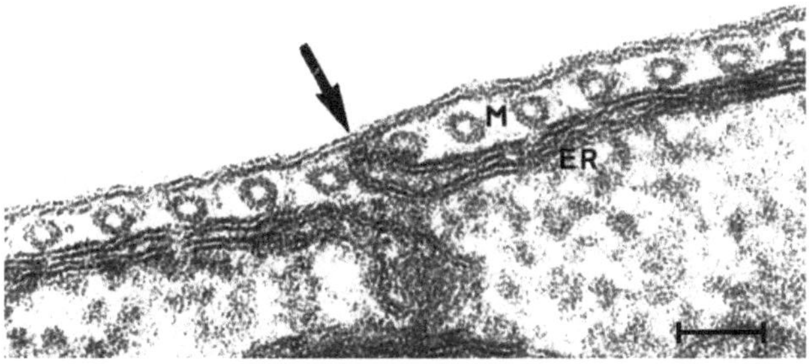

Fig. 2 Cross-section of subpellicular microtubules of *Leishmania mexicana amazonensis*. The *arrow* points to a profile of the endoplasmic reticulum eventually inserting between the regularly spaced microtubules (from Pimenta and De Souza 1985 with permission)

are inserted between previously existing ones in such a way that the distance between the microtubules remains constant. This implies the existence of a mechanism that controls the assembly of new microtubules in close connection with increases in the diameter of the cell, probably in response to the insertion of new plasma membrane components.

As in other organisms, α/β-tubulin heterodimers are the major proteins found in subpellicular microtubules (Bordier et al. 1982; Stieger et al. 1984). COOH-terminal tyrosinilation of α-tubulin has been shown to take place in *T. brucei* (Stieger et al. 1984). In these organisms, the tubulin genes are arranged in a cluster or 13–18 tandemly repeated alternating α/β pairs, although variations may occur (Reviewed in Kohl and Gull 1998).

In the Bodonina suborder, besides the presence of two flagella and a conspicuous set of bended microtubules that constitute the preoral ridge, a set of subpellicular microtubules is also present. However, as observed in detergent-extracted and critical point-dried cells, subpellicular microtubules in these organisms only partially encircle the cell body, forming two subsets of 22-nm thick microtubules: one that partially surrounds the cell body, and another that remains attached to the recurrent flagellum after extraction (Attias et al. 1996).

2.2 In Apicomplexans

Subpellicular microtubules of Apicomplexa were observed for the first time as early as 1942 by Emmel et al. in *Plasmodium*. Then, Bringmann and Holz (1953) prepared whole mounts of lysed parasites in which subpellicular microtubules could be visualized by electron microscopy. Another attempt to study the ultrastructure of *T. gondii* with a transmission electron microscope was made by Meyer and Andrade Mendonça (1955). At that time, ultrathin sectioning was not available and whole tachyzoites fixed with osmium tetroxide were too thick to

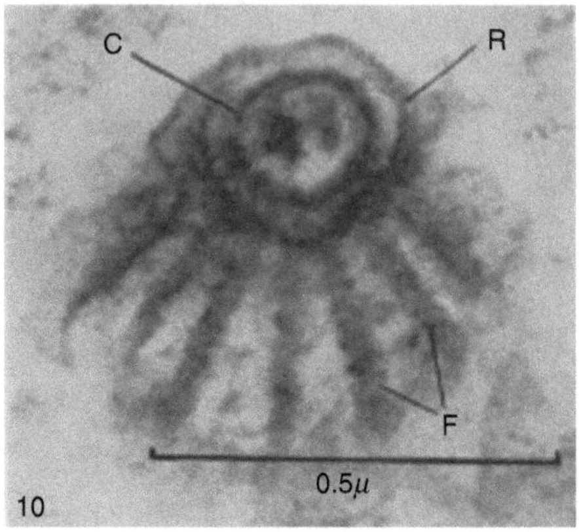

Fig. 3 Ultrathin section of the apical portion of *Toxoplasma gondii* showing the subpellicular microtubules (F) arising from the apical polar ring (R) and, inside it, the conoid (c) (from Garnham et al. 1962 with permission)

allow observation of the inner structure. Later on, in 1957, the same authors published the first electron micrographs of ultrathin sections of *T. gondii* in tissue cultures, but the first mention of subpellicular microtubules, then called "peripheral fibrils," is found in a paper by Garnham et al. (1962) (Fig. 3). "Surface fibrils" were also described by McLaren and Paget (1968) in *Eimeria tenella* and by Sheffield and Melton (1968) in *T. gondii*. The role of the subpellicular microtubules in maintaining the shape of these protists was first proposed by Kikkawa and Gueft (1964). A few years later, in 1972, several publications already made references to the subpellicular microtubules of *T. gondii,* based on observations made on thin sections (Jones et al. 1972; Vivier and Petitprez 1972) or by negative staining of tachyzoites (De Souza 1972).

Apicomplexa zoites lack microtubule-built locomotory structures such as flagella, although these are found in microgametocytes. Despite this absence of flagella, zoites of the Apicomplexa phylum – merozoites, trophozoites, tachyzoites, bradizoites, and sporozoites – are highly motile and rapidly penetrate into host cells. Active invasion depends on parasite motility and the sequential secretion of products from the micronemes and rhoptries, club-shaped secretory organelles that discharge their products through the apical portion of the parasite. Sequential secretion by micronemes and rhoptries and the coordinated movement of the conoid promote the attachment of the infective form to the surface of the host cell and the subsequent invasion of the host cell, where the parasite multiplies within a parasitophorous vacuole (Reviews in De Souza 2006; Ravindran and Boothroyd 2008).

3 Visualization of the Whole Network of Subpellicular Microtubules

Only by electron microscopy techniques, it became possible to observe subpellicular microtubules. However, very early, before thin sectioning and efficient chemical fixation became available, it was possible to observe them in ruptured or detergent-extracted samples negatively stained. Soon, its presence was described in all genera and evolutive forms of Trypanosomatids and Apicomplexans.

3.1 *In Trypanosomatids*

Since the first observations of thin sections, it became clear that subpellicular microtubules are present throughout the trypanosomatid cell body except at the flagellar pocket region, where at one point the microtubule corset is replaced by a flagellum attachment zone (FAZ) formed by a filamentous structure. This structure is also associated with a group of three to four microtubules that are closely associated with the cisternae of the endoplasmic reticulum and originate close to the basal bodies. These microtubules differ from subpellicular microtubules because they are not depolymerized when the cells are incubated in a medium with high salt concentrations (Sherwin and Gull 1989). Additionally, they are the only microtubules stained with the monoclonal antibody 1B41, which recognizes β-tubulin (Gallo and Precigout 1988). These subpellicular microtubules are also found in structures such as the cytostome–cytopharynx complex, a deep invagination of the plasma membrane that may reach the proximities of the nuclear region.

Pioneering attempts to observe the distribution of subpellicular microtubules in whole cells were made by Angelopoulos (1970), who ruptured protozoan cells using surface tension spreading followed by critical point drying (Fig. 4). The second attempt was carried out by De Souza and Benchimol (1984), whose method consisted of adhering the protozoan to a formvar/carbon-coated grid previously coated with poly-L-lysine, followed by a light extraction with Triton-X 100, critical point drying, and observation under a high-voltage electron microscope (1,000 KV) installed at the University of Colorado, Boulder. The extraction of the membrane did not interfere with the general shape of the cell and maintained the integrity of the subpellicular microtubules (De Souza and Benchimol 1984). The third attempt was by Gull and coworkers, who used negative staining to obtain a sharp visualization of microtubule arrangement (Fig. 5) in *Trypanosoma brucei* (Gull 1999). A fourth approach used field emission scanning electron microscopy of detergent-extracted cells (Fig. 6). It is important to point out that none of these methods revealed the presence of a typical microtubule-nucleating center from which the microtubules emerge.

3.2 *In Apicomplexans*

In Apicomplexans, the presence of microtubules located below the inner membrane complex was noticed since the first observations made in thin sections of well-preserved tachyzoites of *T. gondii* (Garnham et al. 1962; Sheffield and Melton 1968). However, a better view of the extension of this system was obtained when

Fig. 4 Distribution of the subpellicular microtubules in whole cells of *Crithidia fasciculata* ruptured and critical point dried. (mc) microtubule curve, (f) flagellum, (pa) posterior apex. *Bar*: 1 μm (from Angelopoulos 1970 with permission)

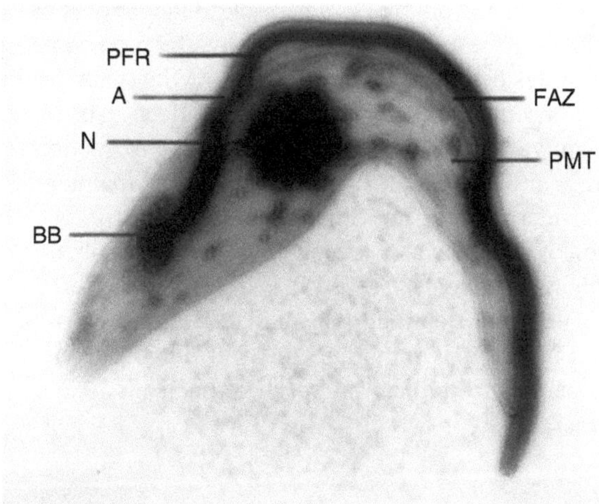

Fig. 5 Negative staining of *Trypanosoma brucei* cytoskeleton after detergent extraction. (FAZ) flagellar attachment zone, (BB) basal body, (PFR) paraflagellar rod, (PMT) subpellicular microtubules, (A) axonema and (N) nucleus (from Gull 1999, with permission)

the whole tachyzoite was slightly disrupted by rinsing with distilled water, deposited on formvar- and carbon-coated grids, negatively stained with phosphotungstic acid, and observed under an electron microscope (Fig. 7) (De Souza 1972;

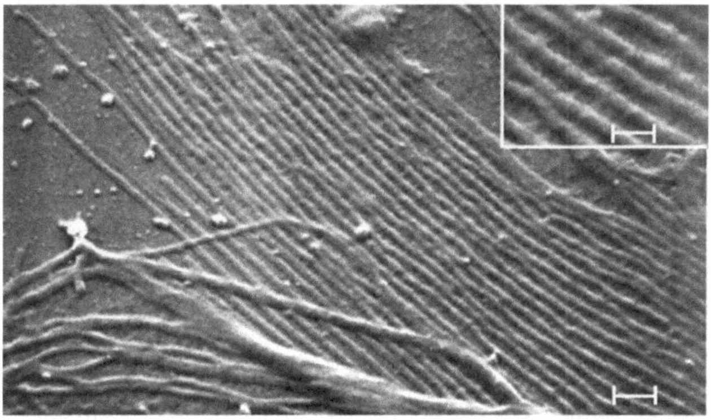

Fig. 6 Subpellicular microtubules of a kinetoplastid observed by field emission scanning electron microscopy after detergent extraction and critical point drying. Spacing of microtubules of various lengths is kept by regularly spaced bridges. *Bar* = 300 nm. *Inset*= 600 nm

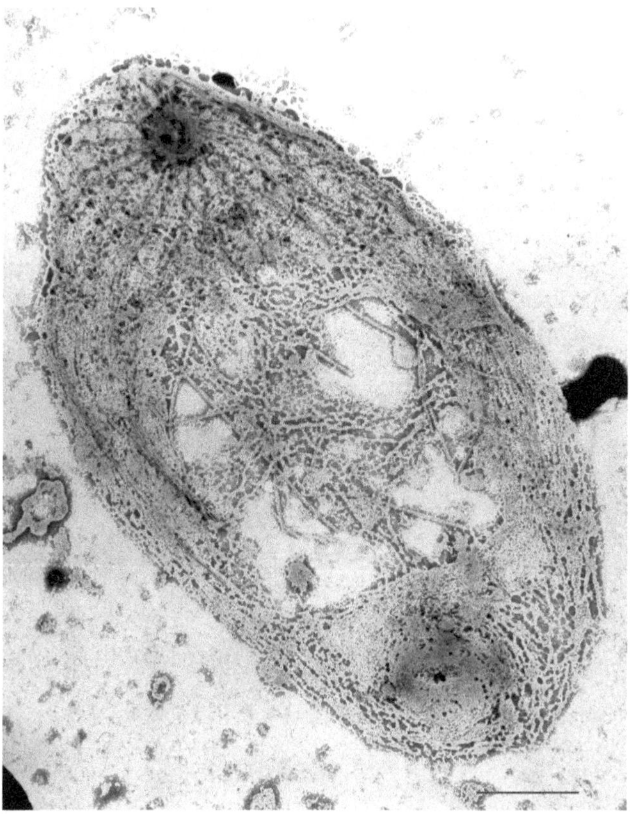

Fig. 7 Negative staining of a tachyzoite after membrane extraction. The 22 subpellicular microtubules around the posterior polar ring can be individualized. The central pair of microtubules is shown by the *white arrowhead*. *Bar* = 1 μm

Vivier and Petitprez 1972), as previously done with sporozoites of *Eimeria ninakohlyakimovae* (Roberts and Hammond 1970). Using this simple approach, it was shown that the subpellicular network remains intact and that the microtubules radiate from a polar ring located at the base of the conoid, a structure also formed by microtubules (De Souza 1972, 1974). Two short intraconoidal microtubules and 22 subpellicular microtubules were observed in all *T. gondii* tachyzoites. In addition, it was shown that the microtubules ended freely, reaching about 70% of the cell length toward the posterior region of the cell. These initial observations were confirmed and extended to other members of the Apicomplexa phylum, with some differences in the number of subpellicular microtubules. For instance, 24 microtubules were counted in both *E. ninakohlyakimovae* (Roberts and Hammond 1970) and *Eimeria acervulina* (Russel and Burns 1984), 15–16 in *Plasmodium berghei* sporozoites (Vanderberg et al. 1967), and 22 in *Sarcocystis neurona* (Speer and Dubey 2001). Subpellicular microtubules are usually evenly distributed around the apical ring, but in *P. falciparum* they form a band instead. It was later shown that the basal polar ring constitutes a microtubule-organizing center (MTOC) from which the microtubules originate, projecting toward the posterior region, which corresponds to the plus end where new tubulin dimers are added for microtubule growth (Russel and Burns 1984). Several genera of Apicomplexa have two other structures constituted by tubulin: the conoid and the microtubules that constitute the mitotic apparatus (Review in Morrissette and Sibley 2002); the conoid will be considered further in this chapter but the mitotic spindle is not within the scope of the present review. Field emission scanning electron microscopy of *T. gondii* tachyzoites, whose membrane pellicle was extracted with detergent before fixation, corroborates the uniform distribution of subpellicular microtubules around the polar ring (Fig. 8). In its resting position, the conoid lies below the polar ring and is surrounded by subpellicular microtubules, making its observation rather difficult; however, in the extruded state, it rises above the polar ring, and the helical assembly of its microtubules can be observed.

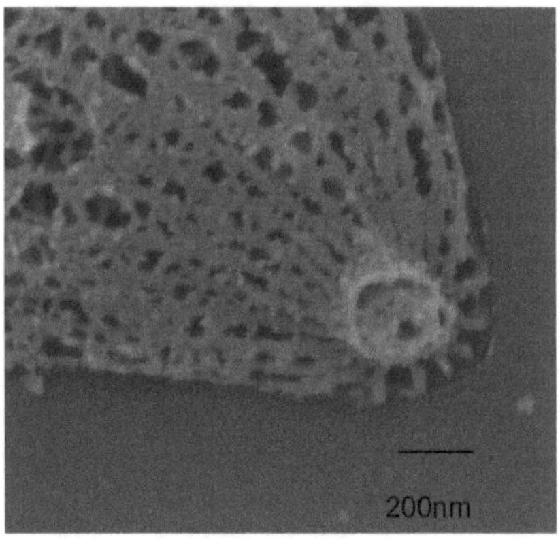

Fig. 8 Field emission scanning electron micrograph of the anterior portion of *Toxoplasma gondii* after extraction with detergent and critical point drying. The conoid arises from the inner side of the polar ring to which subpellicular microtubules are attached. *Bar*: 200 nm

4 High-Resolution Images of the Microtubules

High-resolution images of microtubules were reported as early as in the 1970s. However, in recent years, improvements both in fixation techniques, as freeze-substitution, and in microscopic resources, as electron tomography, brought new insights and information into the constitution of these structures, revealing peculiarities that are neither found in higher eukaryotes nor shared between Trypanosomatids and Apicomplexans.

4.1 In Trypanosomatids

Four approaches were used to obtain high-resolution images of subpellicular microtubules. The first approach was based on previous results by Mizuhira and Futeasaku (1972), who showed that the addition of tannic acid to the glutaraldehyde solution significantly improved the preservation of the microtubules. In addition, some microtubules appeared as though they had been negatively stained so that their internal structure could be visualized. Using this technique, it was observed that the subpellicular microtubules of trypanosomatids are made up of 13 typical protofilaments (Soares and De Souza 1977) (Fig. 9). However, this technique did not reveal further details on the associations of the microtubules with each other or with the plasma membrane. The second approach used quick-freezing, followed by freeze-fracture, deep-etching, and rotary replication (Souto-Padron et al. 1984), resulting in well-preserved microtubules. Indeed, longitudinal views of the microtubules clearly revealed their internal structures (Fig. 10) and showed the presence of filaments that connect the microtubules to each other. It was also clear that these connections are regularly spaced. In addition, the connections of the microtubules with the plasma membrane, the endoplasmic reticulum, and the reservosomes were all visualized. We still do not know the nature of these filaments. Although proteins associated with subpellicular microtubules have been identified, we still have not elucidated a clear topological relationship between structure and composition. The third approach analyzed microtubule structure by optical diffraction and revealed a spacing of 5 nm between protofilaments and a 4-nm axial periodicity corresponding to the tubulin subunits, which were also in freeze-etching replicas. The pith of the shallow left-hand three-start helix is 12°. The fourth approach used the fracture-flip technique, allowing the visualization, at high resolution, of the actual inner surface of the *Leishmania* plasma membrane (Fig. 11). The subpellicular microtubules were seen clearly. Treatment of the sample with trypsin led to the disappearance of the microtubules, leaving behind parallel arrays of particles that may correspond to proteins that link the microtubules to the plasma membrane (Hou et al. 1992).

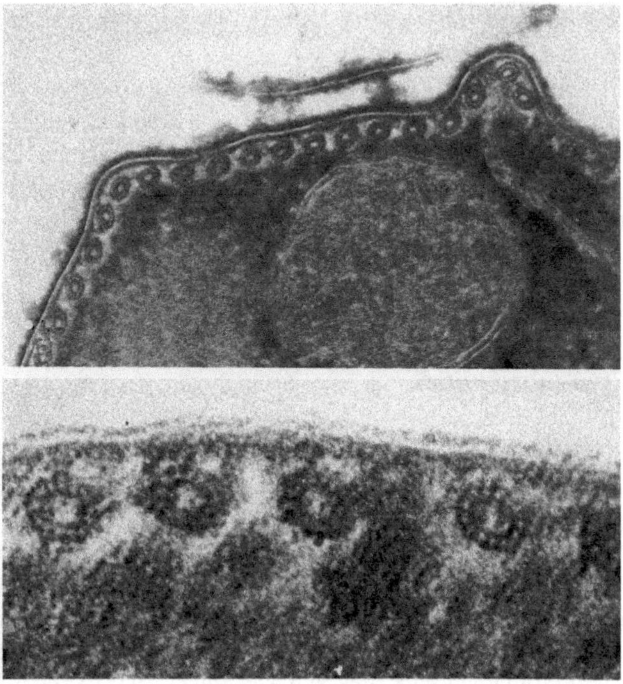

Fig. 9 Cross-section of subpellicular microtubules of *Leptomonas samueli* fixed with glutaraldehyde and tannic acid. (**a**) General view of the assembly of microtubules under the plasma membrane. (**b**) The 13 tubulin subunits of each microtubule can be counted, as well as connections between the microtubules and the plasma membrane. *Bars*: (**a**) 100 nm, (**b**) 20 nm (From Soares and De Souza 1977 with permission)

4.2 In Apicomplexa

As mentioned above, the subpellicular microtubules of Apicomplexa show striations along their length (Fig. 12) (De Souza 1974). Such striations were better visualized following the application of Fourier analysis techniques to the images of isolated, frozen-hydrated microtubules in which a 32-nm periodicity was observed (Fig. 13) (Morrissete et al. 1997). This suggested that the subpellicular microtubules are decorated along their length with a MAP that could be responsible for their great stability under high pressure, cold and treatment with detergents typically used to isolate them. A similar periodicity was also observed in rows of intramembranous particles observed in freeze-fracture replicas of the inner pellicular membrane of *Toxoplasma* that overlaid the subpellicular microtubules (Fig. 14). That was the first undisputable evidence that subpellicular microtubules are linked to the pellicle and could be associated with parasite motility.

It is interesting to point out that Russel and Burns (1984) had been able to see that subpellicular microtubules interact side-on, rather than end-on, with the polar

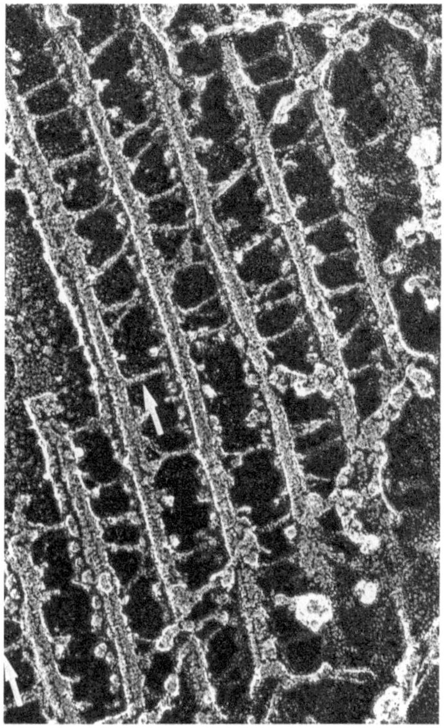

Fig. 10 Replica of quick-freeze, deep-etched, rotary-shadowed epimastigote form of *Trypanosoma cruzi* showing the subpellicular microtubules and the filamentous bridges connecting them to each other side by side

ring, an observation corroborated with the use of cryo-electron tomography (Cyrklaff et al. 2007) (Fig. 15). Besides confirming the lateral association between subpellicular microtubules and the polar ring, cryoelectrontomography revealed that the microtubules on the cell walls of *Plasmodium berghei* sporozoites are extended at the luminal side by an additional 3 nm in comparison to the microtubules of mammalian cells (Fig. 15). Furthermore, Fourier analysis revealed an 8-nm longitudinal periodicity of the luminal constituent, suggesting the presence of a molecule interacting with tubulin dimers, probably stabilizing them. Subpellicular microtubules of *T. gondii* tachyzoites shared the same features but microtubules from pig brain cells did not, indicating a conserved pattern among Apicomplexans.

5 Immunocytochemical Characterization of the Microtubules

Immunocytochemistry arrived as a powerful tool for the identification and localization of several components of the cell. It has been largely applied both for the identification of subtypes of tubulin and for the localization of other proteins associated with it that participate in the maintenance of the assembly and motion of the complex.

Fig. 11 The inner surface of the plasma membrane of *Leishmania major* as well as of a portion of the flagellum (*asterisk*) showing the subpellicular microtubules as revealed by the fracture-flip technique. *Bar*, 250 nm (from Hou et al. 1992)

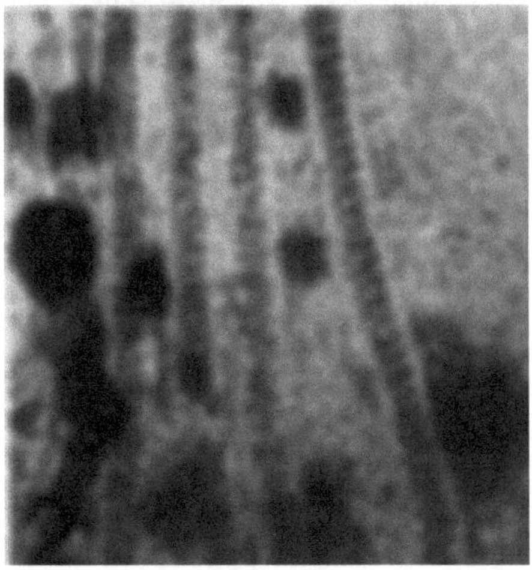

Fig. 12 Striations along the subpellicular microtubules of *Toxoplasma gondii* observed in negatively stained tachyzoites (from De Souza 1974 with permission)

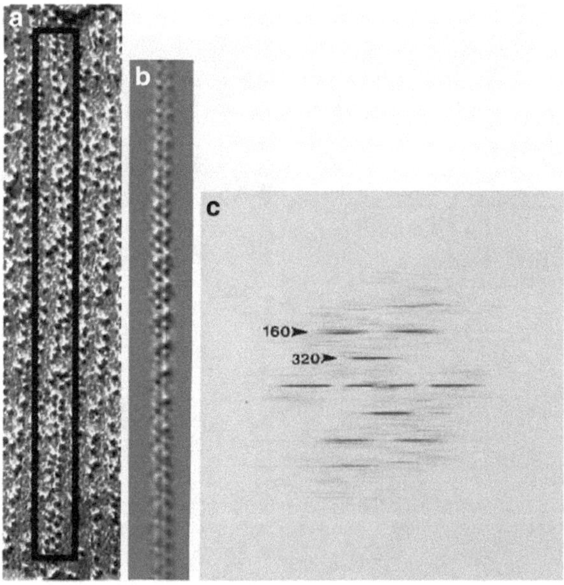

Fig. 13 Analysis of longitudinal double particle rows from freeze-fractured samples (**a**) provides a calculated diffraction pattern (C), which exhibits a pronounced 320 Å periodicity, corresponding to alternate particles in the rows. (**b**) A line-filtered reconstruction of the double particle rows, generated from information found on the strong layer lines. Note that the particles in adjacent rows are pitched relative to row length. When areas of the fracture face containing both single and double rows of IMPs are used to generate computed diffraction patterns (from Morrissette et al. 1997 with permission)

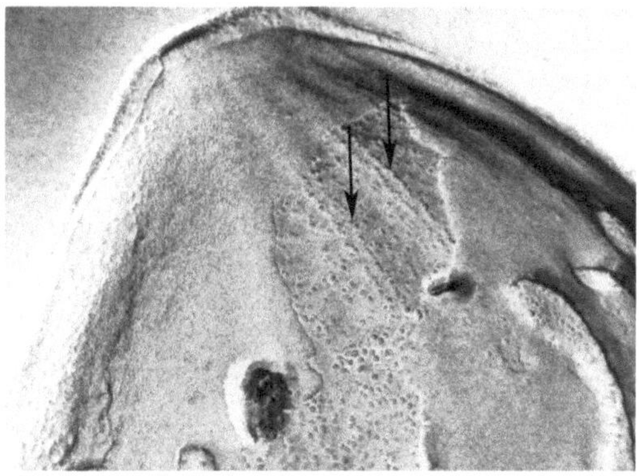

Fig. 14 Freeze-fracture of *Toxoplasma gondii* tachyzoites. P face of the inner pellicular membrane showing the double rows of intramembranous particles positioned according to the subpellicular microtubules (from Cintra and De Souza, 1985a, with permission)

5.1 In Trypanosomatids

Monoclonal antibodies that recognize several tubulin isotypes have been used to characterize the subpellicular microtubules of trypanosomatids. Multiple tubulin genes exist in trypanosomes (Seebeck and Gehr 1983) but the mRNAs of only one

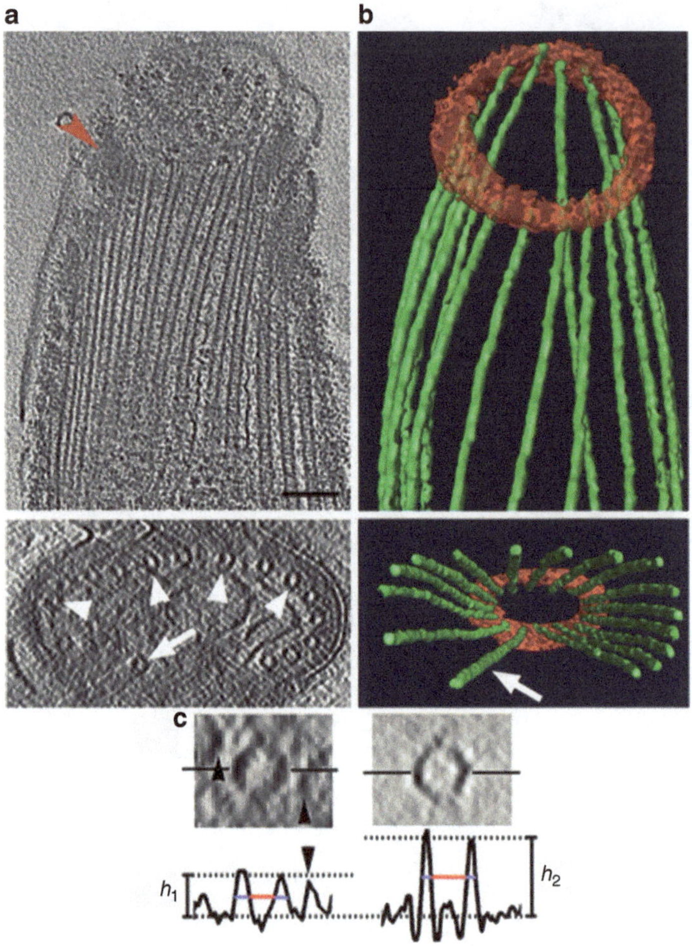

Fig. 15 (**a**) Longitudinal (*top*) and cross (*bottom*) slices through a tomographic reconstruction from the apical part of a *Plasmodium* sporozoite revealing subpellicular microtubules. *Red arrowhead*, polar ring; *white arrowheads*, circumferentially arranged microtubules; *arrow*, the lone microtubule. (**b**) Volume-rendered representations of the microtubules (*green*) and the polar ring (*red*) from the same tomogram. (**c**) Cross-sections of microtubules from a sporozoite (*left*) and from in vitro polymerization (*right*). *Arrowheads* indicate neighboring microtubules. Density distributions along the *horizontal lines* are displayed in the graphs. *Dotted lines* indicate background and maximal heights (h1 and h2). Microtubule widths (*blue-red lines*) were determined at half-heights. *Red line,* diameter of lumen; *blue lines*, microtubule walls. *Bar*, 100 nm (from Cyrklaff et al. 2007, with permission)

type of α-tubulin and one type of β-tubulin have been detected; however, posttranslational modifications such as acetylation, glutamylation, and detyrosination take place (Stieger et al. 1984; Sasse and Gull 1988; Schneider et al. 1997; Westermann et al. 1999). Antibodies recognizing acetylated tubulin, considered by several authors as characteristic of stable microtubules, label subpellicular as well as flagellar microtubules (Sasse and Gull 1988; Souto-Padron et al. 1993). However, acetylated tubulin was also detected in *T. brucei* spindle microtubules, which are labile structures (Sasse and Gull 1988). The use of an antibody recognizing tyrosinated α-tubulin showed that polymerizing microtubules have a high content of tyrosinated tubulin that is subsequently lowered to a basal level. These observations led to the suggestion that this system may provide the cells with a mechanism to discriminate between new and old microtubules (Sasse and Gull 1988). Through the use of several antibodies, it has also been shown that γ-tubulin, usually found in microtubule-organizing centers such as basal bodies, seems to exist in a small subset of subpellicular microtubules (Scott et al. 1997; Libusova et al. 2004).

Robinson et al. (1995) analyzed in detail the polarity of the subpellicular microtubules of *T. brucei*. Using the hook decoration technique, in which detergent-extracted cytoskeletons are incubated in the presence of exogenous tubulin, they observed the decoration of the subpellicular microtubules. In cytoskeletons viewed from the posterior toward the anterior region of the protozoan (relative to the basal bodies and flagellum), 93% of the decorated microtubules had a clockwise hook curvature, thus indicating that the plus end of the microtubules (sites where microtubule growth takes place) is at the posterior region of the cell. The same authors also used the negative staining technique to test polarity by observing points of microtubule growth when the cytoskeleton was incubated in the presence of tubulin. Again, growth was observed in the posterior region. In addition, they took advantage of previous observations that indicated that tyrosinated α-tubulin is a marker for newly assembled microtubules (Kohl and Gull 1998) and incubated the cytoskeleton with the YL 1/2 monoclonal antibody, which recognizes tyrosinated tubulin. Areas of labeling were concentrated in the posterior third of the protozoan cell (Robinson et al. 1995). Another modification found in some microtubules is the presence of tubulin polyglycylation, as occurs in *Giardia lamblia* (Weber et al. 1996; Campanati et al. 2003). No such posttranslational modification of tubulin was found in trypanosomatids (Schneider et al. 1987).

One intriguing question is how the number of microtubules increases during cell division. Available data indicate that the addition of new microtubules into the layer takes place by intercalation between the existing microtubules (Sherwin and Gull 1989). Drugs that interfere with microtubules block the process of cytokinesis, giving rise to cells whose diameter does not reach the values found in control cells; this then triggers the invagination of the plasma membrane so that two new cells are formed (De Souza unpublished observations).

5.2 *In Apicomplexa*

Apicomplexans were shown to have unlinked single copies of genes for α-, β-, and γ-tubulins (Reviewed in Morrissette and Sibley 2002). Both *Plasmodium yoellii* and *P. falciparum* were shown to have two independent genes for α-tubulin (I and II), with the α-tubulin II gene being specifically expressed in the flagella of male gametes (Rawlings et al. 1992). α-tubulin II is a male-specific protein in *P. falciparum*.

Employing the technique of "hook decoration" of *E. acervulina* subpellicular microtubules with tubulin extracted from pig brains, Russel and Burns (1984) demonstrated that the subpellicular microtubules of Apicomplexa grow from the polar ring, a unique microtubule-organizing center, with their plus ends toward posterior end of the parasite. So far, γ-tubulin has not been shown to be associated with the polar ring.

6 Biochemistry

The biochemical nature of several proteins associated with the subpelicular microtubules in Trypanosomatids and their putative roles have been identified. However, the same is not true in what concerns Apicomplexans, may be for the limitations associated with the impossibility of obtention of axenic cultures of any stage.

6.1 *In Trypanosomatids*

We have very little information about the nature of the filaments that connect the subpellicular microtubules of trypanosomatids to each other, to the plasma membrane or to cytoplasmic organelles. As shown in Fig. 2, there is a clear connection between the microtubules and the endoplasmic reticulum, which on some occasions penetrates between the microtubules (Pimenta and De Souza 1985). One of the first suggestions pointed to the possible presence of dynein as a microtubule-associated protein. This hypothesis has not been confirmed.

Biochemical analysis of cytoskeleton preparations, mostly obtained by detergent extraction, pointed to the presence of several proteins associated in some way with the microtubules. First, a 28-kDa protein was identified and localized only in the posterior regions of the cell. This protein, designated as Gb4, is encoded by a large gene consisting of numerous repeated units of 0.6 kb linked (Rindisbacher et al. 1993). Second, there is a family of high-molecular weight proteins that consist of more than 50 nearly identical tandemly repeated highly conserved 38-amino acid units (Affolter et al. 1994; Schneider et al. 1988). One of the proteins, known as MARP-1 (microtubule-associated repetitive protein), was isolated and was shown in vitro to bind to tubulin at sites different from those involved in the binding of the brain MAPs Tau and MAP2 (Hemphill et al. 1992). Under immunofluorescence microscopy, part of the protein in mammalian cells was observed binding to the microtubules. Third, Balaban et al. (1989) isolated the subpellicular microtubules of *T. brucei* using a high-ionic strength salt solution, and identified two peaks of 52 and 53 kDa. The first one corresponded to a tubulin-binding protein. Experiments of in vitro microtubule assembly using brain tubulin showed that, in the presence of the 52-kDa protein, the assembled microtubules formed bundles with regular (7 nm, center-to-center) cross-links. Subsequently, the same group reported the presence of a 15-kDa protein that promotes the bundle formation of microtubules assembled in vitro and is localized along the subpellicular microtubules (Balaban and Goldman 1992). Fourth, Thomas et al. (2000) described the kinetoplastid membrane protein-11 (KMP11) and showed its downregulation in parasites treated with vinblastine. Fifth, Vedrenne et al. (2002) found two related low-molecular weight cytoskeleton-associated proteins, designated as CAP15 and CAP17, in the cytoskeleton of *T. brucei*. Immunocytochemistry observations showed labeling associated with the subpellicular microtubules, especially in the anterior region of the protozoan cell. However, the resolution was not sufficient to determine the exact localization of these two proteins. Again, mammalian cells transfected with a CAP15 plasmid labeled the microtubules, thus confirming that it is a microtubule-associated protein. Sixth, Schneider et al. (1988) reported the presence of a microtubule-binding protein, designated as p41, that carries covalently bound fatty acids. The association of this protein with the microtubules was dependent on the presence of Ca, as it was released when incubation was carried out in the presence of EGTA. Seventh, Kratzerová et al. (2001) used a monoclonal antibody (MA-01) to describe a 210-kDa microtubule-interacting protein that labels the subpellicular microtubules located at the posterior region of *Leishmania* promastigotes. Moreover, this antibody also labeled the flagellum and the mitotic spindle. Eighth, a 60-kDa protein with 10–12 transmembrane domains, designated as NRAMP1, was found to associate with the microtubules as well as with the plasma membrane, thus suggesting that it is involved in the association between these two structures (Kishi et al. 1996). Ninth, a 210-kDa protein detected using a monoclonal antibody was closely associated with the cross-bridges lying between the microtubules (Woods et al. 1992). Finally, and more recently, Baines and Gull (2008) described a highly phosphorylated protein, designated as WCB, recognized by a monoclonal antibody. This protein presents an N-terminal C2 domain characteristic of membrane-associated proteins and a repetitive, charged C-terminal region with the characteristics of a microtubule-binding domain. However, no high-resolution localization of the protein has been reported. With RNA interference depletion of WCB, it was shown that this protein is essential for cell morphogenesis.

7 Drug Sensitivity

Causative agents of several diseases of great impact on human and veterinary health belong to the phylum Apicomplexa – Plasmodium, Toxoplasma, Eimeria. The same is true for Trypanosomatids that include the agents of Chaga's disease, African Trypanosomiasis, and Leishmaniasis. Most of these diseases can be efficiently treated and cured, so there is a permanent concern about finding new drugs for their treatment. In this sense, exclusive organelles such as the apicoplast of Apicomplexa, the glycosomes of Trypanosomatids, and metabolic pathways that are not shared by the hosts are targets with a great potential.

7.1 In Trypanosomatids

The sensitivity of the subpellicular microtubules of trypanosomatids to disrupting agents is very particular. Most microtubules are sensitive to low temperature, high pressure, and several drugs. For instance, drugs such as colchicine and vinblastine efficiently depolymerize the cytoplasmic microtubules of mammalian cells. However, in the case of trypanosomatids, some of these agents do not interfere with the subpellicular microtubules (Filho et al. 1978; Grellier et al. 1999). In contrast, drugs such as oryzalin and trifluralin (Chan and Fong 1994), which have no effect on mammalian microtubules, are potent disruptors of the trypanosomatid cytoskeleton (Seebeck and Gehr 1983; Chan et al. 1991; Bogitsh et al. 1999). Phenothiazines are also highly effective against subpellicular microtubules (Seebeck and Gehr 1983). On the other hand, drugs such as Taxol, which stabilize microtubules, induced significant morphological changes in the tested trypanosomatids (Baum et al. 1981; Webovetz et al. 2003; Dantas et al. 2003; Havens et al. 2000; Kapoor et al. 1999). All this information, in association with the fact that subpellicular microtubules remain intact throughout the cell cycle, clearly indicates that these are not classical microtubules as described in mammalian cells.

7.2 In Apicomplexa

Studies using classical drugs that disrupt the microtubules found in mammalian cells, such as colchicine and vinblastine, showed that the subpellicular microtubules of Apicomplexan protozoa are highly resistant to them. Only at high concentrations (1 mM colchicine) was the shortening of the microtubules observed. However, the microtubules were markedly sensitive to dinitroanilines (trifluralin, oryzalin, and ethafluralin), which interfere with plant microtubules (Stokkermans et al. 1996). This resistance is associated with point mutations in α-tubulin, especially in the M or N loops, which coordinate protofilament interaction in the microtubules and in the core of the α-tubulin (Morrissete et al. 2004; Ma et al. 2007).

8 Microtubule–Microtubule and Microtubule–Plasma Membrane Associations

The regular spacing between each microtubule, and between microtubules and the pellicle, is maintained by linker proteins, most of them still unknown at this time. Proteomic analysis is being currently applied and should quickly broaden the knowledge about them.

8.1 In Trypanosomatids

Electron microscopy analysis carried out to evaluate the isolation of subcellular fractions of trypanosomatids showed that even after disruption of the cells the microtubules remained attached to portions of the plasma membrane (Fig. 16) (De Souza 1976; Hunt and Ellar 1974; Dwyer 1980). Subsequently, several methodologies were developed to isolate membrane fractions. In some protocols, the microtubules remained attached and were even used as a morphological criterion to assess the purity of the fraction. In others, however, the microtubules disappeared (Reviewed in De Souza and Cunha e Silva 2003). More recently, a detailed proteomic analysis of isolated plasma membrane sheets with associated microtubules was carried out in the bloodstream forms of *T. brucei* (Bridges et al. 2008). The authors inferred which proteins were associated with the plasma membrane and which were associated with the cytoskeleton. A large number of proteins were identified as belonging to the cytoskeleton. However, no attempts were made to localize these proteins.

8.2 In Apicomplexa

Connections between subpellicular microtubules and the inner membrane complex have been observed in detergent-extracted *T. gondii*, especially when tannic acid was added to the glutaraldehyde solution used to fix the cells before processing for observation by transmission electron microscopy (Cintra and De Souza 1985b). The connections were regularly spaced at intervals of about 30 nm (Fig. 17).

The use of the freeze-fracture technique revealed the presence of a parallel array of closely apposed intramembranous particles on the P fracture face of the inner membrane complex. The distance between each particle strand was 28 nm, which corresponded to the diameter of the microtubules (Dubremetz and Torpier 1978;

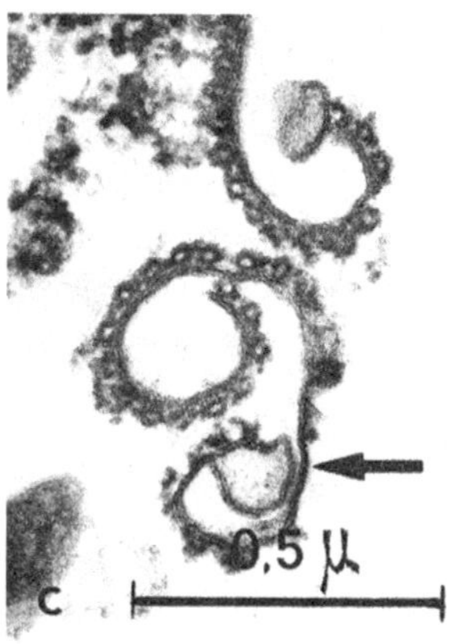

Fig. 16 Subpellicular microtubules remain associated with the plasma membrane even after the disruption of epimastigote forms of *Trypanosma cruzi* (from De Souza 1976 with permission)

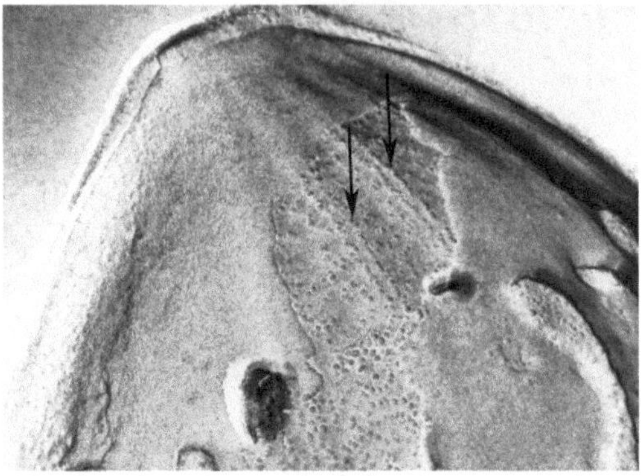

Fig. 17 Tachyzoites of *Toxoplasma gondii* fixed with glutaraldehyde and tannic acid after detergent extraction. Cytoplasm and most of the outer membranes (om) were extracted exposing regularly spaced bridges between the subpellicular microtubules and the inner pellicular membrane (from Cintra and De Souza 1985b with permission)

Porchet and Torpier 1977; Cintra and De Souza 1985a). Similar images were also observed on the fracture faces of the intermediate membrane (Cintra and De Souza 1985a) (Fig. 14). Application of Fourier analysis techniques better revealed this structure, with a 32-nm repeat in the double rows of membrane particles. Drugs known to disrupt cytoskeletal components fail to destroy the integrity of the particle lattice (Morrissete et al. 1997). Although there are no data on the nature of the particles that make up these strands, it is possible that they result from some linkage of the subpellicular microtubules to proteins located in the inner and intermediate membranes via filamentous structures that are not yet identified.

Unlike the subpellicular microtubules of Trypanosomatids, those of Apicomplexans are linked by the minus end to the polar ring, but no lateral bridges connect them laterally. Instead, a subpellicular network of filaments, which are easily observed in tachyzoites following detergent extraction and negative staining, runs parallel to the microtubules (Mann and Beckers 2001). Two proteins were identified in this network: TgIMC1 and TgIMC2. TgIMC1 is similar to articulins, the major cytoskeletal proteins of free-living protists and algae. In *P. falciparum*, a homolog of TgIMC1 with an additional 220 amino acids was identified (Khater et al. 2004). The presence and distribution of plateins, a subfamily of articulins, in *T. gondii* tachyzoites was demonstrated by immunofluorescence by Lemgruber et al. (2009). A diffuse labeling of the whole cell body was observed and electron microscopy of detergent-extracted tachyzoites exposed a network of 10-nm filaments distributed throughout the parasite and labeled by antiplatein antibodies. A predicted IMC protein was identified in the *T. gondii* genome; its sequence had 25% identity and 42% similarity with the platein isoform alpha 1, described in *Euplotes aediculatus*, and 42% identity and 55% similarity with the *Euglena gracilis* homolog, denoting its resemblance to articulins.

8.3 The Conoid in Apicomplexa

As already stated, among the characteristics that define the Apicomplexa, the apical complex is so outstanding that the whole phylum was named after it. In addition to the secretory organelles that discharge their contents through the apical complex, this unique structure is formed by a posterior polar ring to which the subpellicular microtubules attach; it also has two apical rings and a conoid, a hollow cone-shaped structure formed by microtubules in a particular spring-like spiral arrangement, and two intraconoidal microtubules that are shorter than the subpellicular microtubules and may serve to deliver secretory vesicles that participate in host cell invasion (Nichols and Chiappino 1987; Carruthers and Sibley 1997). Early descriptions of the conoid were provided by several authors (Gustafson et al. 1954; McLaren and Paget 1968; Sheffield and Melton 1968; Vivier and Petitprez 1972; de Souza 1974) and its general structural organization was reviewed by Attias and De Souza (in press).

The conoid is not present in all Apicomplexans; for instance, *Plasmodium* spp. lack this structure in their apical complex. On the other hand, it is present in *Toxoplasma*, *Eimeria*, *Gregarina*, and *Sarcocystis*. In *Toxoplasma*, a fine description that included an estimation of the diameter (300–350 nm) and number of fibers (15–23) of the conoid was made by Vivier and Petitprez (1972). By comparing the distances between the polar ring and the conoidal rings as observed by negative staining, De Souza (1974) suggested that the conoid moves up and down (Figs. 4 and 5 from De Souza 1974). The central pair of microtubules inside the conoid was also described in these early papers, but its function and relation to the other conoidal structures remain unknown. Figure 18 depicts the current model of structural organization of the conoid (Attias and De Souza in press; Dubey et al. 1998; Hu et al. 2006). The most consistent contribution on the structural organization of the conoid was provided by Hu et al. (2002), who showed that each conoidal fiber extends by about half of the diameter of the conoid and that there are about six fibers in longitudinal cross-section, confirming the early reports by Vivier and Petitprez (1972). Furthermore, by using transgenic *T. gondii* expressing YFP fused to the N-terminus of tubulin, Hu et al. (2002) confirmed that it is the main structural component of the conoidal spiral fibers. However, labeling of α- and β-tubulin by monoclonal antibodies required vigorous detergent extraction combined with high salt concentrations. Thin sections of parasites fixed in the presence of tannic acid revealed that each of the microtubules in the central pair is composed by a set of 13 protofilaments, as usual. On the other hand, each of the conoidal fibers has only nine protofilaments arranged in an open conformation that resembles a comma (Hu et al. 2002), a very unusual and unique form of tubulin assembly.

The conoid is not present in all Apicomplexans. However, the posterior polar ring is. As the conoid moves upwards, it protrudes above the posterior polar ring, whereas in the nonprotruding state it is hidden by the polar ring and the subpellicular microtubules that irradiate from it (Nichols and Chiappino 1987; Morrissette and Sibley 2002).

Through observations made on the ultrastructure of the cytoskeleton of *E. acervulina*, Russel and Burns (1984) proposed the posterior polar ring as a MTOC. The main evidence for this polarity was the fact that subpellicular microtubules emanate from the polar ring with their plus ends toward the posterior end of

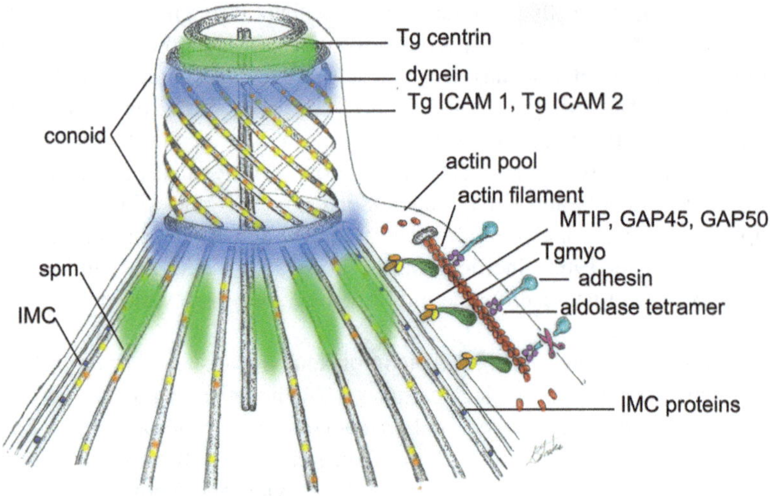

Fig. 18 Scheme of the present knowledge on the molecular and ultrastructural organization on the microtubules and associated proteins in the tachyzoite form of the Apicomplexa *Toxoplasma gondii* summarizing the present knowledge on the structural and molecular organization of the cytoskeleton. Tg centrin, ICAM1 and 2 and Tg centrin have been localized, but its structural interaction with the cytoskeleton has not yet been defined. Tg centrin was localized on the area of the apical polar rings and in patches just under the polar ring, while labeling for dynein was positive at the top and base of the conoid. ICAM 1 and 2 followed the spiral disposition of the conoid fibers, and are sparsely distributed over the subpelicular microtubules (spm). The molecules involved in gliding are represented on the right side of the scheme, where the proportional distance between the plasma membrane and the inner membrane complex (IMC) was disregarded, to allow insertion of all the molecules so far identified (from Attias and De Souza (in press) with permission)

the parasite, as shown by decoration with exogenous tubulin (Heidemann and McIntosh 1980). These results were corroborated for *T. gondii* by Nichols and Chiappino (1987). Later, the presence of γ-tubulin in the apical portion of *P. falciparum* schyzonts was demonstrated by immunofluorescence by Fowler et al. (2001), reinforcing this concept. However, it was not until 2006 that Hu et al. (2006) obtained fractions of *T. gondii* enriched in apical complex cytoskeleton that included the conoid, and identified several of the proteins used to build the conoidal structure. Two of them, TgCAM1 and TgCAM2, have calcium-binding domains and were localized to the conoid area by immunoelectron microscopy. TgDLC is a dynein light chain kinase from *T. gondii* that is 85% identical to human and mouse DLC dynein. It was shown to be present at the spindle poles and in centrioles in addition to the apical portion of the parasites, in which it is distributed above and below the conoid; it is also found on the polar ring at the site of attachment of the subpellicular microtubules. In GFP–DLC-expressing mutants, this protein was also present in the posterior end of the parasites. Two centrin homologs, TgCentrin-2 and TgCentrin-3, were also detected. The first one was predominantly, but not exclusively, found in the conoid-enriched fraction, in the preconoidal rings and in annular patches in the upper third of the cell, below the polar ring. TgCentrin-3 was faintly localized in the conoid, but is also found together with TgCentrin-1 in the centriole. The exact association of these proteins, whether to the conoid or to the posterior polar ring, is yet to be determined.

Conoid motility into and above the posterior polar ring has been pointed out by several authors (De Souza 1974; Monteiro et al. 2001) and recorded on several occasions (Håkansson et al. 1999; Hu et al. 2006). Conoid extrusion can be artificially induced by calcium ionophore stimulation and is blocked by treatment with cytochalasin D (Mondragon and Frixione 1996). On the other hand, jasplakinolide, a membrane-permeable actin-polymerizing and filament-stabilizing drug, was unable to induce conoid extrusion (Shaw and Tilney 1999), although an amazing extension of the apical membrane containing actin filaments was observed. In situ conoid extrusion is observed during recognition and adhesion to host cells and as the parasites egress from the parasitophorous vacuole (Caldas et al. 2010). Once parasites are established in the parasitophorous vacuole inside the host cell, the conoid remains enclosed within the shell formed by the subpellicular microtubules. However, at egress, when the parasites are actively leaving the host cell, the conoid intermittently protrudes beyond the apical end of the microtubules (Hu et al. 2002; Caldas et al. 2010). Extrusion of the conoid above the polar ring is attributed by some authors (Hu et al. 2002; Morrissette 1995) to a torsion of the spiral fibers that compose it; this torsion simultaneously narrows the spiral's diameter and increases its height. However, in measurements made both in ultrathin sections and in field emission scanning electron microscopy of detergent-extracted tachyzoites of *T. gondii*, the diameter and height of the conoid were the same in the resting position as in the extruded state (unpublished data). The localization of DLC proteins around the conoid (Hu et al. 2006) is indicative of a tubulin– dynein interaction resulting in motility.

Conoid motility was demonstrated by Carey et al. (2004) to be independent from gliding motility, since ionomycin-triggered extension of the conoid was successfully inhibited by a small molecule, without interfering with gliding motility. In view of the data gathered to date, the role of calcium in triggering conoid extrusion seems undisputable. Mondragon and Frixione (1996) were first to demonstrate that calcium ionophores A23187 and ionomycin trigger conoid extrusion, as do calcium-ATPase inhibitors such as thapsigargin. On the other hand, when conoid movement was paralyzed either in the extruded or the internalized position, the ability of tachyzoites to invade cells decreased significantly, showing the relevance of motility in this process.

At this point, with the available data, microtubule-associated motor proteins seem to be more likely to be involved in conoid motility than myosin and actin filaments. However, most of the considerations made on Apicomplexa and *Toxoplasma* motility are based on an actin–myosin model.

Although actin and myosin filaments are not the focus of this review, some considerations about these structures are necessary because they may interact with the subpellicular microtubules.

9 In Trypanosomatids

Actin microfilaments were never observed in the cytoplasm of *T. cruzi*. However, cytochalasin, a drug that interferes with actin microfilaments, induces changes in the morphology of bloodstream trypomastigotes and inhibits their movement. In epimastigotes, cytochalasin causes a 48% decrease in peroxidase uptake (Bogitsh et al. 1995). Corrêa et al. (2008) demonstrated that cytochalasin B treatment leads to morphological alterations in the cytoskeletal elements associated with the cytostome–cytopharynx complex, which is responsible for transferrin uptake. Comparative genomic analysis identified a potential role for an actin–myosin system in *T. cruzi*, as this protozoan contains an actin gene as well as an expanded myosin family and a CapZ F-actin capping complex, which are not found in *T. brucei* or *Leishmania* (El-Sayed et al. 2005a, b). The authors of that study suggested that an actin–myosin system might function at the cytostome. Actin and actin-binding proteins have recently been characterized in *T. cruzi* (De Melo et al. 2008). TcActin was observed in several patch-like cytoplasmic structures, spread along the cell body of various *T. cruzi* stages, similar to actin distribution in *Leishmania* (Sahasrabuddhe et al. 2004). In contrast to actin in *Leishmania*, TcActin is not associated with subpellicular microtubules. Although *T. cruzi* actin is similar in structure to the actins of higher eukaryotes, homology modeling has revealed fundamental differences, predominantly in the loops responsible for oligomerization and interaction with actin-binding proteins. Consequently, actin filaments have never been detected in *T. cruzi*. Actin and actin-binding proteins (especially coronin, ADF/cofilin, profilin, formin, and myosin) have been further characterized in *Leishmania* (Reviewed in Sahasrabuddhe and Gupta in press).

10 In Apicomplexans

Since the initial observation of gliding motility in Apicomplexa, it has been suggested that this mechanism relies on an efficient actin–myosin-like system. However, filaments made of these proteins were never visualized in thin sections of zoites. A network of filaments with a diameter compatible with that reported for actin filaments was seen in tachyzoites of *T. gondii* using high-resolution scanning electron microscopy (Schatten et al. 2003). Subsequently, actin was isolated from subpellicular cytoskeleton extracts by binding to DNAse I, and was polymerized in vitro. The filaments formed bound to heavy meromyosin (Parón et al. 2005). Studies using drugs that interfere with actin dynamics showed that they were able to inhibit parasite invasion in the host cells (Mondragon and Frixione 1996; Dobrowolski and Sibley 1996; Morrissette and Sibley 2002). Biochemical analysis showed the presence of actin in these organisms (Field et al. 1993; Webb et al. 1996; Dobrowolski et al. 1997a, b), probably in its globular form (G-actin). Immunofluorescence microscopy of cells incubated in the presence of antibodies recognizing actin, but not with phalloidin, which only binds to filamentous actin, labeled the cells (Dobrowolski et al. 1997a, b). Strong evidence for the presence of actin was provided by the reaction of *T. gondii* tachyzoites when incubated in the presence of jasplakinolide, a drug that induces the polymerization and stabilization of actin filaments; under these conditions, a dramatic assembly of filaments occurred in the apical portion of the cells (Shaw and Tilney 1999). Expression of the actin gene (TgACT1) in baculovirus followed by purification of the expressed actin showed that it rapidly polymerized in vitro into filaments at a critical concentration that was significantly lower than that found for conventional actins. However, the protozoan actin filaments formed were ten times shorter and less stable than mammalian actin formed under the same experimental conditions (Sahoo et al. 2006). These observations led to the suggestion that rapid cycles of assembly and disassembly of actin in Apicomplexa govern the unusual form of gliding motility (Sahoo et al. 2006). Two excellent reviews covering basic aspects of the dynamic of actin participation in gliding motility were published recently (Schuler and Matuschewski 2006a, b).

The first evidence for the presence of myosin in Apicomplexa came from immunofluorescence microscopy using antimyosin antibodies that revealed the presence of myosin in the apical region of the cell (Schwartzman and Pfefferkorn 1983; Webb et al. 1996). Subsequently, it was shown that drugs that inhibit myosin, such as BDM and KT5926, inhibit parasite gliding motility and invasion of host cells (Dobrowolski et al. 1997a, b; Forney et al. 1998; Pinder et al. 1998). These observations were confirmed by the analysis of a conditional knockout of the small myosin motor TgMyoA (Meissner et al. 2002).

A very comprehensive model of the present knowledge on gliding mechanics was provided by Baum et al. (2006). According to this model, microneme-secreted transmembrane proteins of the TRAP family connect on the extracellular side to host cell surface components and are pushed toward the posterior end of the parasite through the aldolase–actin–MyoA machinery. The MyoA tail domain is inserted in the inner membrane complex membrane through MTIP and the whole assembly moves the parasite forward as the membrane flows to the rear end. However, the connection and functional relationship between the TRAP–aldolase–actin–MyoA–MTIP sequence and the subpellicular microtubules and the subpellicular network are not yet established or explained by a convincing model.

11 Functional Data

What are the functions played by subpellicular microtubules in trypanosomatids and in apicomplexans? It is for sure the primary scaffold for maintaining the shape. However, it also must have the plasticity to allow the interconversion between evolutive forms in Trypanosomatids and the articulation necessary for gliding motility and host cell invasion observed in Apicomplexa. Despite any apparent similarity, subpelicular microtubules of Apicomplexa and trypanosomatids do not share many aspects in what concerns its functionality.

11.1 In Trypanosomatids

Available data point to a few functions for subpelicular microtubules in trypanosomatids. The first one is related to the maintenance of cell shape. Indeed, all the treatments that interfere with microtubules led to loss of cell asymmetry, especially in the developmental stages such as trypomastigote and epimastigote forms, with the protozoan acquiring a rounded shape. According to this view, the general shape of the cell is due to the spatial distribution of the microtubules. This distribution is probably maintained by the interaction between the microtubules and other cytoskeletal structures. This system is sensitive to signaling processes taking place during the protozoan life cycle. Indeed, transformation of an amastigote into the trypomastigote form of *T. cruzi* involves changes in the spatial distribution of the microtubules (Meyer and De Souza 1976). When amastigotes stop to divide within the host cell, the flagellar and subpellicular microtubules grow, which could bring about the elongation of the body. A second function that may be attributed to the microtubules is the maintenance of a certain rigidity of the protozoan plasma membrane, thus avoiding the assembly of endocytic vesicles and the fusion of cytoplasmic vesicles with the membrane. In addition, they preclude the direct interaction of other organelles with the plasma membrane, with the exception of the endoplasmic reticulum profiles, which are occasionally seen penetrating between the microtubules (Pimenta and De Souza 1985). Several studies have shown that endocytic and exocytic activities in trypanosomatids only take place in the flagellar pocket, whose membranes are not associated with subpellicular microtubules (Overath and Englster 2004; Field and Carrington 2004). The third possible function involves the association of the microtubules with other cell organelles such as the endoplasmic reticulum. Such connections may play a role in the maintenance of the organelles' shape as well as provide a substrate for organelle movement within the cell.

11.2 In Apicomplexans

Present evidence supports the hypothesis that subpellicular microtubules of Apicomplexa, besides their obvious role in maintaining the cell shape and polarity, participate in the gliding process by anchoring the motor proteins that provide stable support for the insertion of Tg Myo so that it can pull the actin assembly linked to the aldolase-transmembrane adhesion, resulting in the motion of the parasite (Reviewed in Baum et al. 2006). The cell body of Apicomplexa is constricted as they pass along the moving junction during invasion or along the plasma membrane during egress (Caldas et al. 2007). This constriction is most certainly much more a consequence of the passive contraction of adjacent microtubules, as the parasite glides through a space narrower than its diameter, than mediated by microtubule-associated proteins.

The subpellicular microtubules of Apicomplexa do not restrict endo- or exocytic processes because they are not as tightly packed and linked as they are in Trypanosomatids, in order to avoid the secretion of granules; moreover, they do not extend beyond two-thirds of the length of the cell. Also, the micropore, a cytostome-like specialized structure believed to be the site of endocytosis in *T. gondii* (Nichols et al. 1994), is situated in the upper third of the cell body, indicating that the subpellicular microtubules do not restrict the internalization of nutrients at that site. The fact that the subpellicular microtubules of Apicomplexa end free at the posterior (plus) end opens up the possibility that the parasites can elongate by reducing or become shorter by increasing the distance between the subpellicular microtubules.

12 Perspectives

Subpellicular microtubules are present in Trypanosomatids and in Apicomplexans. In both the groups, microtubules are disposed side by side at regular intervals, running parallel under the plasma membrane (Trypanosomatids) and the pellicle (Apicomplexans), encaging the cytoplasmic organelles. This, along with the fact that α/β-tubulin dimers are their main constituent, is the main and almost the only similarity that both types of microtubules share. The regular bridges that link the subpellicular microtubules of trypanosomatids turn them into rigid nutshells unable to bend or constrict the cell body, as is observed in *Toxoplasma*, *Plasmodium* and other Apicomplexa. However, because microtubules in trypanosomatids form helical sheets, they can swell or shrink under osmotic stress, causing the cells to become shorter and spherical or corkscrew-shaped, as observed in *Phytomonas staheli* (Attias et al. 1987). The subpellicular microtubules of Apicomplexa, on the other hand, react to variations in the osmotic pressure of the environment by spreading like the ribs of an umbrella.

At this point, although a lot of information both on the ultrastructure and on the proteomics of subpellicular microtubules has been gathered, the correlation between these data is still missing. Elucidating this correlation will be necessary for the identification of the proteins that link subpellicular microtubules to each other, to the plasma membrane and to cytoplasmic organelles such as mitochondria and the endoplasmic reticulum. The identification of a microtubule-organizing center in Trypanosomatids certainly would be helpful in explaining how they change their shape during their life cycle.

As for Apicomplexa, the challenge is not smaller and several points remain obscure or disregarded. Among them is the question of the actual role played by conoid motion in gliding and invasion. Parasites lacking conoids, such as

Plasmodium, are as competent in invading cells as *Toxoplasma* and *Eimeria*, which have conoids. Moreover, the regulation and mechanics of conoid motion and conoid interaction with motor proteins are not yet understood.

It is known that micronemes and rhoptries secrete their contents at the apical portion of the parasite, but it is not known whether conoid positioning above or below the polar ring is necessary for this secretion to occur. The same applies to the central pair of microtubules: its role and relation to other components of the cytoskeleton and secretory organelles remains obscure. Another point that needs clarification is how adhesion proteins secreted by micronemes are incorporated into the plasma membrane to bridge extracellular molecules to intracellular actin–myosin and the related protein machinery that is involved in gliding.

The subpellicular network is another enigmatic structure; very little is known about its exact structure, location, or function.

The anchoring of the actin–myosin motor to subpellicular microtubules, mediated by the inner pellicle, is another point about which there is much left to elucidate, including the role of these microtubules in cell shape and motility. All these questions are exciting and intriguing puzzles that should receive attention and be cleared out in the years to come through the use of molecular, biochemical, and ultrastructural techniques.

Acknowledgments The work carried out at the authors' laboratory was supported by CNPq/MCT, FINEP, CAPES, DECIT, and FAPERJ.

References

Affolter M, Hemphill A, Roditi I, Muller N, Seebeck T (1994) The repetitive microtubule-associated proteins MARP-1 and MARP-2 of *Trypanosoma brucei*. J Struct Biol 112:241–251

Angelopoulos E (1970) Pellicular microtubules in the family Trypanosomatidae. J Protozool 17:39–51

Attias M, De Souza W (in press) A review of the Apicomplexa cytoskeleton. Trends Mol Cell Biol

Attias M, Bezerra JL, Oliveira DP, Souza W (1987) Ultrastructure of *Phytomonas Staheli* in diseased coconut (Cocos nucifera) and oil palm (Elaeis Guineensis). J Submicrosc Cytol 19:93–100

Attias M, Vommaro RC, Souza W (1996) Computer aided three-dimensional reconstruction of the free-living protozoan Bodo sp (Kinetoplastida Bodonida). Cell Struct Funct 21:297–306

Baines A, Gull K (2008) WCB is a C2 domain protein defining the plasma membrane-sub-pellicular microtubule corset of kinetoplastid parasites. Protist 159:115–125

Balaban N, Goldman R (1992) Isolation and characterization of a unique 15 kilodalton *trypanosome* subpellicular microtubule-associated protein. Cell Motil Cytoskeleton 21:138–146

Balaban N, Waithaka HK, Njogu AR, Goldman R (1989) Isolation of a subpellicular microtubule protein from *Trypanosoma brucei* that mediates crosslinking of microtubules. Cell Motil Cytoskeleton 14:393–400

Baum SG, Wittner M, Nadler JP, Horwitz SB, Dennis JE, Schiff PB, Tanowitz MB (1981) Taxol, a microtubule stabilizing agent, blocks the replication of *Trypanosoma cruzi*. Proc Natl Acad Sci USA 78:4571–4575

Baum J, Papenfuss AT, Baum B, Speed TP, Cowman AF (2006) Regulation of apicomplexan actin-based motility. Nat Rev Microbiol 4:621–628

Bogitsh BJ, Ribeiro-Rodrigues R, Carter (1995) In vitro effect of mannan and cytochalasin B on the uptake of horseradish peroxidase and 14-sucrose by *Trypanosoma cruzi* epimastigotes. J Parasitol 81:144–148

Bogitsh BJ, Middleton OL, Ribeiro-Rodrigues R (1999) Effects of the antitubulin drug trifluralin on the proliferation and metacyclogenesis of *Trypanosoma cruzi* epimastigotes. Parasitol Res 85:475–480

Bordier C, Garavito RM, Armbruster B (1982) Biochemical and structural analyses of microtubules in the pellicular membrane of *Leishmania tropica*. J Protozool 29:560–565

Bridges DJ, Pitt AR, Hanrahan O, Brennan K, Voorheis HP, Herzyk P, de Koning HP, Burchemore RJS (2008) Characterisation of the plasma membrane subproteome of bloodstream form *Trypanosoma brucei*. Proteomics 8:83–99

Bringmann G, Holz J (1953) Toxoplasma gondii in the electronmicroscopical picture. Z Hyg Infektionskr 137:186–191

Caldas LA, DeSouza W, Attias M (2007) Calcium ionophore-induced egress of *Toxoplasma gondii* shortly after host cell invasion. Vet Parasitol 147:210–220

Caldas LA, de Souza W, Attias M (2010) Microscopic analysis of calcium ionophore activated egress of *Toxoplasma gondii* from the host cell. Vet Parasitol 167:8–18

Campanati L, Troester H, Monteiro-Leal LH, Spring H, Trendelenburg MF, De Souza W (2003) Tubulin diversity in trophozoites of Giardia lamblia. Histochem Cell Biol 119:323–331

Carey KL, Westwood NJ, Mitchison TJ, Ward GE (2004) A small-molecule approach to studying invasive mechanisms of Toxoplasma gondii. Proc Natl Acad Sci USA 101:7433–7438

Carruthers VB, Sibley LD (1997) Sequential protein secretion from three distinct organelles of Toxoplasma gondii accompanies invasion of human fibroblasts. Eur J Cell Biol 73:114–123

Casanova M, Crobu L, Blaineau C, Burgeois N, Bastien P, Pagès M (2009) Microtubule-severing proteins are involved in flagellar length control and mitosis in Trypanosomatids. Mol Microbiol 71:1353–1370

Chan MM, Fong D (1994) Plant microtubule inhibitors against trypanosomatids. Parasitol Today 10:448–451

Chan MM, Triemer RE, Fong D (1991) Effect of the anti-microtubule drug oryzalin on growth and differentiation of the parasitic protozoan *Leishmania mexicana*. Differentiation 46:15–21

Cintra WM, De Souza W (1985a) Distribution of intramembranous particles and filipin-sterol complexes in the cell membranes of *Toxoplasma gondii*. Eur J Cell Biol 37:63–69

Cintra WM, De Souza W (1985b) Immunocytochemical localization of cytoskeletal proteins and electron microscopy of detergent extracted tachyzoites of *Toxoplasma gondii*. J Submicroscop Cytol 17:503–508

Corrêa JR, Atella GC, Batista MM, Soares MJ (2008) Transferrin uptake in *Trypanosoma cruzi* is impaired by interference on cytostome-associated cytoskeleton elements and stability of membrane cholesterol but not by obstruction of chlatrin-dependent endocytosis. Exp Parasitol 119:58–66

Cyrklaff M, Kudryashev M, Leis A, Leonard K, Baumeister W, Menard R, Meissner M, Frischknecht F (2007) Cryoelectron tomography reveals periodic material at the inner side of subpellicular microtubules in apicomplexan parasites. J Exp Med 204:1281–1287

Dantas AP, Barbosa HS, De Castro SL (2003) Biological and ultrastructural effects of the anti-microtubule agent taxol against *Trypanosoma cruzi*. J Submicroscop Cytol Pathol 35:287–294

De Melo LD, Sant'Anna C, Reis SA, De Souza W, Cunha e Silva NL (2008) Evolutionary conservation of actin-binding proteins in *Trypanosoma cruzi* and unusual sub-cellular localization of the actin homologue. Parasitology 135:955–965

De Souza W (1972) Mise en évidence et structure du système microtubulaire de *Toxoplasma gondii*. C R Acad Sci 275:2899–2901

De Souza W (1974) Fine structure of the conoid of *Toxoplasma gondii*. Rev Inst Med Trop São Paulo 16:32–38

De Souza W (1976) Associations membrane-microtubules chez *Trypanosoma cruzi*. J Microscop Biol Cell 25:189–190

De Souza W (2006) Secretory organelles of pathogenic protozoa. An Acad Bras Cienc 78:271–291
De Souza W (2008) Electron microscopy of trypanosomes. A historical view. Mem Inst Oswaldo Cruz 103:313–325
De Souza W, Benchimol M (1984) High voltage electron microscopy of critical point dried trypanosomatids. J Submicroscop Cytol 16:237–242
De Souza W, Cunha e Silva NL (2003) Cell fractionation of parasitic protozoa – a review. Mem Inst Oswaldo Cruz 98:151–170
Dobrowolski JM, Sibley LD (1996) *Toxoplasma* invasion of mammalian cells is powered by the actin cytoskeleton of the parasite. Cell 84:933–939
Dobrowolski JM, Carruthers VB, Sibley LD (1997a) Participation of myosin in gliding motility and host cell invasion by *Toxoplasma gondii.* Mol Microbiol 26:163–173
Dobrowolski JM, Niesman IR, Sibley LD (1997b) Actin in the parasite *Toxoplasma gondii* is encoded by a single copy gene, ACT1 and exists primarily in a globular form. Cell Motil Cytoskeleton 37:253–262
Dubey JP, Lindsay DS, Speer CA (1998) Structures of Toxoplasma gondii tachyzoites, bradyzoites, and sporozoites and biology and development of tissue cysts. Clin Microbiol Rev 11:267–299
Dubremetz JF, Torpier G (1978) Freeze-fracture study of the pellicle of an Eimerian sporozoite (Protozoa, Coccidia). J Ultrastruct Res 62:94–109
Dwyer DM (1980) Isolation and partial characterization of surface membranes from *Leishmania donovani* promastigotes. J Protozool 27:176–182
El-Sayed NM, Myler PJ, Bartholomeu DC, Nilsson D, Aggarwal G, Tran AN, Ghedin E, Worthey EA, Delcher AL, Blandin G, Westenberger SJ, Caler E, Cerqueira GC, Branche C, Haas B, Anupama A, Arner E, Aslund L, Attipoe P, Bontempi E, Bringaud F, Burton P, Cadag E, Campbell DA, Carrington M, Crabtree J, Darban H, da Silveira JF, de Jong P, Edwards K, Englund PT, Fazelina G, Feldblyum T, Ferella M, Frasch AC, Gull K, Horn D, Hou L, Huang Y, Kindlund E, Klingbeil M, Kluge S, Koo H, Lacerda D, Levin MJ, Lorenzi H, Louie T, Machado CR, McCulloch R, McKenna A, Mizuno Y, Mottram JC, Nelson S, Ochaya S, Osoegawa K, Pai G, Parsons M, Pentony M, Pettersson U, Pop M, Ramirez JL, Rinta J, Robertson L, Salzberg SL, Sanchez DO, Seyler A, Sharma R, Shetty J, Simpson AJ, Sisk E, Tammi MT, Tarleton R, Teixeira S, Van Aken S, Vogt C, Ward PN, Wickstead B, Wortman J, White O, Fraser CM, Stuart KD, Andersson B (2005a) The genome sequence of Trypanosoma cruzi, etiologic agent of Chagas disease. Science 309:409–415
El-Sayed NM, Myler PJ, Blandin G, Berriman M, Crabtree J, Aggarwal G, Caler E, Renauld H, Worthey EA, Hertz-Fowler C, Ghedin E, Peacock C, Bartholomeu DC, Haas BJ, Tran AN, Wortman JR, Alsmark UC, Angiuoli S, Anupama A, Badger J, Bringaud F, Cadag E, Carlton JM, Cerqueira GC, Creasy T, Delcher AL, Djikeng A, Embley TM, Hauser C, Ivens AC, Kummerfeld SK, Pereira-Leal JB, Nilsson D, Peterson J, Salzberg SL, Shallom J, Silva JC, Sundaram J, Westenberger S, White O, Melville SE, Donelson JE, Andersson B, Stuart KD, Hall N (2005b) Comparative genomics of trypanosomatid parasitic protozoa. Science 309:404–409
Emmel L, Jakob A, Golz H (1942) Elektronenoptische Untersuchungen an Malaria-Sporozoite und Beobachtungen an Kulturformen von Leishmania donovani. Deutsch Tropenmed Z 46:344–348 (in Dubremetz and Fergusonm (2009))
Field MC, Carrington M (2004) Intracellular membrane transport systems in *Trypanosoma brucei.* Traffic 5:905–913
Field SJ, Pinder JC, Clough B, Dluzewski AR, Wilson RJ, Gratzer WB (1993) Actin in the merozoite of the malaria parasite, *Plasmodium falciparum.* Cell Motil Cytoskeleton 25:43–48
Filho SA, de Almeida ER, Gander ES (1978) The influence of hydroxyurea and colchicine on growth and morphology of *Trypanosoma cruzi.* Acta Trop 35:229–237
Forney JR, Vaughan DK, Yang S, Healey MC (1998) Actin-dependent motility in *Cryptosporidium parvum* sporozoites. J Parasitol 84:908–923

Fowler RE, Smith AM, Whitehorn J, Williams IT, Bannister LH, Mitchell GH (2001) Microtubule associated motor proteins of *Plasmodium falciparum* merozoites. Mol Biochem Parasitol 117:187–200

Gallo JM, Precigout E (1988) Tubulin expression in trypanosomes. Biol Cell 64:137–143

Garnham PC, Baker JR, Bird RG (1962) Fine structure of cystic form of Toxoplasma gondii. Br Med J 1:83–84

Grellier P, Sinou V, Garreau-de-Loubresse N, Bylèn E, Boulard Y, Schrével J (1999) Selective and reversible effects of vinca alkaloids on *Trypanosoma cruzi* epimastigote forms: blockage of cytokinesis without inhibition of the organelle duplication. Cell Motil Cytoskeleton 42:36–47

Gull K (1999) The cytoskeleton of Trypanosomatid Parasites. Annu Rev Microbiol 53:629–635

Gustafson PV, Agar HD, Cramer DI (1954) An electron microscope study of Toxoplasma. Am J Trop Med Hyg 3:1008–1021

Håkansson S, Morisaki H, Heuser J, Sibley LD (1999) Time-lapse video microscopy of gliding motility in *Toxoplasma gondii* reveals a novel, biphasic mechanism of cell locomotion. Mol Biol Cell 10:3539–3547

Havens CG, Bryant N, Asher L, Lamocreaux L, Perfetto S, Brendle JJ, Werbovetz KA (2000) Cellular effects of leishmanial tubulin inhibitors on *L. donovani*. Mol Biochem Parasitol 110:223–236

Heidemann SR, McIntosh JR (1980) Visualization of the structural polarity of microtubules. Nature 286:517–519

Hemphill A, Affolter M, Seebeck T (1992) A novel microtubule-binding motif identified in a high molecular weight microtubule-associated protein from *Trypanosoma brucei*. J Cell Biol 117:95–103

Hou W-Y, Pimenta PFP, Ru-Long S, Pinto da Silva P (1992) Stereo views and immunogold labeling of the pellicular microtubules at the inner surface of the plasma membrane of *Leishmania* as revealed by fracture-flip. J Histochem Cytochem 40:1309–1318

Hu K, Roos DS, Murray JM (2002) A novel polymer of tubulin forms the conoid of *Toxoplasma gondii*. J Cell Biol 156:1039–1050

Hu K, Johnson J, Florens L, Fraunholz M, Suravajjala S, DiLullo C, Yates J, Roos DS, Murray JM (2006) Cytoskeletal components of an invasion machine–the apical complex of Toxoplasma gondii. PLoS Pathog 2:e13

Hunt RC, Ellar DJ (1974) Isolation of the plasma membrane of a trypanosomatid flagellate: general characterisation and lipid composition. Biochim Biophys Acta 339:173–189

Jones TC, Yeh S, Hirsch JG (1972) The interaction between Toxoplasma gondii and mammalian cells. I. Mechanism of entry and intracellular fate of the parasite. J Exp Med 136:1157–1172

Kapoor P, Sachdeva M, Madhubala R (1999) Effect of the microtubule stabilising agent taxol on leishmanial protozoan parasites in vitro. FEMS Microbiol Lett 176:429–435

Khater EI, Sinden RE, Dessens JT (2004) A malaria membrane skeletal protein is essential for normal morphogenesis, motility, and infectivity of sporozoites. J Cell Biol 167:425–432

Kikkawa Y, Gueft B (1964) Toxoplasma cysts in the human heart, an electron microscopic study. J Parasitol 50:217–225

Kishi F, Yoshida T, Also S (1996) Location of NRAMP1 molecule on the plasma membrane and its association with microtubules. Mol Immunol 33:1241–1246

Kohl L, Gull K (1998) Molecular architecture of the trypanosome cytoskeleton. Mol Biochem Parasitol 93:1–9

Kratzerová L, Draberová E, Juliano C, Viklicky V, Fiori PL, Cappuccinelli P, Dráber P (2001) Cell cycle-dependent changes in localization of a 210-kDa microtubule-interacting protein in *Leishmania*. Exp Cell Res 266:270–278

Lemgruber L, Kloetzel JA, Souza W, Vommaro RC (2009) *Toxoplasma gondii*: further studies on the subpellicular network. Mem Inst Oswaldo Cruz 104:706–709

Li R, Gundersen GG (2008) Beyond polymer polarity: how the cytoskeleton builds a polarized cell. Nat Rev Mol Cell Biol 9:860–873

Libusova L, Sulimenko T, Sulimenko V, Hozák P, Dráber P (2004) Gamma-Tubulin in *Leishmania*: cell cycle-dependent changes in subcellular localization and heterogeneity of its isoforms. Exp Cell Res 295:375–386

Ma C, Li C, Ganesan L, Oak J, Tsai S, Sept D, Morrissete NS (2007) Mutations in α-tubulin confer Dinitroaniline resistance at a cost to microtubule function. Mol Biol Cell 18:4711–4720

Mann T, Beckers C (2001) Characterization of the sub-pellicular network, a filamentous membrane skeletal components in the parasite *Toxoplasma gondii.* Mol Biochem Parasitol 115:257–268

McLaren DJ, Paget GE (1968) A fine structural study on the merozoite of *Eimeria tenella* with special reference to the conoid apparatus. Parasitology 58:561–571

Meissner M, Schluter D, Soldati D (2002) Role of *Toxoplasma gondii* myosin A in powering parasite gliding and host cell invasion. Science 298:837–840

Meyer H, Andrade Mendonça I (1955) Electron microscopy observations of *Toxoplasma* Nicolle et Manceaux grown in tissue cultures (first note). Parasitology 45:449–451

Meyer H, De Souza W (1976) Electron microscopic study of *Trypanosoma cruzi* periplast in tissue cultures. I. Number and arrangement of the peripheral microtubules in the various forms of the parasite's life cycle. J Protozool 23:385–390

Meyer H, Porter KR (1954) A study of *Trypanosoma cruzi* with the electron microscope. Parasitology 44:16–23

Mizuhira V, Futeasaku Y (1972) New fixation for biological membranes using tannic acid. Acta Histochem Cytochem 5:233–236

Mondragon R, Frixione E (1996) $Ca^{(2+)}$-dependence of conoid extrusion in *Toxoplasma gondii* tachyzoites. J Eukaryot Microbiol 43:120–127

Monteiro VG, de Melo EJ, Attias M, de Souza W (2001) Morphological changes during conoid extrusion in Toxoplasma gondii tachyzoites treated with calcium ionophore. J Struct Biol 136:181–189

Morrissette NS (1995) The apical cytoskeleton of *T gondii*. PhD thesis, University of Pennsylvania, Philadelphia, PA, 191 pp

Morrissete NS, Murray JM, Roos DR (1997) Subpellicular microtubules associate with an intramembranous particle lattice in the protozoan parasite *Toxoplasma gondii.* J Cell Sci 110:35–42

Morrissete NS, Mitra A, Sept D, Sibley LD (2004) Dinitroanilines bind α-tubulin to disrupt microtubules. Nol Biol Cell 15:1960–1968

Morrissette NS, Sibley LD (2002) Disruption of microtubules uncouples budding and nuclear division in *Toxoplasma gondii.* J Cell Sci 115:1017–1025

Nichols BA, Chiappino ML (1987) Cytoskeleton of *Toxoplasma gondii.* J Protozool 34:217–226

Nichols BA, Chiappino ML, PAvesio CE (1994) Endocytosis at the micropore of *Toxoplasma gondii*. Parasitol Res 80:91–98

Overath P, Englster M (2004) Endocytosis, membrane recycling and sorting of GPI-anchored proteins: *Trypanosoma brucei* as a model system. Mol Microbiol 53:735–744

Parón A, Mondragón M, González S, Ambrosio JR, Guerrero AL, Mondragón R (2005) Identification and purification of actin from the subpellicular network of *Toxoplasma gondii* tachyzoites. Mol Biochem Parasitol 35:883–894

Pimenta PF, De Souza W (1985) Fine structure and cytochemistry of the endoplasmic reticulum and its association with the plasma membrane of *Leishmania mexicana amazonensis*. J Submicroscop Cytol 17:413–419

Pinder JC, Fowler RE, Dluzewski AR, Bannister LH, Lavin FM, Mitchell GH, Wilson RJ, Gratzer WB (1998) Actomyosin motor in the merozoite of the malaria parasite. Plasmodium falciparum: implications for red cell invasion. J Cell Sci 111:1831–1839

Porchet E, Torpier G (1977) Etude du germe infectieux de S*arcocystis tenella* ET *Toxoplasma gondii* par la technique du cryodecapage. Z Parasitenkd 54:101–124

Ravindran S, Boothroyd JC (2008) Secretion of proteins into host cells by Apicomplexan parasites. Traffic 9:647–656

Rawlings DJ, Fujioka H, Fried M, Keister DB, Aikawa M, Kaslow DC (1992) Alpha-tubulin II is a male-specific protein in *Plasmodium falciparum*. Mol Biochem Parasitol 56:239–250

Rindisbacher L, Hemphill A, Seebeck T (1993) A repetitive protein from *Trypanosoma brucei* which caps the microtubules at the posterior end of the cytoskeleton. Mol Biochem Parasitol 58:83–96

Roberts WL, Hammond DM (1970) Ultrastructural and cytological studies on sporozoites of four *Eimeria* species. J Protozool 17:76–86

Robinson DR, Sherwin T, Ploubidou A, Byard EH, Gull K (1995) Microtubule polarity and dynamics in the control of organelle positioning, segregation, and cytokinesis in the trypanosome cell cycle. J Cell Biol 128:1163–1172

Russel DG, Burns RG (1984) The polar ring of coccidion sporozoites: a unique microtubule-organizaing center. J Cell Sci 65:193–207

Sahasrabuddhe AA, Gupta CM (2009) Trypanosomatid actins: a new class of eukaryotic actins. Trends Cell Mol Biochem 4:15–23

Sahasrabuddhe AA, Bajpai VK, Gupta CM (2004) A novel form of actin in *Leishmania*: molecular characterization, subcellular localization and association with subpellicular microtubules. Mol Biochem Parasitol 134:105–114

Sahoo N, Beatty W, Heuser J, Sept D, Sibley LD (2006) Unusual kinetic and structural properties control rapid assembly and turnover of actin in the parasite *Toxoplasma gondii*. Mol Biol Cell 17:895–906

Sasse R, Gull K (1988) Tubulin post-translational modifications and the construction of microtubular organelles in *Trypanosoma brucei*. J Cell Sci 90:577–589

Schatten H, Sibley LD, Ris H (2003) Structural evidence for actin-like filaments in *Toxoplasma gondii* using high-resolution lwo voltage field emission scanning electron microscopy. Microsc Microanal 9:330–335

Schneider A, Sherwin T, Sasse R, Russel DG, Gull K, Seebeck T (1987) Subpellicular and flagellar microtubules of *Trypanosoma brucei* are extensively glutamylated. J Cell Biol 104:431–438

Schneider A, Hemphill A, Wyler T, Seebeck T (1988) Large microtubule-associated protein of *T. brucei* has tandemly repeated, near-identical sequences. Science 241:459–462

Schneider A, Plessmann U, Weber K (1997) Subpellicular and flagellar microtubules of *Trypanosoma brucei* are extensively glutamylated. J Cell Sci 110:431–437

Schuler H, Matuschewski K (2006a) Regulation of Apicomplexan microfilament dynamics by a minimal set of actin-binding proteins. Traffic 7:1433–1439

Schuler H, Matuschewski K (2006b) *Plasmodium* motility: actin not actin's like actin. Trends Parasitol 22:146–147

Schwartzman JD, Pfefferkorn ER (1983) Immunofluorescence localization of myosin at the anterior pole of the coccidian *Toxoplasma gondii*. J Protozool 30:657–661

Scott V, Sherwin T, Gull K (1997) Gamma-tubulin in trypanosomes: molecular characterisation and localisation to multiple and diverse microtubule organising centres. J Cell Sci 110:157–168

Seebeck T, Gehr P (1983) Trypanocidal action of neuroleptic phenothiazines in *Trypanosoma brucei*. Mol Biochem Parasitol 9:197–208

Shaw MK, Tilney LG (1999) Induction of an acrosomal process in *Toxoplasma gondii*: visualization of actin filaments in a protozoan parasite. Proc Natl Acad Sci USA 96:9095–9099

Sheffield HG, Melton ML (1968) The fine structure and reproduction of *Toxoplasma gondii*. J Parasitol 54:209–226

Sherwin T, Gull K (1989) The cell division cycle of Trypanosoma brucei brucei: timing of event markers and cytoskeletal modulations. Phil Trans R Soc Lond B 323:573–588

Soares TC, De Souza W (1977) Fixation of trypanosomatids for electron microscopy with the glutaraldehyde-tannic acid method. Z Parasitenkd 53:149–154

Souto-Padron T, De Souza W, Heuser JE (1984) Quick-freeze, deep-etch rotary replication of *Trypanosoma cruzi* and *Herpetomonas megaseliae*. J Cell Sci 69:167–178

Souto-padron T, Cunha e Silva NL, De Souza W (1993) Acetylated alpha-tubulin in Trypanosoma cruzi: immunocytochemical localization. Mem Inst Oswaldo Cruz 88:517–528

Speer CA, Dubey JP (2001) Ultrastructure of schizonts and merozoites of Sarcocystis neurona. Vet Parasitol 95:263–271

Stieger J, Wyler T, Seebeck T (1984) Partial purification and characterization of microtubular protein from *Trypanosoma brucei*. J Biol Chem 259:4596–4602

Stokkermans TJ, Schwartzman JD, Keenan K, Morrissete NS, Tilney LG, Roos DS (1996) Inhibition of *Toxoplasma gondii* replication by dinitroaniline herbicides. Exp Parasitol 84:355–370

Thomas MC, Garcia-Perez JL, Alonso C, Lopez MC (2000) Molecular characterization of KMP11 from *Trypanosoma cruzi*: a cytoskeleton-associated protein regulated at the translational level. DNA Cell Biol 19:47–57

Vanderberg J, Rhodin J, Yoeli M (1967) Electron microscopic and histochemical studies of sporozoite formation in *Plasmodium berghei*. J Protozool 14:82–103

Vedrenne C, Giroud C, Robinson D, Besteiro S, Bosc C, Bringaud F, Baltz T (2002) Two related subpellicular cytoskeleton-associated proteins in *Trypanosoma brucei* stabilize microtubules. Mol Biol Cell 13:1058–1070

Vivier E, Petitprez A (1972) Donnés ultrastructurales complémentaries, morphologiques et cytochimiques, sur *Toxoplasma gondii*. Protistologica VII:199–221

Webb SE, Fowler RE, O'Shaughnessy C, Pinder JC, Dluzewski AR, Gratzer WB, Bannister LH, Mitchell GH (1996) Contractile protein system in the asexual stages of the malaria parasite *Plasmodium falciparum*. Parasitology 112:451–457

Weber K, Schneider A, Muller N, Plessmann U (1996) Polyglycylation of tubulin in the diplomonad *Giardia lamblia*, one of the oldest eukaryotes. FEBS Lett 393:27–30

Webovetz KA, Sackett DL, Delfin D, Bhattacharya G, Salem M, Obrzut T, Rattendi D, Bacchi C (2003) Selective antimicrotubule activity of N1-phenyl-3, 5-dinitro-N4, N4-di-n-propylsulfanilamide (GB-II-5) against kinetoplastid parasites. Mol Pharmacol 64:1325–1333

Westermann S, Schneider A, Horn EK, Weber K (1999) Isolation of tubulin polyglutamylase from *Crithidia*; binding to microtubules and tubulin, and glutamylation of mammalian brain alpha- and beta-tubulins. J Cell Sci 112:2185–2193

Woods A, Baines AJ, Gull K (1992) A high molecular mass phosphoprotein defined by a novel monoclonal antibody is closely associated with the intermicrotubule cross bridges in the *Trypanosoma brucei* cytoskeleton. J Cell Sci 103:665–675

Flagellum Structure and Function in Trypanosomes

Johanna Buisson and Philippe Bastin

Contents

Abstract Trypanosomes are flagellated protozoan parasites responsible for devastating diseases in human and cattle. Recently, they have emerged as new models to study cilia and flagella thanks to powerful reverse genetics approaches coupled to the full sequencing of the genome of several species. In this chapter, we describe the ultra-structural features of the *Trypanosoma brucei* flagellum, revealing evolutionarily conserved aspects of the axoneme or the basal body and specific elements such as the paraflagellar rod or the flagellum attachment zone. We update the numerous functions demonstrated for this organelle, keeping in mind that most data were obtained from cultured parasites. The next challenges will be the determination of

J. Buisson and P. Bastin (✉)
Trypanosome Cell Biology Unit, Institut Pasteur & CNRS, 25 rue du Docteur Roux, 75015 Paris, France
e-mail: philippe.bastin@pasteur.fr

W. de Souza (ed.), *Structures and Organelles in Pathogenic Protists,*
Microbiology Monographs 17, DOI 10.1007/978-3-642-12863-9_3,
© Springer-Verlag Berlin Heidelberg 2010

the role of the flagellum in the complex *T. brucei* life cycle, transiting through tissues of the tsetse fly vector and swimming in the bloodstream of mammals.

1 Introduction

Cilia and flagella are appendages present at the surface of numerous eukaryotic cells. These organelles are composed of an axoneme, a set of microtubules made of nine doublets, circularly disposed around a central pair, wrapped by a specific membrane. They show a remarkable structural conservation throughout evolution from protists to mammals. However, their number, length and position can vary from one organism to another and even from one cell type to another in the same organism. In mammals, they are present in numerous tissues (respiratory epithelium, spermatozoa, eye inner ear, kidney,...) where they are responsible for various functions. Accordingly, defects in these organelles lead to a wide variety of diseases and developmental defects including polycystic kidney disease, primary ciliary dyskinesia and retinal degeneration (Fliegauf et al. 2007).

Many protozoan parasites possess a flagellum that has often been proposed to be essential for their survival and/or for host infection. These parasites are responsible for some devastating human and animal diseases. This review will focus on the flagellum of the African trypanosome *Trypanosoma brucei*, with reference to related Kinetoplastid parasites when available. *T. brucei* is a uniflagellated eukaryotic motile protozoan belonging to the Kinetoplastida order, which is defined by the presence of a single mitochondrion whose highly condensed DNA is made of large and mini-circles and called the kinetoplast (Liu et al. 2005). The trypanosome is transmitted by the bite of the tsetse fly and is responsible for African trypanosomiasis or sleeping sickness in humans. It accounts for at least 100,000 human infections. It can also infect cattle, causing nagana, a disease that restrains economic development in sub-Saharan Africa. If not treated with appropriate drugs, sleeping sickness is always fatal. Trypanosomes proliferate in the bloodstream as extracellular parasites and create multiple symptoms such as fever, swollen lymph glands, aching muscles and joints, headaches and irritability. Patients at early stages of infection are treated with suramine or pentamidine, discovered in 1921 and 1941, respectively, which ensure curing in 90% of the cases. At a late stage of infection, parasites cross the blood–brain barrier and patients need to be treated with arsenic derivatives such as melarsoprol, discovered in 1949. This heavy treatment is efficient but leads to encephalopathy in 5–10% patients and cases of resistance are rising.

During its life cycle, *T. brucei* alternates between two hosts, the tsetse fly and different mammals, with a complex parasitic cycle. A tsetse fly becomes a vector when feeding on an infected mammalian host and taking trypanosomes with the blood meal. The trypanosomes follow the digestive path during which they undergo successive biochemical and morphological changes. They first proliferate in the midgut, differentiate and migrate to the salivary glands where they undergo another phase of multiplication (Vickerman 1985). Being released in the saliva when mature, they are ready to be transmitted to a new host during the next blood meal. Exactly as in mammals, the parasite remains extracellular during its long stay in the fly. Whatever the host and the stage, the trypanosome always possesses a flagellum. However, flagellum length and positioning evolves during the life cycle. The procyclic form (found in the fly midgut) and the bloodstream form (present in the blood of mammals) can be grown in culture. Therefore, most studies have been performed on these two life cycle stages.

Procyclic and bloodstream form trypanosomes are described as trypomastigotes, i.e. the flagellum emerges from the posterior region of the cell and is attached on the cell body for most of its length (20–25 μm). The flagellum induces a propulsive wave that drags the cell forward, hence, they swim with the flagellum first, defining an anterior–posterior axis (Fig. 3). Despite close morphological resemblance, several differences can be noted. The bloodstream form cells exhibit a dense surface coat (12–15 nm) composed of tightly packed variant surface glycoproteins (VSG) (Vickerman 1969; Vickerman and Luckins 1969). The replacement of VSG during antigenic variation allows the parasite to escape the host immune response and is responsible for the typical peaks of parasitemia. Another difference is the extreme posterior positioning of the basal body compared to the procyclic form (Figs. 3 and 4). Endocytosis is much more active in bloodstream cells, and clathrin-coated vesicles are more prominent in that life cycle stage (Natesan et al. 2007) (Fig. 2a). Significant differences exist at the level of metabolism and surface protein expression, but will not be covered here.

2 The Structure of the Flagellum

Four distinct regions can be defined in the trypanosome flagellum of procyclic (Fig. 1) and bloodstream forms (Fig. 2) (1) the basal body (Figs. 1b and 2b) that is rooted in the cell body; (2) the transition zone (Figs. 1c, d and 2c, d) located in the flagellar pocket; (3) the axoneme alone, starting in the flagellar pocket (Figs. 1e and 2e), emerging at the surface of the cell and (4) the axoneme accompanied by the paraflagellar rod (PFR), both running along the cell body (Figs. 1f and 2f).

3 The Basal Body

The basal body is the microtubule organising centre of the flagellum. In procyclic and bloodstream trypanosomes, it is located at the posterior tip of the cell. Its positioning can vary during the life cycle, as encountered in several stages of development in the tsetse fly where the basal body is found in an anterior position relative to the nucleus. Such stages are termed ‘epimastigotes’ and also exist in the related parasite *Trypanosoma cruzi*, responsible for Chagas disease. The basal body

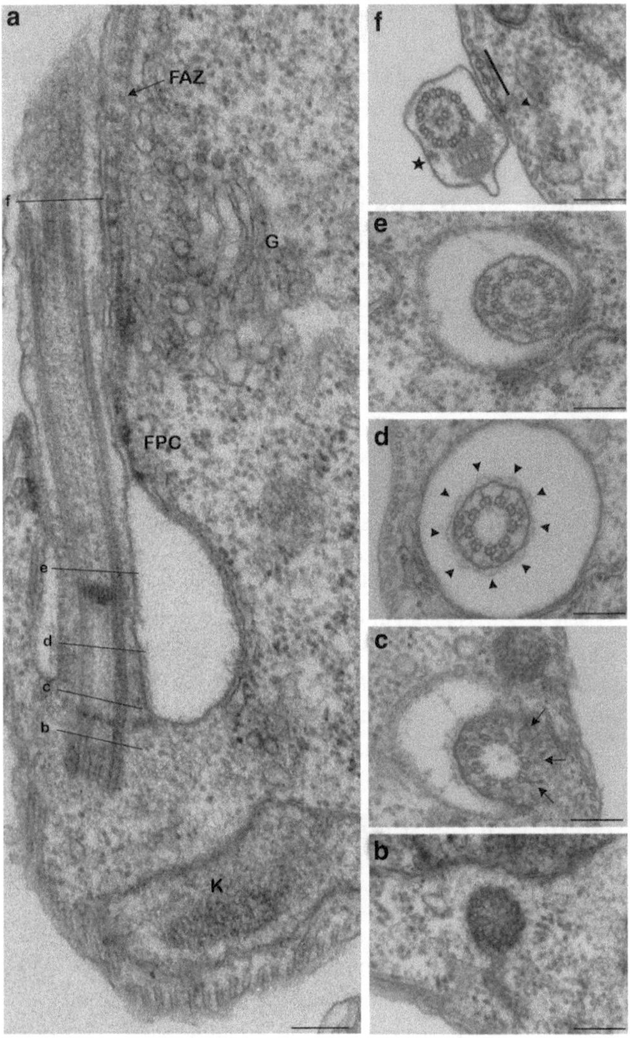

Fig. 1 Ultra-structure of the flagellum in culture procyclic form of *T. brucei*. (**a**) Longitudinal section through the flagellar pocket and the base of the flagellum, including the mature basal body. The pro-basal body is not visible on this section. (**b–f**) Cross-sections through the flagellum at the indicated areas of (**a**). (**b**) Basal body. (**c**) Base of the transition zone. Notice the transitional fibres (*arrows*). (**d**) Transition zone. *Arrowheads* indicate the collarette. (**e**) Axoneme within the flagellar pocket lumen. Electron-dense material is frequently observed between peripheral doublets and the membrane. (**f**) The flagellum contains the axoneme and the PFR after its exit from the flagellar pocket. The *bar* indicates the position of the four microtubules associated to the FAZ and the *arrowhead* points at the FAZ filament. The *star* indicates an IFT particle. *FPC* flagellar pocket collar, *G* Golgi apparatus, *K* kinetoplast. *Scale bars* are 100 nm

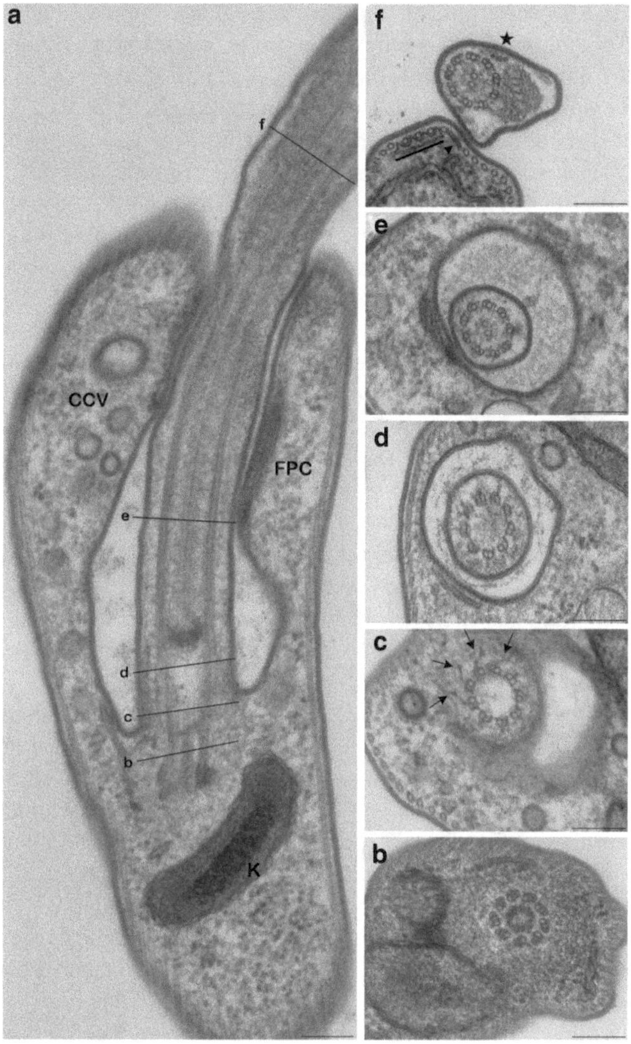

Fig. 2 Ultra-structure of the flagellum in cultured bloodstream form of *T. brucei*. (**a**) Longitudinal section through the flagellar pocket and the base of the flagellum, including the mature basal body. The pro-basal body is not visible on this section. (**b–f**) Cross-sections through the flagellum at the indicated areas of (**a**). (**b**) Basal body. (**c**) Base of the transition zone. Notice the transitional fibres (*arrows*). (**d**) Transition zone. (**e**) Axoneme within the flagellar pocket lumen. Electron-dense material is frequently observed between peripheral doublets and the membrane. (**f**) The flagellum contains the axoneme and the PFR after its exit from the flagellar pocket. The *bar* indicates the position of the four microtubules associated to the FAZ and the *arrowhead* points at the FAZ filament. Notice the dense surface coat that decorates the membrane, including that of the flagellar pocket. *CCV* clathrin-coated vesicles, *FPC* flagellar pocket collar, *K* kinetoplast. *Scale bars* are 100 nm

shares a similar organisation with the centrioles. It is a complex made of a mature basal body subtending the axoneme and a shorter adjacent pro-basal body found in a perpendicular position (arrow on Fig. 6a). The proximal part of a mature basal body is built up with nine triplets of microtubules (A, B and C) organised in a barrel-like structure around a central axis (Figs. 1a and 2a for longitudinal sections and Figs. 1b and 2b for cross-sections). The pro-basal body is restricted to that proximal part. At the distal part of the mature basal body, the C microtubules end while A and B continue to form the microtubule doublets of the transition zone. Electron-dense material is present in the lumen and a cartwheel-like structure is present at the proximal part (Lacomble et al. 2009). Subtle differences in basal body structures may exist between bloodstream and procyclic forms but further comparative studies will be required to figure out their extent. Several proteins have been localised to the basal body, such as gamma-tubulin (Scott et al. 1997), a leucine-rich protein (Morgan et al. 2005), centrins (He et al. 2005; Selvapandiyan et al. 2007) and intraflagellar transport (IFT) proteins (Absalon et al. 2008b). All these proteins are conserved throughout evolution and the quest for trypanosome-specific basal body proteins is still open. One candidate is KMP11, a short protein that localises not only to the basal body but also to the flagellum and appears to participate in the control of basal body segregation and flagellar adhesion (Li and Wang 2008). However, it is not clear at this stage whether this function is ensured by KMP-11 at the basal bodies or (and) at the flagellum.

The basal body is physically linked to the kinetoplast (mitochondrial DNA or kDNA) by a network of filaments extending from the proximal end to the kDNA through the mitochondrial membranes, the tripartite attachment complex (TAC) (Ogbadoyi et al. 2003; Robinson and Gull 1991). A single molecular component has been identified so far: the p166 protein that has an amino-terminal mitochondrial targeting signal and is predicted to be a transmembrane protein. It is essential for segregation of kinetoplast DNA but is not involved in its duplication nor in the duplication of basal bodies (Zhao et al. 2008). The TAC filaments are also recognised by a monoclonal antibody, termed MAb22, that was obtained after immunisation with purified skeletons of trypanosome flagellum (Bonhivers et al. 2008a). Unfortunately, this antibody does not blot and so far the antigen it detects has not been identified. It is nevertheless frequently used as a marker of the basal body/kinetoplast.

3.1 The Transition Zone

The transition zone is located in the flagellar pocket and is the region between the distal ends of the C-tubules and the start of the central pair of microtubules of the axoneme at the basal plate (Figs. 1c, d and 2c, d). It is also there that the flagellum membrane starts (Figs. 1a, c and 2a, c). At the base of the transition zone, electron-dense projections appear to connect each outer microtubule doublet to the flagellar membrane (arrows on Figs. 1c and 2c). These radial fibres are forming a necklace that could be responsible for a physical separation of the flagellum from the cytoplasm. They are thought to serve as a docking site for the proteins that enter the flagellum (Deane et al. 2001). In the procyclic form, an additional structure called the collarette shows material around the flagellum membrane in the flagellar pocket lumen at the proximal part of the transition zone. Originally described in longitudinal section (Vickerman 1973), transversal sections show a well-organised structure mimicking the nine microtubule doublets, each microtubule doublet having its corresponding replica. These double-tube-like units are linked to one another forming a continuous necklace around the flagellum (Lacomble et al. 2009). The function of such an arrangement remains enigmatic.

The tubulin folding co-factor C (or RP2) is the only protein that has been unambiguously localised to the transition zone fibres, where it could act as a 'quality control gateway' for tubulin before incorporation in the flagellar compartment (Stephan et al. 2007). The protein makes a ring around the mature basal body but does not localise to the pro-basal body. The typical 'double-spot' images frequently reported in many papers would correspond to viewing the structure from the side (Stephan et al. 2007). A Trypanosomatid-specific coiled-coil rich protein (TBBC) has been localised to the basal body (Dilbeck et al. 1999) and only stains the mature basal body by indirect immunofluorescence (Absalon et al. 2008a), suggesting it is associated to the transition zone. In its absence, flagella get detached from the cell body and basal body migration is aborted (Absalon et al. 2007). Construction of the transition zone is associated with that of the flagellum, as inactivation of IFT (see below) results in failure to elongate the transition zone properly, albeit without interfering with basal body duplication (Absalon et al. 2008a) (Fig. 5e, f).

3.2 *The Axoneme*

The axoneme is in continuity with the transition zone and starts when the flagellum is still in the flagellar pocket. It exhibits the canonical circular arrangement of nine doublets of microtubules (A and B) surrounding a central pair apparatus of singlet microtubules (Figs. 1e and 2e). The central pair initiates at the basal plate and extends to the distal tip of the flagellum. The singlet microtubules of the central pair are connected to one another and exhibit several projections (Branche et al. 2006). Here too, further electron microscopy studies will be necessary to determine the exact structures of these projections and to allow comparison with other organisms. The outer microtubule doublets are linked to one another via a nexin link and radial spokes project from the A microtubule towards the central pair. They display on microtubule A inner and outer dynein arms responsible for the flagellar movement and thus for cell motility. Comparative genomic and proteomic studies revealed that numerous proteins belonging to the dynein arms, central pair and radial spoke family are well conserved in trypanosomes and functional analyses have confirmed that the conservation extends beyond the structural level (Baron et al. 2007a, b; Branche et al. 2006; Broadhead et al. 2006; Dawe et al. 2007; Ralston et al. 2006). Over recent years, trypanosomes have emerged as potent models to study the role of genes involved in axoneme motility (Baron et al. 2007b; Dawe et al. 2005, 2007; Hutchings et al. 2002), especially those whose defects are associated to genetic diseases such as primary ciliary dyskinesia (our unpublished data).

3.3 The Paraflagellar Rod

When the flagellum emerges from the flagellar pocket, the PFR, an additional lattice-like structure, is present along the axoneme until its distal tip (Figs. 1f and 2f). The PFR has a similar diameter as the axoneme and is restricted to Euglenoids (Bastin et al. 1996; Cachon et al. 1988; Maga and LeBowitz 1999). Transmission cross-sections show three domains, defined by their relative position to the axoneme. The closest domain of the axoneme is called the proximal domain and is physically linked to the axoneme via fibres connecting to the doublets four and seven. The distal domain has a very similar pattern as the proximal one, showing stacks of plates whereas the intermediate domain is constituted of filaments perpendicular to the plates (Farina et al. 1986). The connection between the PFR and the axoneme is very strong and resists to a variety of treatments, thus making PFR purification difficult (Russell et al. 1983). Recent proteomic studies reveal that the PFR comprises at least 20 proteins, most of them being specific to Trypanosomatid species (Portman et al. 2009). PFR1 and PFR2 are coiled-coil rich proteins and constitute the main components of the PFR (Deflorin et al. 1994; Schlaeppi et al. 1989). Their genes are conserved in other Trypanosomatid species (Maga and LeBowitz 1999) and in *Euglena* (Ngô and Bouck 1998) but absent in the genomes of species without a PFR. Absence of PFR1 or PFR2 leads to a disrupted PFR, with an accumulation of the other protein at the distal tip of the flagellum (Bastin et al. 1998, 1999b; Maga et al. 1999; Santrich et al. 1997). *T. brucei* and *Leishmania mexicana* mutants with a disrupted PFR show strongly diminished motility, although their axoneme looks normal. Consequently, PFR appears necessary for motility but its exact contribution remains enigmatic. The PFR is extremely reduced in certain Trypanosomatid species that bear endosymbionts, yet these do not appear to swim less well compared to the closest species with a normal PFR (Gadelha et al. 2005). Interpretation of the motility phenotype of the PFR mutants described above remains difficult as this could reflect flagellum structural alterations and possibly indirect perturbation of axoneme function rather than a specific contribution of the PFR to flagellum beating.

Finally, it is amazing to observe that some evolutionarily conserved proteins are specifically localised to the PFR, such as several adenylate kinases (Pullen et al. 2004) and calmodulin (Ridgley et al. 2000), suggesting involvement of nucleotides and calcium in assembly and function of this intriguing structure.

4 Flagellar Positioning and Adhesion

The flagellum emerges from the flagellar pocket at the posterior end of the cell and is attached to the cell body for most of its length (with the exception of the distal tip) and follows a left-handed helical path towards the anterior end (Fig. 3a, c). The site of attachment defines a specialised region of the cell body that has been called the flagellum attachment zone (FAZ) (Kohl and Gull 1998). The FAZ is the association of two distinct structures (Figs. 1f and 2f):

(1) The FAZ filament is positioned in a gap between two microtubules of the sub-pellicular corset. It is composed of a row of punctuated electron-dense structures (Fig. 3b, d) with a periodicity of ~95 nm when looking at a longitudinal section. In transversal cross-sections, it looks like an electron-dense area found in a gap between two microtubules of the corset (arrowhead on Figs. 1f and 2f). FAZ1 is a large protein containing 36–70 repetitions of a unique 14 amino acid domain and is specifically localised in the FAZ filament. However, it does not appear necessary for filament assembly but could contribute to cell cycle progression (Vaughan et al. 2008). (2) The FAZ also comprises a subset of four specialised microtubules that have greater biochemical stability upon high-salt treatment (Sherwin and Gull 1989) and that are always associated with the smooth endoplasmic reticulum (bar on Figs. 1f and 2f). Knowledge of the molecular nature of these microtubules remains limited to tubulins: gamma-tubulin (Scott et al. 1997) and possibly a specific isoform of beta-tubulin (Gallo et al. 1988).

Inside the trypanosome flagellum, a filamentous structure seems to connect the region of the membrane associated with the FAZ to the proximal domain of the PFR (Figs. 1f and 2f) (Sherwin and Gull 1989), and electron-dense material fills the inter-space between the flagellum membrane and the cell membrane. FLA1 is a Trypanosomatid-specific trans-membrane glycoprotein that is essential for flagellum attachment in *T. cruzi* (Cooper et al. 1993) and in *T. brucei* (LaCount et al. 2002), and has been proposed to act as a linker between the cell surface and the flagellum. It should be noted that a FAZ filament and the four microtubules are well present in *Leishmania* species, but their length is much shorter compared to trypanosome species. Accordingly, genes encoding FLA1 and FAZ1 are present in all three sequenced genomes of *Leishmania* subspecies (Kohl and Bastin 2005).

Despite its privileged location, there is no direct evidence that the FAZ is really involved in flagellum attachment. FAZ1 RNAi mutants show partially detached flagella (Vaughan et al. 2008) but this is relatively modest and multiple RNAi mutants exhibit a stronger phenotype (Absalon et al. 2007; Hutchings et al. 2002;

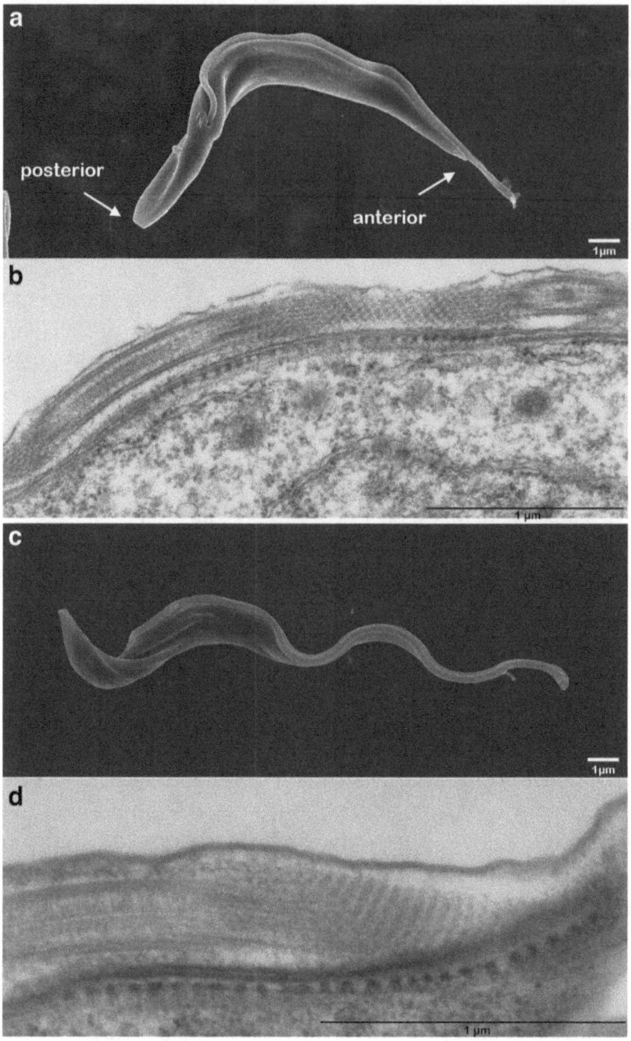

Fig. 3 Flagellum disposition at the cell surface of cultured procyclic and bloodstream trypanosomes. (**a**, **c**) Scanning electron micrographs of a procyclic (**a**) and a bloodstream (**c**) trypanosome. The posterior end is at the left side of the image and the cells would swim from left to right. (**b**, **d**) Longitudinal section through the flagellum attachment zone in a procyclic (**b**) and a bloodstream (**d**) trypanosome. The repetitive nature of FAZ electron-dense structures found immediately underneath the plasma membrane is easily recognised

Li and Wang 2008; Rodgers et al. 2007; Selvapandiyan et al. 2007). The FAZ has been proposed to play a pivotal role in the definition of cell division (Robinson et al. 1995), possibly providing structural information required for the positioning of the cleavage furrow during cytokinesis. This hypothesis is strengthened by the fact that formation of a shorter new flagellum results in a shorter FAZ and in formation of abnormally small cells (see below) (Kohl et al. 2003).

The flagellum is always positioned in a flagellar pocket, a surface depression of the plasma membrane that adopts an inverted flask shape and that is the only site of vesicle exchange between the cytoplasm and the surface (Field et al. 2007; Overath and Engstler 2004). This collar of the flagellar pocket contains cytoskeletal material where a single protein has been identified so far. BILBO1 is a Trypanosomatid-specific protein that contains EF-hand domains and localises exclusively to the collar. In its absence, the flagellar pocket is not made anymore and vesicular trafficking is dramatically halted. However, the flagellum is still assembled but it is shorter and not attached to the cell body. The FAZ is not made and the basal body migrates to the far posterior end of the cell (Bonhivers et al. 2008b). In the case where the flagellum is absent, formation of the flagellar pocket is altered, resulting in reduced endocytosis activity and mis-localisation of flagellar pocket proteins (Absalon et al. 2008a). The flagellum, therefore, contributes to the formation of a normal flagellar pocket.

5 Flagellum Construction

5.1 *Trypanosome Cell Cycle*

Trypanosomes, like all the Trypanosomatids, replicate by binary fission, without disassembling their cytoskeleton. The first detectable event in the cell cycle is the maturation of the pro-basal body elongating to become competent for nucleating a new flagellum. This maturation process is concomitant with the formation of a new pro-basal body alongside each mature basal body (Sherwin and Gull 1989). Such a cell possesses two basal body complexes, the pre-existing one that is already carrying its flagellum and the newly matured one. In procyclic cells, the new flagellum is assembled in the same flagellar pocket but curiously is positioned in an anterior position (Fig. 5b, c) (Absalon et al. 2008b; Briggs et al. 2004; Grasse 1961). However, as the new flagellum elongates and exits from the flagellar pocket, it is found in the right posterior position (Fig. 4b, f). When emerging at the cell surface, the new flagellum possesses its own flagellar pocket (Figs. 5d and 6a). When viewed from the posterior end, the new flagellum is always positioned on the left side of the old one. The already existing flagellum remains in place when the new one is built, unlike in *Chlamydomonas* and metazoa where flagella are disassembled at every cell cycle for the basal bodies to be used as centrioles at the spindle pole for mitosis before nucleating the new flagellum (Cavalier-Smith 1974). The new flagellum keeps on elongating and cytokinesis is initiated from the anterior end of the cell (Fig. 4d, h). In procyclic trypanosomes, the cell inheriting the new flagellum is smaller than the cell retaining the pre-existing one (Fig. 4d), the flagellum continues to grow until definitive size is reached and the cell can undergo another duplication cycle (Farr and Gull 2009).

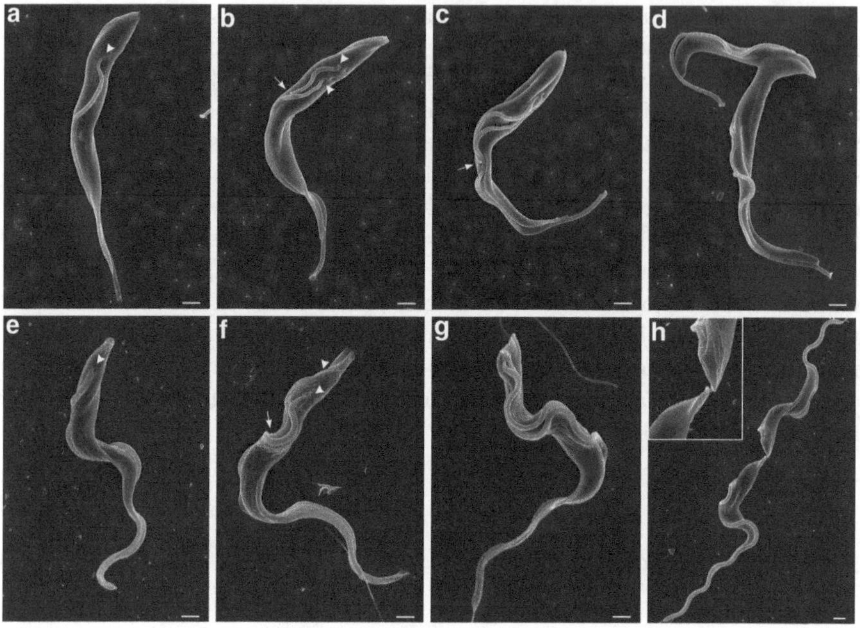

Fig. 4 The cell cycle of cultured procyclic (**a–d**) and bloodstream (**e–h**) trypanosomes. (**a**, **e**) Cells at the G1 stage possess a single flagellum emerging from the flagellar pocket (*arrowhead*) at the posterior end. (**b**, **f**) A new flagellum is assembled, emerges from its own flagellar pocket (*arrowhead*) and is always found in a posterior position relative to the old flagellum. Its tip elongates towards the anterior end of the cell and is in close proximity of the old flagellum (*arrow*). (**c**, **g**) The new flagellum further elongates with its distal tip physically connected by the FC only in the procyclic form (**c**, *arrow*). (**d**, **h**) Cell division takes place, initiating through the anterior end. Notice the different sizes of the two daughter cells in the procyclic form (**d**), which is far less obvious in bloodstream cells (**h**). The *inset* in (**h**) shows the magnified area of the midbody. *Scale bar* is 1 μm

In procyclic trypanosomes, an unusual transmembrane complex, the flagellar connector (FC), forces this positioning (Moreira-Leite et al. 2001) (Fig. 6b). The FC is an amazingly stable structure that is resistant to detergent or high-salt treatment and is capable to withstand the force induced by the vigorous beating of the new flagellum whose movement is not synchronised with that of the old one. This structure is mobile since it migrates along the old flagellum as the new flagellum elongates (Briggs et al. 2004; Kohl et al. 2003). The FC is composed of three distinct zones, each composed of electron-dense plates on the internal side of each flagellum, of filaments connecting the opposite plates in the interflagellum space and of filaments connecting the plates to specific outer microtubule doublets in each flagellum (Fig. 6b). The FC is formed very early in the cell cycle, before the new flagellum exits the pocket and probably coincident with the formation of the new flagellar axoneme. Such a structure is not found at the tip of the mature flagellum. There is some flexibility in the time point for the presumed dissociation of the FC,

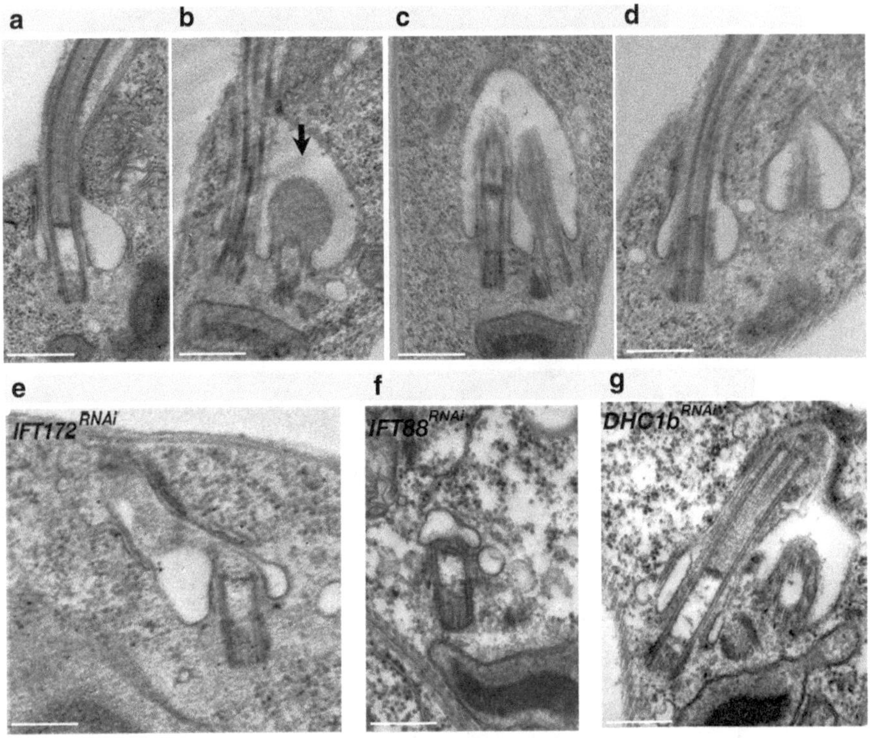

Fig. 5 The new flagellum is assembled in the existing flagellar pocket in procyclic trypanosomes and its construction is inhibited in the absence of IFT. Sections through the flagellar pocket of wild-type (**a**, **d**) or *IFT172*^{RNAi} cells induced for 48 h (**e**) or of *IFT88*^{RNAi} (**f**), *DHC1b*^{RNAi} (**g**) induced for 72 h. Notice the large amount of material only in very short flagella of wild-type cells (**b**). The *arrow* on (**b**) indicates the new flagellum. It grows in the same flagellar pocket as the old one (**b**, **c**) and elongates until two separate flagellar pockets are recognised (**d**). *IFT172*^{RNAi} and *IFT88*^{RNAi} mutants fail to elongate a flagellum and exhibit bald basal bodies (**e**, **f**). Notice the shorter transition zone compared to wild-type cells or *DHC1b*^{RNAi} cells where formation of short flagella occurs, with accumulation of IFT material (**g**). *Bars* are 500 nm. Modified from Absalon et al. (2008b)

allowing detachment of the tip of the flagellum. It can occur at anytime after initiation and early progression of the cleavage furrow. The FC has not been found in bloodstream form. Nevertheless, the tip of the new flagellum always stays close to the side of the mature flagellum during elongation (Briggs et al. 2004) and little is known on how the correct tracking of the new flagellum is ensured in that life cycle.

The growth of the new flagellum is accompanied by the migration of the two basal body complexes and kinetoplast segregation. This relationship is bimodal, with a slow separation of the basal bodies at early stages of flagellum elongation, followed by a rapid migration at later stages (Robinson et al. 1995). Strikingly, the turning point corresponds to an arrest of FC migration (Absalon et al. 2007;

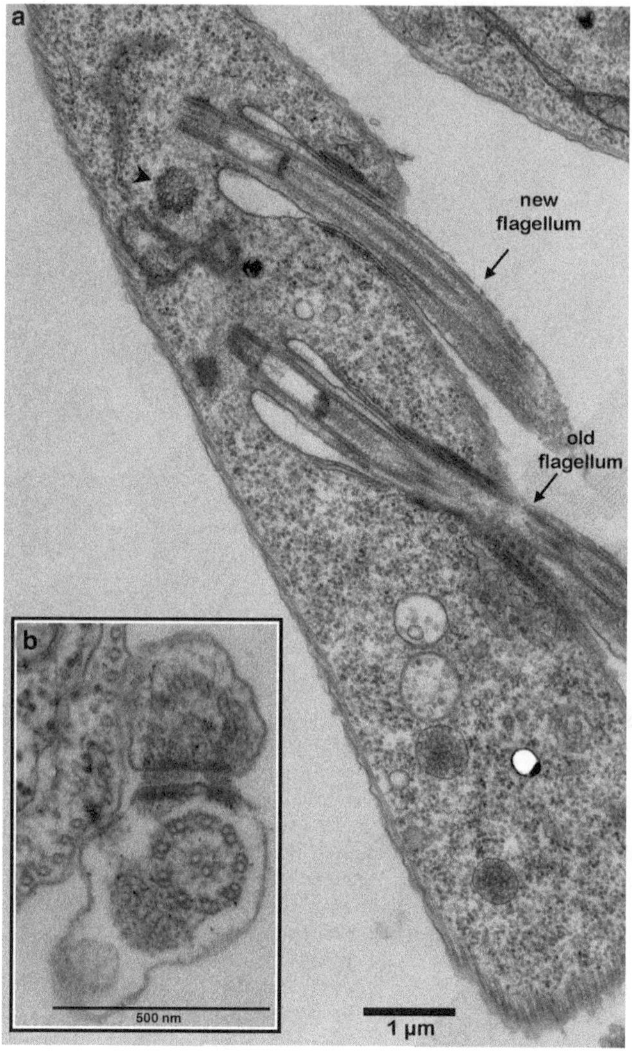

Fig. 6 Duplication of the flagellum in procyclic cells. (**a**) Sections through a wild-type cell that possesses two flagella as indicated. The basal body, the transition zone, the axoneme and the flagellar pocket with its collar for each flagellum are clearly visible. The pro-basal body (arrowhead) of the new flagellum, sectioned at the level of its cartwheel, is found perpendicular to the mature basal body. (**b**) Section through the tip of the new flagellum showing the connection with the old flagellum. The flagellar connector is a complex structure composed of three connection zones between the opposing flagellar membranes. Some electron-dense plates are visible on both sides of the old and the new flagellum

Davidge et al. 2006), leading to the proposal that in addition to its role in positioning of the new flagellum and its associated FAZ structures, the FC could be involved in exhaustive basal body migration. This is in agreement with the limited migration of basal bodies in *Crithidia*, *Leishmania* or *T. cruzi*, i.e. species that lack an FC. Blocking formation, base-to-tip motility or connection of the new flagellum all result in aborted basal body migration (Absalon et al. 2007).

Positioning of the new flagellum and FAZ defines the cytokinesis axis that is initiated at the anterior end of the FAZ and proceeds following a helical pattern towards the posterior end of the cell producing two daughter cells (Robinson et al. 1995). Construction of a shorter new flagellum is related with the formation of a shorter FAZ and results in the birth of shorter cells, with a direct relationship with flagellum length (Kohl et al. 2003). Remarkably, a similar situation is encountered during the life cycle when parasites differentiate in the tsetse fly (Sharma et al. 2008). The complete absence of a new flagellum is accompanied by the formation of a shorter new FAZ and the cleavage furrow is initiated in an asymmetric way leading to one normal cell retaining the old pre-existing flagellum and a small daughter cell with no flagellum. This non-flagellated cell is smaller (half the size), loses its polarity but can continue nuclear and kinetoplast DNA duplication. However, it completely fails to construct a FAZ and is not able to divide leading to multinucleated cell and eventually to cell death (Kohl et al. 2003). The flagellum is, therefore, a key actor in the trypanosome cell cycle.

5.2 Intraflagellar Transport

Building cilia or flagella is a huge commitment, as this requires the correct production and assembly of a large number of proteins, both in time (right moment of the cell cycle) and in space (in a separate compartment of the cell). Proteomic analysis showed that the skeleton of the *T. brucei* flagellum, i.e. the axoneme and the PFR without the membrane and the matrix, is composed of at least 330 different peptides (Broadhead et al. 2006). A similar fraction contains 223 proteins in the ciliate *Tetrahymena* whereas a complete *Chlamydomonas* flagellum, including the membrane and matrix fractions, is made of 510 proteins (Pazour et al. 2005). Since the flagellum does not possess any ribosomes, all the components needed for the construction and maintenance of this organelle are synthesised in the cytoplasm of the cell body and must be imported in the flagellum where they have to be transported to the site of incorporation, the distal tip (Bastin et al. 1999a). This process is mediated by IFT, first discovered in *Chlamydomonas* (Kozminski et al. 1993).

IFT is a bidirectional motility process where molecular motors associated to complex cargoes travel along the axoneme from the base to the tip of the flagellum (Fig. 7). The motor for anterograde transport is a heterotrimeric complex, belonging to the kinesin-II family, able to walk on microtubules towards the plus end. Then, the anterograde motor and the associated IFT complex are recycled to the cell body. The motor ensuring retrograde transport is an isoform of cytoplasmic dynein, cytoplasmic dynein 1b, now called cytoplasmic dynein 2. This complex is composed of at least four different subunits, a heavy intermediate, light intermediate and light chain. The current model proposes that during flagellum construction IFT

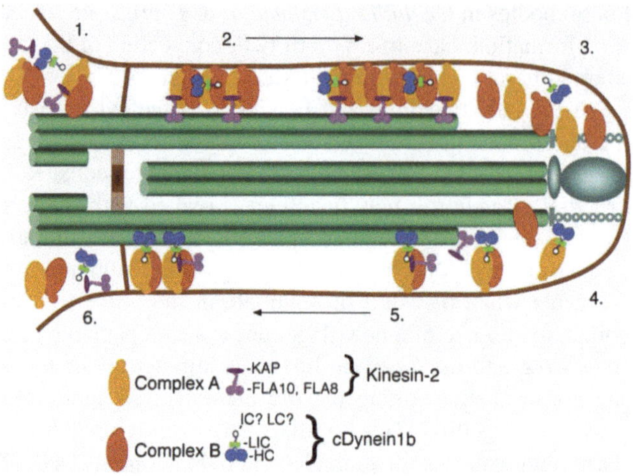

Fig. 7 Model for intraflagellar transport inferred from functional studies in the green algae *Chlamydomonas*. Step 1: gathering of IFT particles and motors in the peribasal body region. Step 2: kinesin-2-mediated anterograde transport of IFT complexes A and B and inactive cDynein1b, which is attached to complex B. Active kinesin-2 is associated with complex A. Step 3: release of complexes A and B and inactive cDynein1b into the flagellar tip compartment, followed by dissociation of complexes A and B and release of cDynein1b from complex B. Step 4: complex A binds to active cDynein1b via the LIC subunit, complex B then associates with complex A, and kinesin-2 binds to active cDynein1b. Step 5: active cDynein1b transports everything back to the cell body. Step 6: IFT components are recycled to the cell body. Reproduced with permission from Pedersen et al. (2006)

transports axoneme precursors to the distal tip for incorporation in the growing structure. Recycled products would be brought back to the cell body by retrograde IFT.

IFT particles are visible by electron microscopy (Fig. 1F, arrow and have originally been isolated from purified *Chlamydomonas* flagella on a sucrose density gradient (Cole et al. 1998). Two separate complexes (A and B) were identified and the amino acid sequences of these IFT polypeptides are characterised by an abundance of WD repeats, TPR repeat and coiled-coil motifs that are involved in protein–protein interaction. This fact is coherent with a role of the particles in the transport of flagellum precursors and turnover products. The exact way of interaction with motors, precursors and turnover product is still unknown as well as their regulation. Almost all IFT proteins and motors are conserved in Trypanosomatids with a few exceptions, such as the presence of two different genes for DHC1b (Adhiambo et al. 2005) or the absence of the KAP, suggesting that their kinesin II is hetero- or homodimeric (Julkowska and Bastin 2009). All IFT proteins investigated so far are essential for flagellum formation in *T. brucei* (Absalon et al. 2007, 2008b; Adhiambo et al. 2009; Davidge et al. 2006; Kohl et al. 2003) and *L. mexicana* (Adhiambo et al. 2005), as in most other species (Hao and Scholey 2009). Absence of the anterograde motor or of any protein of the B complex prevents flagellum elongation, leaving only a bald basal body (Fig. 5e, f) whereas inhibition of retrograde transport or of any IFT protein of the A complex results in abortive flagellum construction, with accumulation of other IFT proteins (Fig. 5g) (Absalon et al. 2008b).

In *T. brucei*, IFT proteins are found at the basal body and in the flagellum (Absalon et al. 2008b; Adhiambo et al. 2009). Life cell analysis revealed the continuous movement of IFT particles, which move along the axoneme at about 2 μm/s in the anterograde direction and 3.5 μm/s in the retrograde movement (Absalon et al. 2008b). This is not restricted to the new flagellum as IFT also takes place in the old flagellum, suggesting an important function in maintenance. IFT could also play a critical role in the exchange of material between the flagellum compartment and the cell body, possibly in relationship with sensory functions (Rotureau et al. 2009). Understanding the dynamics of IFT in trypanosomes, both in vitro and in vivo, could yield exciting results given the spectacular modulation of flagellum length, structure and positioning during the parasite cycle (Kohl et al. 2003; Sharma et al. 2008).

6 Functions of a Motile Flagellum

6.1 Characteristics of Flagellum Motility

The flagellum is responsible for trypanosome motility. It beats in a helicoidal way (Hill 2003), and two different types of waves are easily detected (Branche et al. 2006). The main one initiates from the distal tip of the flagellum and propagates towards the base with low amplitude and high frequency, inducing forward movement of the cell. The other one starts from the base of the flagellum and propagates towards the distal tip with high amplitude and low frequency. It is responsible for the reorientation of the cell (Branche et al. 2006) and can also induce backward motility (Baron et al. 2007a). The trypanosome is, therefore, able to swim in the three dimensions of the space. These features are conserved in other Trypanosomatid species (Gadelha et al. 2007). Outer dynein arms are the main motors for forward motility (Baron et al. 2007a; Branche et al. 2006) and inner dynein arms are likely to govern reverse movement, although mutants lacking inner arms have still not been described. Central pair and radial spokes control dynein activity (Branche et al. 2006; Dawe et al. 2007; Ralston et al. 2006), as well as the dynein regulatory complex whose very first molecular subunit was discovered and characterised in *T. brucei* (Hutchings et al. 2002). The role of dynein arms, central pair projections or radial spokes in flagellum beating is conserved during evolution and has been demonstrated in trypanosomes, green algae and mammals. Despite this central role of the axoneme, trypanosomes possess the PFR whose ablation dramatically reduces motility. Several hypotheses have been put forward to explain the contribution of the PFR to flagellum beating, yet the reasons for presence of such a large structure remain elusive (Ralston et al. 2009).

6.2 Flagellum Motility Contributes to Completion of Cell Division

Formation of the new flagellum plays essential functions in cell morphogenesis, contributing to basal body positioning, flagellar pocket formation and definition of the cytokinesis axis. Flagella exert yet another role at the end of the cell cycle, once cytokinesis is almost complete when the two future daughter cells face each other (Fig. 4d, h). The flagella point in opposite direction and appear to drag the cells apart. This force is necessary to tear the midbody (Fig. 4h) and to allow cell separation. Many paralysed mutants at the procyclic form fail at this stage, except if the cultures are shaken indicating that this action is purely mechanical (Branche et al. 2006; Ralston et al. 2006). However, mutants with reduced motility do not exhibit such a severe phenotype and are able to divide almost normally. In contrast, even slight alterations of motility have dramatic consequences in the bloodstream stage of the parasite (Branche et al. 2006; Broadhead et al. 2006; Ralston and Hill 2006). Cells fail cytokinesis but keep on duplicating their nuclei and kinetoplasts and assemble multiple flagella, resulting in ‘monster cells’ that ultimately die. The role of motility in final separation at the end of the cell cycle is conserved throughout evolution, not only with flagellar/ciliary movement (Brown et al. 1999) but also with amoeboid movement (Tuxworth et al. 1997).

6.3 Motility and Clearance of Surface Proteins

At a given time during infection, the trypanosome surface is covered by a dense coat made of ten million identical copies of a defined VSG (Cross 1975). Antibodies produced against VSG will progressively lead to trypanosome destruction, but a new *VSG* gene can be activated and these cells will lead to the emergence of a novel population. Clearance of surface-bound antibodies allows the parasite to resist for longer to the immune system. VSG and bound antibodies are captured by endocytosis, transit through early endosomes where acidification leads to dissociation of the antibody–VSG couple. The VSG proteins are rapidly returned to the surface whereas the antibody is sent to the lysosome for destruction. Careful quantification demonstrated that the whole VSG surface pool was turned over in only 12 min and the internal pool ($\sim$10%) in only 1 min (Engstler et al. 2004). However, the subpellicular corset of microtubules prevents endocytosis at the surface of the trypanosome cell, with the exception of the flagellar pocket. This implies that only a small surface, located at the posterior end of the cell, has to deal with such huge amounts of material (Fig. 8a). Strikingly, forward motility is required for the efficient transfer of VSG and bound antibodies from the cell surface to the posterior end where the flagellar pocket is located (Engstler et al. 2007). In the outer dynein arm mutant that can only swim backward, bound antibodies accumulate at the anterior end of the cell (Fig. 8b), demonstrating the role of motility in the dynamics of surface molecules (Engstler et al. 2007).

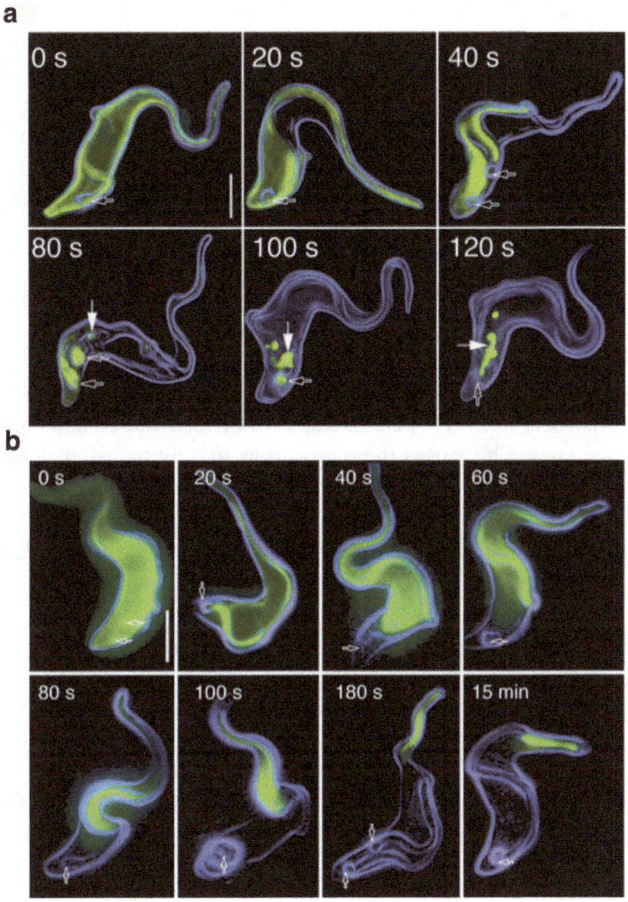

Fig. 8 Forward motility is required for efficient transfer of surface-bound material to the flagellar pocket. (**a**) Visualization of antibody removal. Cells were surface labelled with blue-fluorescent dye and incubated for 10 min on ice with anti-VSG-specific IgG (*green*). Following 0–3 min of incubation at 37°C, cells were fixed and permeabilised. *Open arrows* indicate the position of the flagellar pocket, and *filled arrows* point to the lysosome. *Scale bar*, 3 μm. (**b**) The swimming direction determines the route of movement of IgG-VSG on the cell surface. Trypanosomes deprived of dynein arm intermediate chain (DNAI1) show backward motion (Branche et al. 2006) and reversed direction of IgG-VSG movement. RNAi against DNAI1 was induced for 10 h. Reproduced with permission from Engstler et al. (2007)

7 Conclusions and Perspectives

The flagellum is a prominent, yet complex, organelle in African trypanosomes. Over the last few years, great progress has been achieved in understanding the molecular composition and the function of individual elements, unravelling an array of essential and sometimes unexpected functions during the parasite cell cycle. Despite the high sequence and functional conservation of axoneme and IFT proteins, peculiar adaptations have been achieved to fulfil trypanosome-specific functions. The next challenges reside in the determination of flagellum functions and remodelling in vivo during infection of mammals and of tsetse flies. Motility likely participates to trypanosome migration from one tissue to the other in the insect, leaving the peritrophic space of the midgut to reach the proventriculus and then the salivary glands (Peacock et al. 2007). Another fascinating process is parasite attachment to insect tissues that takes place via the flagellum and is a common feature of almost all Trypanosomatid species (Kollien et al. 1998; Tetley and Vickerman 1985; Vickerman 1973). In such situations, the flagellum membrane elongates and spreads on the surface or around microvillae of insect epithelial cells, develops novel structures resembling hemi-desmosomes and a motility organelle becomes responsible for anchoring the parasite to a defined location. These interactions only occur at very specific stages of differentiation and are restricted to certain insect tissues. This raises the question as to how parasites detect an appropriate surface for adhesion [reviewed in (Bates 2008)]. The recent discoveries of exhaustive sensory functions for primary cilia (Singla and Reiter 2006) and for motile cilia (Shah et al. 2009) in mammals and algae argue for a similar role of the parasite flagellum. This becomes especially significant when one takes into account the fact that trypanosomes swim with their flagellum leading, placing the tip of the organelle at the direct interface with the host (Rotureau et al. 2009). The flagellum being essential for survival at the bloodstream stage, it represents a promising drug target for the future as about two-thirds of its components are specific to Trypanosomatids (Broadhead et al. 2006).

Acknowledgments Work in the authors' laboratory is funded by the Institut Pasteur and the CNRS. We thank the Plateforme de Microscopie Ultrastructurale for providing access to their equipment. J.B. is funded by an MNRT fellowship.

References

Absalon S, Kohl L, Branche C, Blisnick T, Toutirais G, Rusconi F, Cosson J, Bonhivers M, Robinson D, Bastin P (2007) Basal body positioning is controlled by flagellum formation in Trypanosoma brucei. PLoS ONE 2:e437

Absalon S, Blisnick T, Bonhivers M, Kohl L, Cayet N, Toutirais G, Buisson J, Robinson D, Bastin P (2008a) Flagellum elongation is required for correct structure, orientation and function of the flagellar pocket in Trypanosoma brucei. J Cell Sci 121:3704–3716

Absalon S, Blisnick T, Kohl L, Toutirais G, Dore G, Julkowska D, Tavenet A, Bastin P (2008b) Intraflagellar transport and functional analysis of genes required for flagellum formation in trypanosomes. Mol Biol Cell 19:929–944

Adhiambo C, Forney JD, Asai DJ, LeBowitz JH (2005) The two cytoplasmic dynein-2 isoforms in Leishmania mexicana perform separate functions. Mol Biochem Parasitol 143: 216–225

Adhiambo C, Blisnick T, Toutirais G, Delannoy E, Bastin P (2009) A novel function for the atypical small G protein Rab-like 5 in the assembly of the trypanosome flagellum. J Cell Sci 122:834–841

Baron DM, Kabututu ZP, Hill KL (2007a) Stuck in reverse: loss of LC1 in Trypanosoma brucei disrupts outer dynein arms and leads to reverse flagellar beat and backward movement. J Cell Sci 120:1513–1520
Baron DM, Ralston KS, Kabututu ZP, Hill KL (2007b) Functional genomics in Trypanosoma brucei identifies evolutionarily conserved components of motile flagella. J Cell Sci 120: 478–491
Bastin P, Matthews KR, Gull K (1996) The paraflagellar rod of kinetoplastida: solved and unsolved questions. Parasitol Today 12:302–307
Bastin P, Sherwin T, Gull K (1998) Paraflagellar rod is vital for trypanosome motility. Nature 391:548
Bastin P, MacRae TH, Francis SB, Matthews KR, Gull K (1999a) Flagellar morphogenesis: protein targeting and assembly in the paraflagellar rod of trypanosomes. Mol Cell Biol 19:8191–8200
Bastin P, Pullen TJ, Sherwin T, Gull K (1999b) Protein transport and flagellum assembly dynamics revealed by analysis of the paralysed trypanosome mutant snl-1. J Cell Sci 112:3769–3777
Bates PA (2008) Leishmania sand fly interaction: progress and challenges. Curr Opin Microbiol 11:340–344
Bonhivers M, Landrein N, Decossas M, Robinson DR (2008a) A monoclonal antibody marker for the exclusion-zone filaments of Trypanosoma brucei. Parasit Vectors 1:21
Bonhivers M, Nowacki S, Landrein N, Robinson DR (2008b) Biogenesis of the trypanosome endo-exocytotic organelle is cytoskeleton mediated. PLoS Biol 6:e105
Branche C, Kohl L, Toutirais G, Buisson J, Cosson J, Bastin P (2006) Conserved and specific functions of axoneme components in trypanosome motility. J Cell Sci 119:3443–3455
Briggs LJ, McKean PG, Baines A, Moreira-Leite F, Davidge J, Vaughan S, Gull K (2004) The flagella connector of Trypanosoma brucei: an unusual mobile transmembrane junction. J Cell Sci 117:1641–1651
Broadhead R, Dawe HR, Farr H, Griffiths S, Hart SR, Portman N, Shaw MK, Ginger ML, Gaskell SJ, McKean PG, Gull K (2006) Flagellar motility is required for the viability of the bloodstream trypanosome. Nature 440:224–227
Brown JM, Hardin C, Gaertig J (1999) Rotokinesis, a novel phenomenon of cell locomotion-assisted cytokinesis in the ciliate Tetrahymena thermophila. Cell Biol Int 23:841–848
Cachon J, Cachon M, Cosson MP, Cosson J (1988) The paraflagellar rod: a structure in search of a function. Biol Cell 63:169–181
Cavalier-Smith T (1974) Basal body and flagellar development during the vegetative cell cycle and the sexual cycle of Chlamydomonas reinhardii. J Cell Sci 16:529–556
Cole DG, Diener DR, Himelblau AL, Beech PL, Fuster JC, Rosenbaum JL (1998) Chlamydomonas kinesin-II-dependent intraflagellar transport (IFT): IFT particles contain proteins required for ciliary assembly in Caenorhabditis elegans sensory neurons. J Cell Biol 141:993–1008
Cooper R, de Jesus AR, Cross GA (1993) Deletion of an immunodominant Trypanosoma cruzi surface glycoprotein disrupts flagellum-cell adhesion. J Cell Biol 122:149–156
Cross GA (1975) Identification, purification and properties of clone-specific glycoprotein antigens constituting the surface coat of Trypanosoma brucei. Parasitology 71:393–417
Davidge JA, Chambers E, Dickinson HA, Towers K, Ginger ML, McKean PG, Gull K (2006) Trypanosome IFT mutants provide insight into the motor location for mobility of the flagella connector and flagellar membrane formation. J Cell Sci 119:3935–3943
Dawe HR, Farr H, Portman N, Shaw MK, Gull K (2005) The Parkin co-regulated gene product, PACRG, is an evolutionarily conserved axonemal protein that functions in outer-doublet microtubule morphogenesis. J Cell Sci 118:5421–5430
Dawe HR, Shaw MK, Farr H, Gull K (2007) The hydrocephalus inducing gene product, Hydin, positions axonemal central pair microtubules. BMC Biol 5:33
Deane JA, Cole DG, Seeley ES, Diener DR, Rosenbaum JL (2001) Localization of intraflagellar transport protein IFT52 identifies basal body transitional fibers as the docking site for IFT particles. Curr Biol 11:1586–1590

Deflorin J, Rudolf M, Seebeck T (1994) The major components of the paraflagellar rod of Trypanosoma brucei are two similar, but distinct proteins which are encoded by two different gene loci. J Biol Chem 269:28745–28751

Dilbeck V, Berberof M, Van Cauwenberge A, Alexandre H, Pays E (1999) Characterization of a coiled coil protein present in the basal body of Trypanosoma brucei. J Cell Sci 112(Pt 24): 4687–4694

Engstler M, Thilo L, Weise F, Grunfelder CG, Schwarz H, Boshart M, Overath P (2004) Kinetics of endocytosis and recycling of the GPI-anchored variant surface glycoprotein in Trypanosoma brucei. J Cell Sci 117:1105–1115

Engstler M, Pfohl T, Herminghaus S, Boshart M, Wiegertjes G, Heddergott N, Overath P (2007) Hydrodynamic flow-mediated protein sorting on the cell surface of trypanosomes. Cell 131:505–515

Farina M, Attias M, Souto-Padron T, De Souza W (1986) Further studies on the organization of the paraxial rod of Trypanosomatids. J Protozool 33:552–557

Farr H, Gull K (2009) Functional studies of an evolutionarily conserved, cytochrome b5 domain protein reveal a specific role in axonemal organisation and the general phenomenon of post-division axonemal growth in trypanosomes. Cell Motil Cytoskeleton 66:24–35

Field MC, Natesan SK, Gabernet-Castello C, Koumandou VL (2007) Intracellular trafficking in the trypanosomatids. Traffic 8:629–639

Fliegauf M, Benzing T, Omran H (2007) When cilia go bad: cilia defects and ciliopathies. Nat Rev Mol Cell Biol 8:880–893

Gadelha C, Wickstead B, de Souza W, Gull K, Cunha-e-Silva N (2005) Cryptic paraflagellar rod in endosymbiont-containing kinetoplastid protozoa. Eukaryot Cell 4:516–525

Gadelha C, Wickstead B, Gull K (2007) Flagellar and ciliary beating in trypanosome motility. Cell Motil Cytoskeleton 64:629–643

Gallo JM, Precigout E, Schrevel J (1988) Subcellular sequestration of an antigenically unique beta-tubulin. Cell Motil Cytoskeleton 9:175–183

Grasse PP (1961) La reproduction par induction du blepharoplaste et du flagelle de Trypanosoma equiperdum. C R Acad Sci 252:3917–3921

Hao L, Scholey JM (2009) Intraflagellar transport at a glance. J Cell Sci 122:889–892

He CY, Pypaert M, Warren G (2005) Golgi duplication in Trypanosoma brucei requires Centrin2. Science 310:1196–1198

Hill KL (2003) Biology and mechanism of trypanosome cell motility. Eukaryot Cell 2:200–208

Hutchings NR, Donelson JE, Hill KL (2002) Trypanin is a cytoskeletal linker protein and is required for cell motility in African trypanosomes. J Cell Biol 156:867–877

Julkowska D, Bastin P (2009) Tools for analysing intraflagellar transport in trypanosomes. Meth Cell Biol 93:59–80

Kohl L, Bastin P (2005) The flagellum of Trypanosomes. In: International review of cytology, Vol 244. Academic Press, New york, pp 227–285

Kohl L, Gull K (1998) Molecular architecture of the trypanosome cytoskeleton. Mol Biochem Parasitol 93:1–9

Kohl L, Robinson D, Bastin P (2003) Novel roles for the flagellum in cell morphogenesis and cytokinesis of trypanosomes. EMBO J 22:5336–5346

Kollien AH, Schmidt J, Schaub GA (1998) Modes of association of Trypanosoma cruzi with the intestinal tract of the vector Triatoma infestans. Acta Trop 70:127–141

Kozminski KG, Johnson KA, Forscher P, Rosenbaum JL (1993) A motility in the eukaryotic flagellum unrelated to flagellar beating. Proc Natl Acad Sci U S A 90:5519–5523

Lacomble S, Vaughan S, Gadelha C, Morphew MK, Shaw MK, McIntosh JR, Gull K (2009) Three-dimensional cellular architecture of the flagellar pocket and associated cytoskeleton in trypanosomes revealed by electron microscope tomography. J Cell Sci 122:1081–1090

LaCount DJ, Barrett B, Donelson JE (2002) Trypanosoma brucei FLA1 is required for flagellum attachment and cytokinesis. J Biol Chem 277:17580–17588

Li Z, Wang CC (2008) KMP-11, a basal body and flagellar protein, is required for cell division in Trypanosoma brucei. Eukaryot Cell 7:1941–1950

Liu B, Liu Y, Motyka SA, Agbo EE, Englund PT (2005) Fellowship of the rings: the replication of kinetoplast DNA. Trends Parasitol 21:363–369
Maga JA, LeBowitz JH (1999) Unravelling the kinetoplastid paraflagellar rod. Trends Cell Biol 9:409–413
Maga JA, Sherwin T, Francis S, Gull K, LeBowitz JH (1999) Genetic dissection of the Leishmania paraflagellar rod, a unique flagellar cytoskeleton structure. J Cell Sci 112:2753–2763
Moreira-Leite FF, Sherwin T, Kohl L, Gull K (2001) A trypanosome structure involved in transmitting cytoplasmic information during cell division. Science 294:610–612
Morgan GW, Denny PW, Vaughan S, Goulding D, Jeffries TR, Smith DF, Gull K, Field MC (2005) An evolutionarily conserved coiled-coil protein implicated in polycystic kidney disease is involved in basal body duplication and flagellar biogenesis in Trypanosoma brucei. Mol Cell Biol 25:3774–3783
Natesan SK, Peacock L, Matthews K, Gibson W, Field MC (2007) Activation of endocytosis as an adaptation to the mammalian host by trypanosomes. Eukaryot Cell 6:2029–2037
Ngô HM, Bouck GB (1998) Heterogeneity and a coiled coil prediction of trypanosomatid-like flagellar rod proteins in Euglena. J Eukaryot Microbiol 45:323–333
Ogbadoyi EO, Robinson DR, Gull K (2003) A high-order trans-membrane structural linkage is responsible for mitochondrial genome positioning and segregation by flagellar basal bodies in trypanosomes. Mol Biol Cell 14:1769–1779
Overath P, Engstler M (2004) Endocytosis, membrane recycling and sorting of GPI-anchored proteins: Trypanosoma brucei as a model system. Mol Microbiol 53:735–744
Pazour GJ, Agrin N, Leszyk J, Witman GB (2005) Proteomic analysis of a eukaryotic cilium. J Cell Biol 170:103–113
Peacock L, Ferris V, Bailey M, Gibson W (2007) Dynamics of infection and competition between two strains of Trypanosoma brucei in the tsetse fly observed using fluorescent markers. Kinetoplastid Biol Dis 6:4
Pedersen LB, Geimer S, Rosenbaum JL (2006) Dissecting the molecular mechanisms of intra-flagellar transport in chlamydomonas. Curr Biol 16:450–459
Portman N, Lacomble S, Thomas B, McKean PG, Gull K (2009) Combining RNA interference mutants and comparative proteomics to identify protein components and dependences in a eukaryotic flagellum. J Biol Chem 284:5610–5619
Pullen TJ, Ginger ML, Gaskell SJ, Gull K (2004) Protein targeting of an unusual, evolutionarily conserved adenylate kinase to a eukaryotic flagellum. Mol Biol Cell 15:3257–3265
Ralston KS, Hill KL (2006) Trypanin, a component of the flagellar Dynein regulatory complex, is essential in bloodstream form African trypanosomes. PLoS Pathog 2:e101
Ralston KS, Lerner AG, Diener DR, Hill KL (2006) Flagellar motility contributes to cytokinesis in Trypanosoma brucei and is modulated by an evolutionarily conserved dynein regulatory system. Eukaryot Cell 5:696–711
Ralston KS, Kabututu ZP, Melehani JH, Oberholzer M, Hill KL (2009) The Trypanosoma brucei flagellum: moving parasites in new directions. Annu Rev Microbiol 63:335–362
Ridgley E, Webster P, Patton C, Ruben L (2000) Calmodulin-binding properties of the paraflagellar rod complex from Trypanosoma brucei. Mol Biochem Parasitol 109:195–201
Robinson DR, Gull K (1991) Basal body movements as a mechanism for mitochondrial genome segregation in the trypanosome cell cycle. Nature 352:731–733
Robinson DR, Sherwin T, Ploubidou A, Byard EH, Gull K (1995) Microtubule polarity and dynamics in the control of organelle positioning, segregation, and cytokinesis in the trypanosome cell cycle. J Cell Biol 128:1163–1172
Rodgers MJ, Albanesi JP, Phillips MA (2007) Phosphatidylinositol 4-kinase III-beta is required for Golgi maintenance and cytokinesis in Trypanosoma brucei. Eukaryot Cell 6:1108–1118
Rotureau B, Morales MA, Bastin P, Spath GF (2009) The flagellum-MAP kinase connection in Trypanosomatids: a key sensory role in parasite signaling and development? Cell Microbiol 11(5):710–718

Russell DG, Newsam RJ, Palmer GC, Gull K (1983) Structural and biochemical characterisation of the paraflagellar rod of Crithidia fasciculata. Eur J Cell Biol 30:137–143

Santrich C, Moore L, Sherwin T, Bastin P, Brokaw C, Gull K, LeBowitz JH (1997) A motility function for the paraflagellar rod of Leishmania parasites revealed by PFR-2 gene knockouts. Mol Biochem Parasitol 90:95–109

Schlaeppi K, Deflorin J, Seebeck T (1989) The major component of the paraflagellar rod of Trypanosoma brucei is a helical protein that is encoded by two identical, tandemly linked genes. J Cell Biol 109:1695–1709

Scott V, Sherwin T, Gull K (1997) gamma-tubulin in trypanosomes: molecular characterisation and localisation to multiple and diverse microtubule organising centres. J Cell Sci 110 (Pt 2):157–168

Selvapandiyan A, Kumar P, Morris JC, Salisbury JL, Wang CC, Nakhasi HL (2007) Centrin1 is required for organelle segregation and cytokinesis in Trypanosoma brucei. Mol Biol Cell 18:3290–3301

Shah AS, Ben-Shahar Y, Moninger TO, Kline JN, Welsh MJ (2009) Motile cilia of human airway epithelia are chemosensory. Science 325(5944):1131–1134

Sharma R, Peacock L, Gluenz E, Gull K, Gibson W, Carrington M (2008) Asymmetric cell division as a route to reduction in cell length and change in cell morphology in trypanosomes. Protist 159:137–151

Sherwin T, Gull K (1989) The cell division cycle of Trypanosoma brucei: timing of event markers and cytoskeletal modulations. Philos Trans R Soc Lond B Biol Sci 323:573–588

Singla V, Reiter JF (2006) The primary cilium as the cell's antenna: signaling at a sensory organelle. Science 313:629–633

Stephan A, Vaughan S, Shaw MK, Gull K, McKean PG (2007) An essential quality control mechanism at the eukaryotic basal body prior to intraflagellar transport. Traffic 8: 1323–1330

Tetley L, Vickerman K (1985) Differentiation in Trypanosoma brucei: host-parasite cell junctions and their persistence during acquisition of the variable antigen coat. J Cell Sci 74:1–19

Tuxworth RI, Cheetham JL, Machesky LM, Spiegelmann GB, Weeks G, Insall RH (1997) Dictyostelium RasG is required for normal motility and cytokinesis, but not growth. J Cell Biol 138:605–614

Vaughan S, Kohl L, Ngai I, Wheeler RJ, Gull K (2008) A repetitive protein essential for the flagellum attachment zone filament structure and function in Trypanosoma brucei. Protist 159:127–136

Vickerman K (1969) On the surface coat and flagellar adhesion in trypanosomes. J Cell Sci 5:163–193

Vickerman K (1973) The mode of attachment of Trypanosoma vivax in the proboscis of the tsetse fly Glossina fuscipes: an ultrastructural study of the epimastigote stage of the trypanosome. J Protozool 20:394–404

Vickerman K (1985) Developmental cycles and biology of pathogenic trypanosomes. Br Med Bull 41:105–114

Vickerman K, Luckins AG (1969) Localization of variable antigens in the surface coat of Trypanosoma brucei using ferritin conjugated antibody. Nature 224:1125–1126

Zhao Z, Lindsay ME, Roy Chowdhury A, Robinson DR, Englund PT (2008) p166, a link between the trypanosome mitochondrial DNA and flagellum, mediates genome segregation. EMBO J 27:143–154

The Flagellar Pocket of Trypanosomatids: A Critical Feature for Cell Morphogenesis and Pathogenicity

Paul G. McKean and Keith Gull

Contents

P.G. McKean (✉)
Division of Biomedical and Life Sciences, School of Health and Medicine, Lancaster University, Lancaster LA1 4YQ, UK
e-mail: p.mckean@lancaster.ac.uk

K. Gull
Sir William Dunn School of Pathology, University of Oxford, Oxford OX1 3RE, UK
e-mail: keith.gull@path.ox.ac.uk

W. de Souza (ed.), *Structures and Organelles in Pathogenic Protists*,
Microbiology Monographs 17, DOI 10.1007/978-3-642-12863-9_4,

© Springer-Verlag Berlin Heidelberg 2010

Abstract Trypanosomatids possess a highly ordered array of sub-pellicular microtubules that restrict all vesicular traffic to the flagellar pocket (FP); a small invagination of the plasma membrane located at the base of the flagellum. Although the FP is not an adaptation to parasitism per se, it is without question a key pathogenicity feature that has enabled parasitic trypanosomatid species to exploit a diversity of host environments. In this chapter, we focus on the FP of the African trypanosome *Trypanosoma brucei* and consider recent advances in our understanding of the physical architecture of the FP and the dynamics of FP operation. We conclude with a brief discussion that the trypanosomatid FP represents an example of "extreme biology", i.e. a normal but exaggerated example of the cell biology present at the flagellum base in proliferating flagellated eukaryotic cells.

1 Introduction

Although the characteristic morphologies of trypanosomatid cells vary enormously (both between species and different life cycle stages of the same species), all are constructed from a highly ordered array of sub-pellicular microtubules (Gull 1999). These microtubules, which are extensively cross-linked to each other and the overlying plasma membrane, prohibit all vesicular trafficking via the generalised cell body plasma membrane. As a result, all endocytosis and exocytosis in trypanosomatid cells are focussed to a small bulbous invagination of the plasma membrane at the base of the flagellum; the flagellar pocket (FP). The FP is exquisitely designed for this purpose and although the FP membrane is estimated to comprise less than 5% of the cell surface (Overath and Engstler 2004) the rates of endocytosis in trypanosomes exceed that observed in many eukaryotes (Overath et al. 1997).

Since the FP is a characteristic feature of all trypanosomatids, including free living species, it is not an "adaptation" to parasitism per se but is without question a key pathogenicity feature. The creation of the FP provides an "internal" privileged site that enables trypanosomatid parasites to sequester specific surface proteins away from constant host exposure. This has enabled successful exploitation of such diverse and challenging environments as the mammalian bloodstream (*Trypanosoma brucei*), the cytoplasm of host cells (*Trypanosoma cruzi*) and the parasitophorous vacuole of macrophages (*Leishmania* spp.), as well as various niches within insect vectors.

In this chapter, we intend to focus on the FP of the African trypanosome, *T. brucei*, and consider recent advances in our understanding of the physical architecture of the FP and dynamics of operation. However, as appropriate we will also consider the FP biology of other trypanosomatid species. Finally, we will conclude with a brief deliberation on whether the FP is another example of "extreme biology" in trypanosomatid cells and hence provides new insight into biological principles operating in other ciliated/flagellated eukaryotic cells.

2 Physical Architecture of the Flagellar Pocket

Precise positioning of the FP within trypanosomatid cells varies considerably; indeed, positioning relative to the cell nucleus is central to the nomenclature defining trypanosomatid cell types (Hoare and Wallace 1966). In trypomastigotes, the nucleus is anterior to an assemblage consisting of the FP, basal body and kinetoplast but is posterior to this assemblage in epimastigote forms. More subtle differences in FP positioning are also evident; in *T. brucei* procyclic (insect stage) trypomastigotes, the FP is positioned midway between the cell midpoint and posterior but in bloodstream trypomastigotes is at the extreme posterior pole. The biological significance of these variations are not always apparent but were also noted by David Bruce in his early descriptions of the African trypanosome (Bruce 1911). As can be seen in the original drawings of bloodstream form trypanosomes in Fig. 1, Bruce's (1911) description of the FP varies subtly in intracellular positioning and also in size; the FP in long slender bloodstream form trypanosomes is appreciably smaller than the FP in short stumpy forms.

Viewed in longitudinal section by transmission electron microscopy, the FP is a distinctive electron-translucent compartment immediately internal to the flagellum exit point from the cell body (Fig. 2a). The mature basal body subtending the flagellum is positioned between the base of the FP and the kinetoplast; with a probasal body lying immediately alongside. The flagellum invades the FP of the procyclic trypanosome with a distinct asymmetry with the FP lumen being significantly larger on one side of the flagellum than the other. This FP bulge is invariantly located coincident with the region of the cell that contains the probasal body and Golgi apparatus.

Long Slender

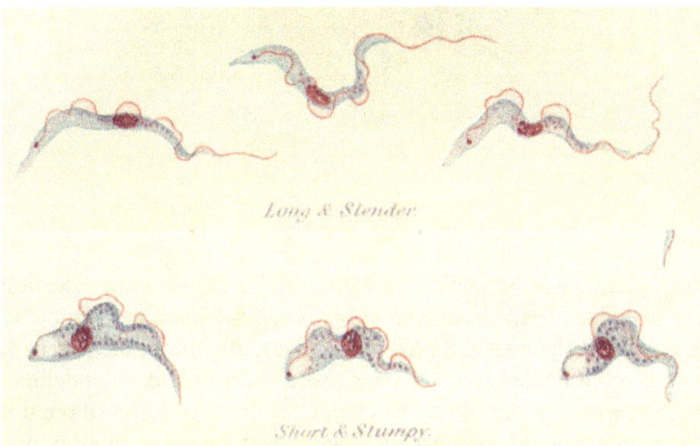

Short Stumpy

Fig. 1 Long slender and short stumpy bloodstream form *Trypanosoma gambiense* as drawn by Colonel David Bruce in 1911. Although it was not remarked upon by Bruce in his original article, it is evident from these drawings that he had noted a marked disparity in the size of the flagellar pocket in different life cycle stages of the African trypanosome. Drawings taken from Bruce (1911)

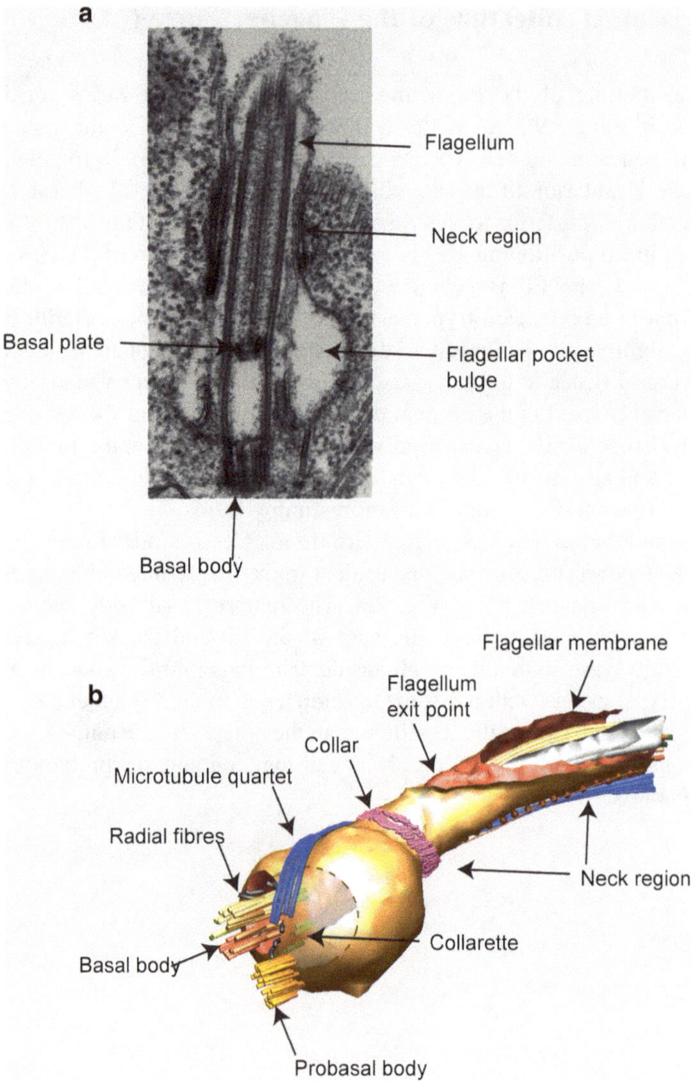

Fig. 2 Architecture of the flagellar pocket. (**a**) Thin-section transmission electron micrograph of the *T. brucei* flagellar pocket. In the *longitudinal section*, the distinctive asymmetry of the flagellar pocket is clearly observed. (**b**) Model tomogram showing the major membrane and cytoskeletal compenents of the flagellar pocket. The major cytoskeletal structures that define FP morphology are clearly visible in this model; the collar, which defines the flagellum exit point from the FP, and the radial fibres and collarette, which mark the flagellum entry point into the FP

The flagellum emerges from the FP through a tightly constricted aperture; at this point, membranes of the flagellum and cell body are tightly apposed and hence the FP lumen is apparently insulated from the external environment. In some trypanosomatids (e.g. *T. brucei*), after the flagellum emerges from the FP, it remains attached to the cell body via a flagellum attachment zone (FAZ); however, in others (e.g. *Leishmania* promastigotes), the flagellum remains unattached. It is worth noting that in trypanosomatid species which form an amastigote stage, which only forms a rudimentary flagellum, the flagellum still extends to occlude the neck of the FP but does not extend much further into the external environment. The relevance of forming a rudimentary flagellum in the amastigote FP is discussed in greater detail later.

2.1 New Insights Provided by Electron Tomography

Significant new insights into the physical three-dimensional architecture of the *T. brucei* FP have been provided recently by electron microscope tomography studies (Lacomble et al. 2009). A 3D model generated by this study (using 435 tomographic slices encompassing 80% of the procyclic form FP) is shown in Fig. 2b. Note that the FP modelled in this study was associated with a new flagellum of a late stage dividing trypanosome and so contains many features that typify the trypanosomatid FP.

This tomographic model reinforces our view of the FP as an asymmetric compartment and confirms that the bulge in the FP membrane in procyclic forms is closely apposed to the Golgi, which is invariantly positioned adjacent to the neck of the FP. Also visible in this model are four "specialised microtubules", termed the microtubule quartet (MtQ), which are nucleated between the basal and probasal body. The MtQ extends around the FP membrane and traverses the FP collar to insert into the sub-pellicular array of microtubules at the junction point between the FP and plasma membrane. This MtQ creates a unique seam in the cytoskeletal corset since they are arranged with an opposite polarity to all other MTs in the sub-pellicular cytoskeleton, namely, plus ends oriented towards the anterior pole of the cell. Although the MtQ has long been referred to as a component of the FAZ, their contribution to flagellum attachment has never been satisfactorily explained and evidence is now accumulating that the MtQ may play critical role(s) in allowing access of macromolecules to the FP (see below).

Examination of the model also reveals two cytoskeletal boundaries that define the 3D architecture of the FP; the flagellum "entry point" into the FP and the flagellum "exit point" into the external environment. The site of flagellum entry is defined by three structural features; transitional fibres, radial fibres and the collar-ette. The transitional fibres radiate from the mature basal body and contact the flagellar membrane at the base of the FP, thus defining the boundary between the cytoplasm of the cell body and the environmentally distinct flagellum compartment. These transitional fibres also serve as a platform for recruitment of proteins destined for import into the flagellum by intra-flagellar transport (IFT), the evolutionarily conserved transport process used to construct most eukaryotic cilium/flagellum (Pedersen and Rosenbaum 2008).

The collarette is also positioned around the basal body but on the external face of the flagellar membrane. Originally described by Keith Vickerman in 1973 (Vickerman 1973), EM tomography reveals a hitherto unappreciated complexity to this structure, which is created from nine regularly spaced double tube-like units connected to each other by plate-like fibrils positioned on the flagellar membrane. It has been speculated that the collarette may act to constrain the flagellar membrane into a discrete tube. It is certainly apparent that the flagellar membrane exhibits no significant curvature until it passes through the collarette but afterwards spreads out to form the base of the pocket. While both the transitional fibres and collarette were described in the literature prior to the electron tomography study of Lacomble et al. (2009), the radial fibres are a novel description. The radial fibres are positioned inside the flagellar membrane around the proximal zone of the basal body and appear to be an internal reflection of the external collarette. The flagellum exit point from the FP is defined by an electron-dense structure, termed the collar, positioned on the cytoplasmic face of the membrane. In electron micrographs, the neck of the FP is tightly constricted and it is suggested that the collar may operate as a cytoplasmic valve regulating the entry (and exit) of macromolecules into (and out of) the lumen of the FP.

The concept of the FP as an enclosed external microenvironment into which macromolecular access and egress is highly regulated is central to its perceived importance in trypanosomatid biology. Recent experiments have provided new insights into how the architecture of the FP may act to regulate macromolecular traffic.

3 Physical Organisation of the Flagellar Pocket and Dynamics of Operation

The studies by Lacomble et al. (2009) provide an unrivalled description of the 3D organisation of the FP. However, the FP is a highly dynamic compartment and it is essential to ask how FP architecture influences endocytosis. Note that in this chapter we do not intend to focus on endocytosis per se [for excellent reviews on this subject, see Morgan et al. (2002a, b), Field and Carrington (2004), Overath and Engstler (2004)] but rather we concentrate on how new understanding of the 3D architecture of the FP alters our view of the dynamics of FP operation.

3.1 Entry of Macromolecules into the Flagellar Pocket: The "Neck Channel"

It has been suggested by several studies that the process of endocytosis is compartmentalised within the FP, being spatially restricted to the anterior portion of the FP

(Engstler et al. 2005). This spatial restriction would seem fitting given that the organelles associated with endocytosis (as well as recycling and secretion) are mainly positioned between the anterior portion of the FP and the nucleus (Overath and Engstler 2004). However, this view of compartmentalised endocytosis within the FP is not supported by recent work on the FP of bloodstream form *T. brucei* (Gadelha et al. 2009). In this study, isothermal fixation techniques and electron tomography were used to generate a 3D view of endocytosis while freeze-fracture EM techniques investigated whether the FP membranes show any evidence of functional compartmentalisation. These studies observed that clathrin-coated pits – unlike many eukaryotes, all endocytosis in trypanosomatids is clathrin-dependent (Balber 1990; Overath and Engstler 2004) – were present throughout the entire FP and found no evidence that the FP membranes were organised into distinct domains, as evidenced by a uniform distribution of intra-membrane particles (IMPs). Taken together, this data suggests that much of the entire FP is likely to be competent for, or involved in, the various stages of recognition and endocytosis of exogenous molecules.

To further investigate macromolecular uptake into the FP, Gadelha et al. used gold-conjugated tomato lectin (TL) to specifically label glycoproteins enriched in the FP and endosomes (Nolan et al. 1999; Atrih et al. 2005). Although the extremely high rates of endocytosis in bloodstream form *T. brucei* makes these studies challenging, incubation of cells in cold media results in endocytosis being reversibly blocked (Brickman et al. 1995). Under these cold conditions, macromolecules enter the FP but are not internalised and Gadelha et al observed gold-conjugated TL accumulating on FP membrane directly adjoining the MtQ. A similar localisation was observed when gold-conjugated wheat-germ agglutinin (a lectin with sugar specificity similar to TL) and bovine serum albumin (a marker of fluid-phase endocytosis in trypanosomes (Coppens et al. 1987)) were used. This suggests that membrane-associated macromolecules may preferentially locate to or move along tracks established by the underlying MtQ. Previous studies carried out by Brickman and Balber in 1990 also provided indications that ricin-binding glycoproteins specifically associate with membrane apposed to the FAZ, MtQ and FP, adding further credence to the argument that these regions form functionally discrete membrane microdomains (Brickman and Balber 1990).

Within the FP neck, the membranes of the flagellum and FP are tightly apposed, potentially sealing the FP lumen from the external environment. However, within this region gold particles were found to reside within a small luminal space created by the membrane area associated with the MtQ (Fig. 3). This space, termed the "neck channel", creates a direct connection between the FP lumen and the external environment which could facilitate traffic of macromolecules across the constricted FP neck. However, if all of the FP is endocytosis-competent, then macromolecules accessing the FP via this discrete channel must redistribute across the entire FP membrane after entering the FP.

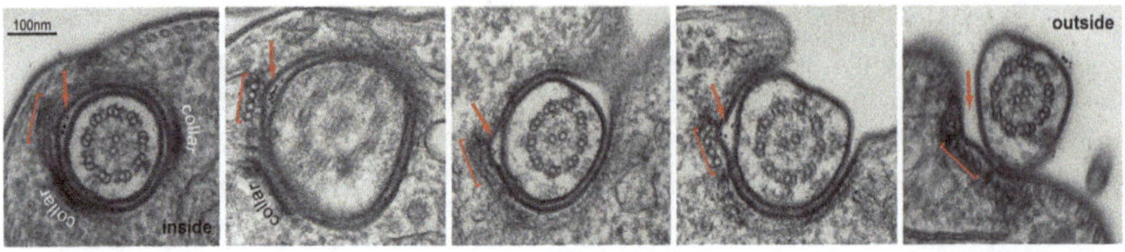

Fig. 3 The MtQ is associated with a channel connecting the lumen of the FP with the external environment. Thin-section electron micrographs through the neck region of bloodstream form trypanosomes which had been previously incubated on ice with tomato lectin conjugated to 5-nm gold. Gold particles are clearly visible and show preferential accumulation in a gap between the membrane of the flagellum and the FP neck (*arrowed*). The neck channel provides a conduit for macromolecular traffic into and out of the FP. Figures taken from Gadelha et al. (2009)

This study provides insights into how macromolecules may access the FP but raises a number of important questions including:

1. Does the neck channel act as a passive conduit or is there an active mechanism moving surface membrane bound material across the FP neck?
2. How does the neck channel select macromolecules that enter and leave the FP?
3. How is the neck channel established by the underlying MtQ?

If entry into the FP via the neck channel is passive it would still be necessary to explain why (and how) surface-bound molecules specifically associate with membrane overlying the MtQ. There are a number of possible explanations ranging from lack of mobility from the 4 MtQ zone to other FP zones in the cold, through to maintenance of a residual much higher density of binding molecules in this region because it is incompetent for endocytosis. It is more probable therefore that entry into the FP is an active process reliant on the underlying MtQ. Surface-bound molecules presumably associate with transmembrane proteins which in turn contact molecular motors associated with the MtQ. A large number of unusual and as yet uncharacterised molecular motors that could facilitate movement of surface-bound material into/out of the FP are certainly encoded in the trypanosome genome (Berriman et al. 2005; Wickstead and Gull 2006, 2007).

The concept of the FP as a privileged "external" microenvironment is well established; however, since trypanosomes allow unconjugated gold particles into the FP lumen it is difficult to see how selection at the FP neck is based on molecular recognition alone. Given the restricted dimensions of the neck channel, it is possible that molecular size plays an important part in the selection process as a number of studies, including that of (Gadelha et al. 2009), have reported that 20-nm gold particles enter the FP but that larger gold particles are excluded. It is clear that we still have much to learn about the mechanisms regulating molecular trafficking into/out of the FP.

Is the neck channel a general feature of the trypanosomatid FP? Although extensive supporting evidence is still lacking, early studies on *Leishmania collosoma* (a trypanosomatid that possesses a free flagellum) have identified a deep, longitudinal fold in the FP membrane directly adjacent to a group of rootlet microtubules (Linder and Staehelin 1977). This suggests that the neck channel and associated MtQ may be a common structural (and functional) feature of the trypanosomatid FP.

3.2 *Membrane Domains and Boundaries Within the Flagellar Pocket*

Although freeze-fracture studies have found no evidence of FP membrane compartmentalisation, it is evident that trypanosomatid surface membranes are organised into four distinct domains – the cell body, the FP neck, the FP and the flagellar membrane. These domains are characterised by distinctive densities of IMPs, and the FP cytoskeletal structures identified in the 3D tomogram are present at two of these boundaries, namely, (1) the collarette at the junction between the flagellum and FP membranes and (2) the collar at the junction between the FP and neck membranes. These membrane boundaries are characterised by distinctive accumulations of IMPs and are likely to be important in restricting the movement of membrane proteins (Linder and Staehelin 1977; Vickerman and Tetley 1990; Yoshikawa et al. 1990). A greater understanding of how these membrane boundaries are established is critical since they are likely to be central players acting to maintain the FP as a distinct microenvironment; selectively retaining receptors within the FP while allowing other proteins (such as the Variant surface glycoprotein; VSG) onto the cell surface.

The question of how the MtQ leads to the formation of the neck channel is a critical question. One would anticipate that connections between the MtQ and the overlying FP membrane cause local distortion of the FP membrane and the formation of a channel. However, examination of the region between the MtQ and the FP membrane does not clarify the nature of these potential linkages, although a cloud of electron-dense, amorphous material is observed to fill the gap between the MtQ and the overlying FP membrane. It will be interesting to determine whether trypanosomatids are able to modulate interactions between the MtQ and overlying FP membrane to open and close the neck channel in response to environmental conditions.

4 The Role of the Flagellar Pocket in Immune Evasion

The entire cell surface of bloodstream *T. brucei* is covered by a dense monolayer of VSG homodimers which are tethered to the plasma membrane by a GPI anchor (Ferguson 1999). Although VSG creates an impenetrable barrier protecting the underlying plasma membrane from immune-mediated damage, this protection is transient and trypanosomes are susceptible to antibody-dependent, complement-mediated lysis (McLintock et al. 1993). To establish persistent infection in the mammalian bloodstream, trypanosomes have evolved specific mechanisms to overcome protective host responses. The key to bloodstream survival is the trypanosome's ability to undergo antigenic variation which involves periodic switching of the VSG surface coat (for reviews see Donelson (2003), Garcia-Salcedo et al. (2004), Horn (2004), Pays (2005), Taylor and Rudenko (2006), Navarro et al. (2007), Stockdale et al. (2008)). Switching occurs at a low frequency and trypanosomes that express a new antigenically distinct VSG (different VSGs share only 15–20% amino acid sequence identity) selectively proliferate in the face of antibody responses elicited to "old" VSG coats. This process can be repeated numerous times as trypanosomes are able to select new VSG coats for expression from a vast genomic repertoire (Berriman et al. 2005).

Although high concentrations of anti-VSG antibody mediate immune clearance (hence the fluctuating parasitaemias observed in the mammalian bloodstream), trypanosomes survive low-moderate antibody concentrations by removing antibody–VSG complexes from the cell surface (O'Beirne et al. 1998). As the VSG molecule is anchored onto the surface by a GPI anchor, it has no interaction with the underlying MT cytoskeleton and can move freely within the membrane. Recent studies by Engstler et al. (2007) demonstrate that antibody–VSG complexes on the surface are propelled towards the posterior of the cell (and consequently the FP) by hydrodynamic forces created by the anterior movement of trypanosome cells in response to flagellar motility. This is an extremely rapid process, with IgG–VSG complexes being internalised via the FP within 2 min. Following uptake into the FP, IgG–VSG complexes are endocytosed into lysosomes, with anti-VSG antibodies being degraded (O'Beirne et al. 1998; Pal et al. 2003) and VSG molecules recycled to the cell surface. Precisely, how trypanosomes are able to degrade immunoglobulin without subjecting internalised VSG molecule to proteolytic cleavage is uncertain. The rapid internalisation of immune complexes via the FP appears to provide an efficient mechanism to prevent complement-mediated lysis of bloodstream form cells and may explain why bloodstream form trypanosomes position the FP at the extreme posterior pole of the cell.

RNAi mutants have been used to investigate the three distinct steps involved in the surface clearance of Ig–VSG complexes (Fig. 4). A requirement for anterior-directed movement was demonstrated by depletion of a dynein arm intermediate chain, which leads to reversal of flagellar beat (Branche et al. 2006). These mutant trypanosomes swim backwards (i.e. posterior pole leading) and so Ig–VSG complexes accumulate at the anterior pole well away from the FP. Ablation of clathrin heavy chain which prevents active endocytosis (see later for more detail) results in the concentration of Ig–VSG at the posterior pole near the FP since active uptake is prevented. The importance of flagellar motility was shown by ablating FLA1 which results in flagellar detachment (LaCount et al. 2002). In the FLA1 mutant Ig-VSG remains distributed over the entire cell surface, apart from the region immediately surrounding the FP. This clearance from around the FP may indicate the existence of localised directional trafficking of membrane-bound Ig–VSG into the FP by MtQ-dependent transport mechanisms.

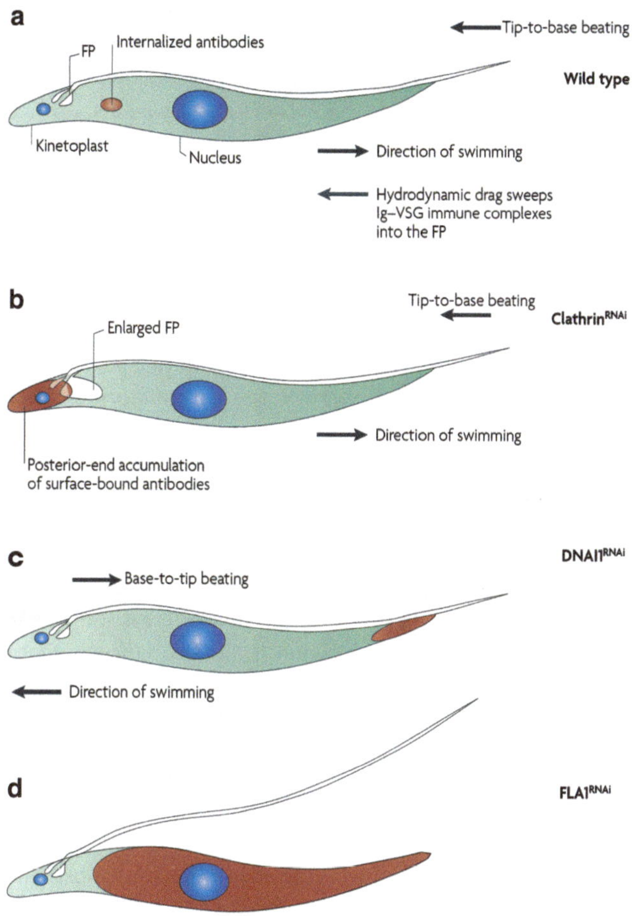

Fig. 4 Cartoon showing the effect of the RNAi-mediated ablation of the clearance of anti-VSG antibodies from the cell surface. (**a**) In wild type bloodstream form, *T. brucei* tip-to-base flagellar beating moves surface-bound anti-VSG antibodies towards the posterior pole for internalisation via the FP; (**b**) When uptake via the FP is blocked, due to RNAi-mediated ablation of clathrin, anti-VSG antibodies accumulate at the posterior pole but are not internalised; (**c**) RNAi-mediated depletion of dynein intermediate chain 1 results in the reversal of flagellar beat and accumulation of anti-VSG antibodies at the anterior pole; (**d**) RNAi-mediated depletion of FLA1 causes flagellar detachment and results in anti-VSG antibodies remaining on the cell surface apart from a region surrounding the FP. Figure taken from Ginger et al. (2008) and based on data in Engstler et al. (2007)

5 Life Cycle Differences in Size and Function of the Flagellar Pocket

Rates of endocytosis via the FP differ markedly between life cycle stages in *T. brucei*, with bloodstream form trypanosomes showing an approximately 10-fold higher rate of endocytosis than insect stage procyclic forms (Morgan et al. 2001;

Allen et al. 2003; Field and Carrington 2004). Note that both long slender and short stumpy bloodstream forms of *T. brucei* exhibit similar rates of endocytosis (Natesan et al. 2007). While the overall architecture of the FP does not differ between life cycle stages, the size of the FP shows significant disparity. For instance, the FP of both procyclic and short stumpy bloodstream forms are noticeably larger than the FP of long slender bloodstream form cells. But this does not correlate with the competency for endo- and exocytosis since the rate of endocytosis in procyclic forms (with a relatively large FP) is 10-fold less than that observed in long slender bloodstream forms which has a smaller FP. Rather, it appears that the trypanosomes capacity for endocytosis is governed by the modulation of endocytic pathways as an adaptation for survival in the mammalian bloodstream.

Natesan et al. (2007) studied endocytic activity during in vitro differentiation of short stumpy bloodstream forms to procyclics by monitoring the expression of clathrin and Rab11; markers for endosomal vesicles and recycling endosomes, respectively. The progress of differentiation was monitored by analysing VSG and CAP5.5 expression, the latter being a stage-regulated cytoskeletal protein expressed in procyclics (Hertz-Fowler et al. 2001). These studies showed that VSG was lost by 4 h and CAP5.5 expression acquired by 8 h after induction of differentiation. The levels of clathrin and Rab11 both decreased, but with different kinetics; 50% of clathrin was still present after 24 h whereas Rab11 was reduced to less than 30% of short stumpy levels within 6 h. This indicates that as *T. brucei* moves between the mammalian bloodstream and the midgut of the tsetse vector the endocytic recycling pathway is rapidly down-regulated. This is perhaps surprising since differentiation involves considerable remodelling of the parasite surface involving removal of the VSG coat and replacement with the GPI-anchored protein procyclin.

Analysis of tsetse-derived metacyclic trypanosomes (found in the salivary gland and responsible for transmitting infection to the mammalian host) showed that in the preparation for re-entry into the mammalian host, endocytic activity is increased 5-fold (compared to midgut procyclics) (Natesan et al. 2007). The most likely explanation as to why endocytosis rates are significantly higher in the mammalian bloodstream relates to the critical role played by endocytosis via the FP in VSG recycling and immune evasion (see above).

6 Molecules Associated with the Flagellar Pocket

While a number of proteins have been localised to the FP, it is critical to distinguish between trypanosomatid proteins transiently associating with the FP (and which also localise to the endocytic pathway or the cell surface) and proteins that form an integral part of the membrane or cytoskeletal framework resident in or used in the construction of the FP. Of course, specificity of location of a membrane protein to the FP may not be absolute. Rather, it could be that although there is widespread belief that some receptors are specifically located to the pocket there is the possibility that this could be merely a reflection of a much higher concentration in that area in contrast to a low (i.e. undetectable by normal surface labelling) on the cell body. Whilst not critical for much of the debate that follows, this issue of sensitivity of detection protocols does speak to the underlying issue of localisation via concentration or absolute specificity by restriction.

6.1 Proteins Involved in the Cytoskeletal Architecture of the Flagellar Pocket

The catalogue of proteins that actively contribute to the functional organisation of the FP is rather limited; however, a critical advance has been made recently with the characterisation of BILBO1 (Bonhivers et al. 2008). BILBO1 specifically localises to the *T. brucei* FP (as shown by in vivo expression of GFP-tagged BILBO1 and anti-BILBO1 specific antibodies), resolving as a distinct horseshoe-shaped structure at the FP neck (Fig. 5a). Immunogold EM confirms that BILBO1 specifically localises to the collar; the defining cytoskeletal feature at the flagellum exit point from the FP (Fig. 5b). The observation that BILBO1 resolves as a horseshoe configuration rather than a complete ring structure is consistent with our 3D understanding of the collar structure provided by electron tomography (Lacomble et al. 2009).

BILBO1 has a critical role in the formation of the FP collar since RNAi-mediated ablation of this protein results in trypanosomes that lack the collar structure at the base of the new flagellum. These mutant trypanosomes are unable to endocytose material from the base of the new flagellum confirming the indispensable requirement for a functional FP for endocytosis in trypanosomatids.

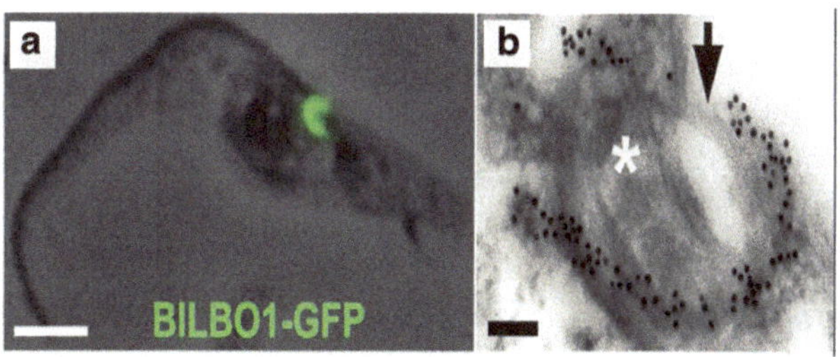

Fig. 5 Localisation of BILBO1 to the flagellar pocket collar. (**a**) Fluorescence image of procyclic *T. brucei* expressing BILBO1:eGFP merged with phase-contrast image of the whole cell. Note that BILBO1:eGFP localises to a distinctive horseshoe-shaped collar at the exit point of the flagellum from the FP. *Scale bar* = 2.5 μm. (**b**) Immunogold localisation of BILBO1 to the FP collar (*arrowed*) using an anti-BILBO1 polyclonal antibody followed by anti-mouse secondary antibody conjugated to 10-nm gold particles. The *asterisk* marks the flagellum transition zone. *Scale bar* = 100 nm. Figure taken from Bonhivers et al. (2008)

Although a new flagellum forms in procyclic cells, indicating that loss of the FP has no effect on flagellum elongation, the flagellum is not attached to the cell body. The effect of BILBO1 ablation on bloodstream forms is even more dramatic as these cells lose all morphological integrity and rapidly round up into spherical shaped cells. But, interestingly, despite these gross morphological effects in bloodstream forms, the new flagellum remains attached to the cell body. Why does the loss of the FP specifically affect flagellum attachment in procyclic forms? It may be significant that outgrowth of the new flagellum in procyclics (but not bloodstream forms) is influenced by the flagellar connector; a mobile trans-membrane junction that attaches the distal tip of the new flagellum to the old flagellum (Moreira-Leite et al. 2001; Davidge et al. 2006). As the FC guides outgrowth of the new flagellum, critical organisational information inherent in the old MT cytoskeleton (including positioning of the FAZ) is imparted to the new cell. The FC forms early during new flagellum elongation within the FP (Davidge et al. 2006). However, following ablation of BILBO1 in procyclic cells the FC is not formed (or at least is defective and unable to maintain robust connection between the new and old flagellum) suggesting that the enclosed FP microenvironment may be needed for FC construction and subsequent FAZ formation (Bonhivers et al. 2008).

TbMORN1 (membrane occupation and recognition nexus repeat protein) has also been localised around the neck of the FP, partially overlapping with the FP collar (as labelled by BILBO1) and other FP related structures such as a centrin-positive bilobed structure that lies adjacent to the Golgi and the posterior end of the FAZ (Morriswood et al. 2009). Although it has been suggested that TbMORN1 serves to anchor FP components to the FP membrane, no structural defects are observed following TbMORN1 ablation. The precise role of TbMORN1 in FP organisation thus remains to be elucidated. Nevertheless, ablation of TbMORN1 does result in severe motility defects in procyclic cells (Broadhead et al. 2006) and is lethal in bloodstream forms (Morriswood et al. 2009) indicating that this FP located protein has important roles in flagellum and flagellum pocket biogenesis.

TbMORN1 and BILBO1 are encoded in the genomes of other trypanosomatids suggesting possible conservative roles in FP biology; however, homologues of TbMORN1 also exist in Apicomplexan genomes indicating further functions for this protein in addition to those associated with the trypanosomatid FP (Morriswood et al. 2009).

6.2 *Specific Receptors Located Within the Flagellar Pocket*

Restricted localisation of specific proteins to within the protected FP microenvironment serves a critical function for trypanosomatid parasites as it prevents, or at least hinders, host immune recognition of important receptors. Two receptors with important roles in the mammalian bloodstream are located within the FP, the Haptoglobin–Hemoglobin receptor (TbHPHbR) and the transferrin receptor (TfR), and both warrant detailed consideration here.

6.2.1 TbHpHbR and Susceptibility of *T. brucei* to Lysis Human Serum

Humans display innate resistance to infection by many species of trypanosome; indeed, resistance to lysis by normal human serum (NHS) is a key feature distinguishing the morphologically identical human-infective *T. b. rhodesiense* and *T. b. gambiense* from non-human infective *T. b. brucei* (Radwanska et al. 2002). Trypanolytic activity is associated with two high-density lipoprotein (HDL) complexes present in human serum; HDL3 and another multi-protein complex containing IgM (Pays and Vanhollebeke 2009). However, the identity of the specific molecule(s) and their mechanism(s) of trypanolysis have only recently been unravelled. Both complexes share three critical components: apolipoprotein A-1 (apoA1), haptoglobin-related protein (Hpr) and apolipoprotein L-1 (apoL1). Although the role of these individual components has been a matter of intense debate, recent studies have shown that ApoL1 is the key component of Trypanosome lytic factor (TLF). Evidence in favour of this is also provided from studies using human serum from patients lacking either Hpr or apoL1; Hpr-negative serum displays full trypanolytic activity whereas serum from patients lacking apoL1 is ineffective (Vanhollebeke et al. 2007a, b). However, it has been noted that apoL1 alone is not as efficient in lysing trypanosomes as native HDL complexes, and it has been suggested that Hpr may act synergistically with apoL1 to exert full trypanolytic effects (Shiflett et al. 2005; Widener et al. 2007; Harrington et al. 2009).

ApoL1 kills trypanosomes by forming anionic pores in the lysosomal membrane causing an influx of chloride ions from the cytoplasm resulting in uncontrolled expansion and lysis (Perez-Morga et al. 2005; Vanhollebeke et al. 2007a, b). This requires prior uptake of apoL1, which is mediated (at least in part) by a GPI-anchored haptoglobin (Hp)–haemoglobin (Hb) receptor (TbHpHbR) located in the FP of bloodstream form *T. brucei* (Vanhollebeke et al. 2008). The principal role of this receptor is to provide trypanosomes with haem (derived from haemoglobin) which promotes growth and aids resistance to oxidative damage. In NHS, TbHpHbR also binds the trypanolytic HDLs (which contain apoL1) via the recognition of Hpr–Hb proteins that are present in these complexes. Since Hpr facilitates apoL1 uptake, this may explain the synergistic action observed between Hpr and apoL1 in trypanolytic function.

Although *T. b. rhodesiense* is resistant to lysis by NHS, this resistant character can be lost following serial passage in mice (van Meirvenne et al. 1976). The trypanosome gene that confers resistance to human serum is termed SRA (serum resistance associated) and encodes a truncated version of a VSG protein which is GPI-anchored but expressed in the FP and endosomal system rather than on the cell surface (Oli et al. 2006). The molecular basis of SRA-mediated resistance is believed to result from an interaction between SRA and apoL1 which neutralises apoL1 before it reaches the lysosome. As SRA is encoded at a single VSG expression site (De Greef et al. 1989) and confers no selective advantage in non-human hosts, it is easy to see how the SRA gene can be lost from the trypanosome genetic repertoire following extended passage in mice. Resistance of *T. b. gambiense* to lysis by human serum does not involve SRA (De Greef et al. 1992) but the mechanisms (and molecules) underpinning human serum resistance in this sub-species are likely to be resolved in the near future. This may or may not entail a separate receptor.

6.2.2 The Transferrin Receptor

T. brucei scavenges the iron transport protein transferrin (Tf) from human plasma using a GPI-anchored TfR, this is specifically located within the FP. TfR is a heterodimeric protein encoded by two genes, ESAG6 and ESAG7 (Expression Site Associated Genes), which are located within the VSG expression sites (Ligtenberg et al. 1994; Salmon et al. 1994). Different versions of ESAG6 and ESAG7 are encoded within various expression sites, and the Tf-R they encode exhibit distinctive binding affinities for Tf derived from diverse mammalian hosts. It has been suggested that expression of distinct TfRs may explain the noticeable growth differences observed when *T. brucei* infects different mammalian hosts (Bitter et al. 1998; Gerrits et al. 2002), but this interpretation has been disputed (Salmon et al. 2005). Following internalisation via the FP Tf–TfR, complexes are dissociated due to a reduction in pH within the endosomal system (Maier and Steverding 1996); the GPI-anchored TfR is then recycled to the FP (in Rab11-positive recycling endosomes) while Tf is degraded following lysosomal fusion and secreted (Pal et al. 2003).

The mechanisms by which FP restricted receptors are prevented from exiting the FP is a critical issue since trafficking to the cell surface could possibly expose proteins to immune recognition by the host. It has been suggested that the extremely high rate of endocytosis observed in trypanosomes may act to restrict low-abundance receptors (such as TfR and TbHpHbR) to the FP, while highly abundant surface proteins (such as VSG) pass onto the cell surface, but this remains a matter of debate (Field and Carrington 2009).

However, it has also been suggested that the nature of plasma membrane attachment may influence localisation of trypanosome proteins. For example, ESAG6 and ESAG7 have a high degree of sequence similarity (Carrington and Boothroyd 1996) but only ESAG6 is GPI-anchored. Consequently TfR is tethered to the FP membrane by a single GPI anchor and so it is suggested that GPI valency may be important for the sub-cellular localisation of proteins. VSG, which has two GPI anchors, has a cell surface localisation; TfR with a single GPI anchor is retained within the FP while GPI-minus VSG is degraded (Schwartz et al. 2005). Since over-expression results in the TfR receptor "spilling out" onto the cell surface (Mussmann et al. 2003), normal mechanisms acting to retain singly GPI-anchored proteins within the confines of the FP are clearly saturable, or easily disrupted by elevated expression levels. TfR localisation is also disrupted following RNAi ablation of the FP collar protein BILBO1, with TfR once again being observed to spread out over the cell surface (Bonhivers et al. 2008). While the dramatic phenotype associated with BILBO1 ablation clearly confirms that FP architecture is critical for normal endo- and exocytic functions, it does not reveal how FP proteins are restricted to the pocket. Although this has not been investigated to date, one would envisage that the discrete membrane boundaries normally established within the FP (see above for more detail) may be severely disrupted in BILBO1 FP mutants. The FP collar is proposed to act as a physical barrier to the diffusion of membrane-associated macromolecules and so the loss of this structure could have a significant impact on the distinctive lipid and protein composition of the flagellum membrane (Tyler et al. 2009).

6.2.3 Cysteine-Rich Repetitive Acidic Transmembrane Protein

In contrast to the two FP proteins discussed above, the *c*ysteine-rich *r*epetitive *a*cidic trans*m*embrane protein (CRAM) is a type 1 transmembrane protein. CRAM is abundantly expressed in procyclic *T. brucei*, but at a relatively low level in bloodstream form cells (Lee et al. 1990; Liu et al. 2000; Yang et al. 2000). The CRAM protein comprises a putative N-terminal signal peptide, an extracellular cysteine-rich repeat, a hydrophobic transmembrane domain and a 41 amino acid C terminal hydrophilic extension. Surface expression of CRAM is restricted to the FP where it is suggested to function as a lipoprotein receptor, although this function is non-essential as CRAM null mutants display no growth or development abnormalities in vitro. The CRAM protein is targeted to the FP via a clathrin-mediated sorting pathway (Hung et al. 2004), with specific transport signals mapping to a region between amino acids −5 and −14 of the C terminus since deletion of this region results in the CRAM protein localising to the endoplasmic reticulum. Larger C terminal deletions (removal of the last 40 amino acids) result in the CRAM locating to the ER, the FP and the surface of the cell body and flagellum; suggesting that the C terminal region contains both FP targeting and retention signals (Yang et al. 2000). The delineation of FP targeting (and retention) motifs is clearly important to advancing our understanding of how trypanosomatids can establish the FP as a distinct microenvironment; we still have much to learn in this regard. With regards to mechanisms acting to retain proteins within the FP, it is interesting to note that over-expressed CRAM does not relocate onto the cell surface (Qiao et al. 2006). This is in marked contrast to over-expression of GPI-anchored TfR, which spills out onto the cell surface, indicating that different mechanisms may act to retain trans-membrane and GPI-anchored proteins within the FP.

Before we conclude this deliberation on FP restricted proteins, it is worth noting that FP receptors are often stage-specific and so the proteomic composition of the FP must differ considerably between developmental stages. TfR and TbHpHbR, for instance, are restricted to the FP of bloodstream form trypanosomes, while the expression of CRAM is significantly elevated in procyclics. Differences in receptor abundance/expression reflect the disparate environmental challenges faced by trypanosomatid cells within the mammalian and insect hosts. To fully appreciate the role of the FP in parasite survival, a greater understanding of these life cycle dependent variations in FP proteomic complexity is required.

7 Molecular Aspects of Endo/Exocytosis: Lessons from RNAi Mutants

Our molecular understanding of the FP and its role in endo- and exocytosis has gained much from the analysis of defined RNAi mutant cell lines and several of these are discussed below.

7.1 Clathrin

The critical role of clathrin in receptor-mediated endocytosis was demonstrated by RNAi ablation of the clathrin heavy chain (TbCLH) (Allen et al. 2003). Clathrin is three-legged triskelion-shaped protomer (comprising three heavy chains and three light chains) that plays a major role in the formation of transport vesicles delivering cargo from the plasma membrane to early endosomes.

In bloodstream form cells, induction of TbCLH RNAi rapidly leads to the formation of a massively enlarged FP that expands to fill the majority of the cell. The formation of this grossly enlarged FP (referred to as a "big-eye" phenotype) is assumed to result from the normal delivery of membrane components to the FP but perturbation of membrane removal. Allen et al further suggest that the specific enlargement of the FP is a consequence of restrictions imposed by the sub-pellicular MT cytoskeleton so that all abnormal membrane accumulation is limited to the FP. Bloodstream form cells exhibiting a big-eye phenotype are extremely fragile and susceptible to lysis, presumably as a consequence of increased intracellular pressure generated by the enlarged FP. It is noticeable, however, that although grossly enlarged the FP retains a spherical appearance and so it is suggested that the gel-like matrix that occupies the lumen of the FP must exert a "compensatory pressure" opposing the pressure created by the expanding FP within the cytoplasm (Allen et al. 2003). Significant phenotypic differences are apparent following ablation of clathrin in procyclic form and bloodstream form trypanosomes. In procyclics, loss of clathrin does not lead to the formation of a "big-eye" but rather the abnormal intracellular accumulation of transport vesicles; indicating transport vesicles are prevented from docking with the FP membrane.

Gross enlargement of the FP is observed following RNAi ablation of a diverse array of proteins, including Rab5, Rab11, actin, dynamin and numerous proteins that have an effect on flagellum motility. While detailed information is provided below on several of these proteins, it is worth pointing out that significant differences exist between the procyclic and bloodstream phenotypes. This serves to highlight that endocytosis (and other recycling and sorting processes) differ significantly between these two trypanosome life cycle stages.

7.2 *Rab Proteins*

Rab proteins are small GTPases that play critical roles in intracellular trafficking in all eukaryotes. A bioinformatic interrogation of the *T. brucei* genome predicts that 16 RAB or RAB-like genes are present, more than in yeast but significantly fewer than in humans (Ackers et al. 2005). Most trypanosome Rab proteins have now been ascribed a function in endocytosis, exocytosis or protein recycling; however; only two Rab proteins will be considered here: Rab5 and Rab11.

T. brucei, in fact, encodes two Rab5 proteins, Rab5A and Rab5B. Both Rab5 proteins are essential for the survival of bloodstream form trypanosomes and their ablation results in morphological abnormalities and enlarged FPs, a phenotype consistent with a block in endocytosis (Fig. 6). While both Rab5A and Rab5B are involved in the early stages of endocytosis in bloodstream form cells, they localise to different compartments. Rab5A localises to endosomes containing VSG, internalised immunoglobulin and transferrin, whereas Rab5B localises to endosomes containing the trans-membrane invariant surface glycoprotein, ISG100. In bloodstream forms, the two Rab5 proteins thus appear to have distinct functions in clathrin-mediated endocytosis. Intriguingly, in procyclics, Rab5A and Rab5B co-localise to the same compartment, suggesting that differences exist in the trafficking of proteins internalised via the FP in different stages of the *T. brucei* life cycle (Hall et al. 2004).

Rab11 is a marker for recycling endosomes, and in bloodstream form cells, TbRab11 partially co-localises with Rab5A, VSG and TfR and so appears to have a critical role in recycling ligands to the FP (Jeffries et al. 2001). The critical role of Rab11 in ligand recycling was confirmed by over-expression of wild type and dominant-positive TbRab11 proteins. The dominant-positive TbRab11 mutant protein contained a single amino acid substitution within the GTP-binding site (Q to L)

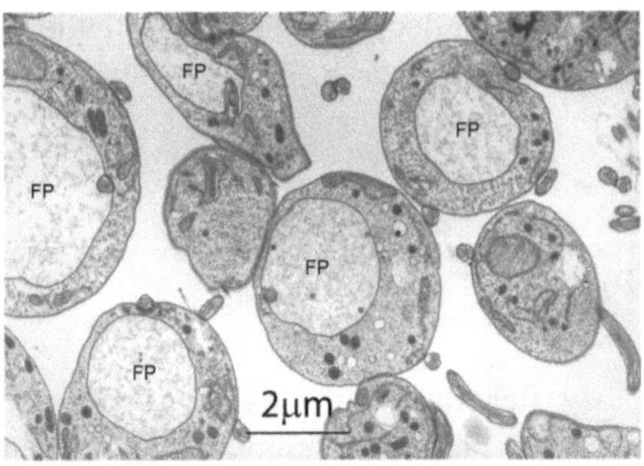

Fig. 6 Bloodstream form trypanosomes showing massive enlargement of the FP pertrubation of endocytosis. Thin-section electron micrograph showing bloodstream form *T. brucei* in which Rab5A has been ablated by RNAi. Induced cells display the "big-eye" phenotype in which the FP is grossly enlarged. *Scale bar* = 2 μm. Electron micrograph kindly provided by Dr Mark Field, University of Cambridge

which resulted in reduced GTPase activity but normal affinity to GTP. The over-expression of either wild type or dominant-positive TbRab11 had no effect on endocytosis but accelerated exocytosis (Pal et al. 2003). RNAi-mediated ablation of TbRab11 is lethal in both procyclic and bloodstream forms with cells rapidly rounding up and becoming non-viable (Hall et al. 2005). However, once again, significant phenotypic differences are observed with regards to FP morphology; bloodstream form cells exhibit a grossly enlarged FP while in procyclics FP morphology remains normal. The ablation of TbRab11 in bloodstream form cells had no apparent effect on fluid-phase endocytosis or internalisation of Tf but affected post-endosomal transport processes (as revealed by lectin endocytosis assays). However, in procyclics ablation of TbRab11 caused a block to both fluid-phase endocytosis and the receptor-mediated uptake of proteins. This suggests that TbRab11 is essential to both bloodstream form and procyclic cells but the principal role of TbRab11 varies during the life cycle.

7.3 Actin

Actin is also intimately involved in endocytosis in trypanosomes, although this role varies during the life cycle. In bloodstream forms, depletion of actin results in the arrest of vesicular trafficking (the formation of clathrin-coated vesicles at the FP is blocked) and the FP becomes grossly enlarged (Garcia-Salcedo et al. 2004). Cell division is also prevented and bloodstream form trypanosomes die as a consequence of these combined morphogenetic defects. In contrast, actin depletion in procyclic forms has no effect on cell division (although cells are smaller and stumpier) and FP morphology is normal. There is, however, a noticeable distortion to the trans-Golgi network and an aberrant accumulation of electron-lucent vesicles around the nucleus. The rationale for this life cycle difference is far from clear as is the precise role of actin in trypanosome endocytosis as many of the proteins that link the actin cytoskeleton to endocytic processes in other eukaryotes are absent from the trypanosome genome.

7.4 Flagellum Proteins

As mentioned previously, flagellum paralysis has severe effects on the ability of bloodstream form *T. brucei* to clear VSG-Immunoglobulin complexes from the cell surface. Prolonged flagellum paralysis or malfunction (resulting from the ablation of both axonemal and PFR proteins) also results in severe morphogenetic defects as bloodstream form cells are unable to divide (Broadhead et al. 2006). The resulting multinucleate "monsters" exhibit hugely enlarged FPs, but once again, in our hands, this phenotype is stage-specific since flagellum paralysis in procyclics does not lead to FP enlargement or cell division defects. Why does flagellar paralysis have such drastic effects on bloodstream form cells? There may be subtle differences depending on the location of the ablated protein, but it is suggested that flagellum dysfunction feeds through to a primary morphogenetic defect (the inability to pull apart at fully execute division) and that this subsequently results in abnormal morphogenetic axes being established and abnormal membrane–cytoskeletal imbalance at the FP. Although several FPs in these monstrous cells are active, there is significant variation in the capacity of individual FPs for antibody and lectin uptake. This suggests that a profound disturbance to the highly polarised trafficking pathway normally established within trypanosome cells results from the failure to complete cytokinesis.

7.5 Dynamin

Unlike most eukaryotes, *T. brucei* encodes only two tandemly linked dynamin-like proteins (TbDLP); these are 97% identical and likely to be functionally equivalent (Morgan et al. 2004; Chanez et al. 2006). Proteins belonging to the evolutionary conserved dynamin GTPase superfamily are responsible for various membrane remodelling events, including scission of clathrin-coated vesicles. However, the role of trypanosome dynamin in endocytosis is a matter of some debate since there are conflicting reports on the effect of TbDLP RNAi-mediated ablation in bloodstream and procyclic form *T. brucei*. Although depletion of TbDLP blocks mitochondrial fission in both life cycle stages (Morgan et al. 2004; Chanez et al. 2006), endocytosis is only affected in procyclic forms. Within 24 h of induction of dynamin RNAi, 50% of procyclic cells exhibit an enlarged FP (indicating perturbation of endocytotic pathways) and the uptake of surface proteins is significantly reduced (Chanez et al. 2006). The generation of a "big-eye" phenotpye in procyclics is unusual as this phenotype is more commonly associated with perturbation of endocytosis in bloodstream trypansomes. The explanation for this discrepancy and also the varying role of TbDLP in endocytosis remains to be resolved.

8 The Flagellar Pocket: A Portal for Secretion of Large Polymeric Proteins

The FP of bloodstream form *T. brucei* contains a number of glycoconjugates that bind the β-D-galactose-specific lectin ricin. These have been purified using ricin-agarose affinity chromatography and the glycan moieties characterised as large (yet, relatively simple) poly-*N*-acetyllactosamine (poly-LacNAc) structures (Atrih et al. 2005). Although the affinity purified preparations were undoubtedly contaminated with lysosomal and endosomal material, this study provides some insight into the composition of FP proteins. It is suggested that the very large poly-LacNAc glycans identified may contribute to the gel-like matrix present in the lumen of the FP.

Modification with poly-*N*-acetyllactosamine has also been proposed as a novel sorting signal within the trypanosome endocytic pathway (Nolan et al. 1999) and various FP-directed proteins [including type-2 VSG (Zamze et al. 1991)] are modified by LacNAc repeats, but precisely how this modification could provide the basis for FP targeting is unclear.

8.1 The Secretion of Proteophosphoglycans from the Flagellar Pocket of **Leishmania**

Until now, we have primarily focussed on the FP of the African trypanosome and host–parasite interactions. However, the FP in other trypanosomatids is equally important as exemplified by *Leishmania* spp. where the FP actively secretes proteophosphoglycan molecules which are important for colonisation of the sandfly vector and virulence in the mammalian host (Rogers et al. 2004). Secretion of PPGs into the sandfly midgut by *Leishmania* promastigotes causes an obstruction and hinders sandfly feeding activity which results in regurgitation of *Leishmania* parasites and the PPG into the skin. The presence of the PPG in the skin exacerbates cutaneous *L. mexicana* infection in several ways (1) stimulates macrophage recruitment to the site of infection; (2) promotes alternative macrophage activation; and (3) increases macrophage arginase expression (Rogers et al. 2009). PPG is thus able to modulate macrophage responses to promote favourable conditions for parasite growth.

How are these large polymeric phosphoglycan molecules assembled and secreted from *Leishmania*? While the extrusion of secreted acid phosphatase (a filamentous phosphoglycoprotein polymer produced by *Leishmania* promastigotes) can be visualised at the neck of the FP by immunofluorescence and immunoelectron microscopy (Stierhof et al. 1994), these filaments (which reach up to 2 μm long) are not observed within the intracellular transport vesicles. It has been suggested therefore that the FP may act as a "polymerisation vessel" for the conversion of monomeric or oligomeric PPG subunits into the long polymeric filamentous structures secreted from the FP (Ilg 2000). Thomas Ilg estimated the concentration of PPG as being 1,000-fold higher in the FP than the surrounding medium (Ilg 2000); an observation that perplexed Ilg since he remarked that the "flagellar reservoir is not a closed compartment". However, the flagellum clearly occludes the neck of the FP creating a relatively closed microenvironment even in amastigote stages. It is intriguing that a flagellum forms at all the amastigote stages, as the flagellum provides no motility function for the cell. However, if the microenvironment created within the FP lumen is indispensable for the polymerisation of PPG filaments, amastigotes may need to form a rudimentary flagellum to selectively seal the FP. It is likely that the rudimentary flagellum in amastigotes also fulfils other critical functions in trypanosomatid cells, including organising the formation of FP architecture, kinetoplast division processes (Robinson and Gull 1991) and may serve a sensory function, analogous to the role of primary cilia in mammalian cells (Gull 2009).

9 The Flagellar Pocket: An Example of Extreme Biology

The FP is an intriguing part of the cyto-architecture of trypanosomatid cells and is clearly pivotal for a parasitic life style. The FP allows parasitic trypanosomatids to cluster and sequester important receptors, enabling critical pathogenicity functions and restricting them from immediate adaptive immune recognition. But, as can be seen in the case of *T. brucei*, the FP also plays a pivotal role in innate immune evasion. But is the trypanosomatid FP exceptional or simply an example of "extreme biology" i.e. a normal but exaggerated feature of the cell biology present at the flagellum/cilium base in all proliferating flagellated/ciliated eukaryotic cells.

Ciliated cells such as *Paramecium*, for instance, also have specialised sites for endo- and exocytosis at the base of the cilium, called parasomal sacs and trichocysts (Allen et al. 1992; Plattner and Kissmehl 2003). Parasomal sacs contain clathrin-coated pits and are associated with endocytic vesicle formation, while trichocysts are exocytic structures that secrete long, filamentous material that may perform either a defensive function or trap prey prior to ingestion. Thus, the association of endo- and exocytosis at the base of cilia (flagella) is not unique to trypanosomatids but perhaps the FP membrane and cytoskeletal architecture has simply become extreme in its elaborations in these organisms. The polarised delivery of material to the base of flagellum/cilium also appears to be a universal feature of ciliated/flagellated cells. For example, proteins required to construct the axoneme and intra-flagellar structures are targeted to transitional fibres radiating from the mature basal body (Rosenbaum and Witman 2002; Stephan et al. 2007), while ciliary membrane proteins are delivered to specific membrane domains located close to the base of the flagellum/cilium (Rogalski and Bouck 1982; Vieira et al. 2006). Similarly, many, if not all, ciliated or flagellated cells have a mechanism for membrane differentiation to allow for specialisation of flagellar and cell body membranes.

Thus, we conclude that the FP of trypanosomatids can be considered as a normal albeit exaggerated aspect of the cell biology of a proliferating flagellated eukaryotic cell. However, this further example of the extreme biology of trypanosomes has proven useful in their adaption to a parasitic lifestyle.

References

Ackers JP, Dhir V et al (2005) A bioinformatic analysis of the RAB genes of *Trypanosoma brucei*. Mol Biochem Parasitol 141:89–97

Allen RD, Schroeder CC et al (1992) Endosomal system of *Paramecium*: coated pits to early endosomes. J Cell Sci 101:449–461

Allen CL, Goulding D et al (2003) Clathrin-mediated endocytosis is essential in *Trypanosoma brucei*. Embo J 22:4991–5002

Atrih A, Richardson JM et al (2005) *Trypanosoma brucei* glycoproteins contain novel giant poly-N-acetyllactosamine carbohydrate chains. J Biol Chem 280:865–871

Balber AE (1990) The pellicle and the membrane of the flagellum, flagellar adhesion zone, and flagellar pocket: functionally discrete surface domains of the bloodstream form of African trypanosomes. Crit Rev Immunol 10:177–201

Berriman M, Ghedin E et al (2005) The genome of the African trypanosome *Trypanosoma brucei*. Science 309:416–422

Bitter W, Gerrits H et al (1998) The role of transferrin-receptor variation in the host range of *Trypanosoma brucei*. Nature 391:499–502

Bonhivers M, Nowacki S et al (2008) Biogenesis of the trypanosome endo-exocytotic organelle is cytoskeleton mediated. PLoS Biol 6:e105

Branche C, Kohl L et al (2006) Conserved and specific functions of axoneme components in trypanosome motility. J Cell Sci 119:3443–3455

Brickman MJ, Balber AE (1990) *Trypanosoma brucei* rhodesiense bloodstream forms: surface ricin-binding glycoproteins are localized exclusively in the flagellar pocket and the flagellar adhesion zone. J Protozool 37:219–224

Brickman MJ, Cook JM et al (1995) Low temperature reversibly inhibits transport from tubular endosomes to a perinuclear, acidic compartment in African trypanosomes. J Cell Sci 108:3611–3621

Broadhead R, Dawe HR et al (2006) Flagellar motility is required for the viability of the bloodstream trypanosome. Nature 440:224–227

Bruce D (1911) The Morphology of Trypanosoma gambiense (Dutton). Proc R Soc Lond Ser B 84:327–332

Carrington M, Boothroyd J (1996) Implications of conserved structural motifs in disparate trypanosome surface proteins. Mol Biochem Parasitol 81:119–126

Chanez AL, Hehl AB et al (2006) Ablation of the single dynamin of *T. brucei* blocks mitochondrial fission and endocytosis and leads to a precise cytokinesis arrest. J Cell Sci 119:2968–2974

Coppens I, Opperdoes FR et al (1987) Receptor-mediated endocytosis in the bloodstream form of *Trypanosoma brucei*. J Protozool 34:465–473

Davidge JA, Chambers E et al (2006) Trypanosome IFT mutants provide insight into the motor location for mobility of the flagella connector and flagellar membrane formation. J Cell Sci 119:3935–3943

De Greef C, Imberechts H et al (1989) A gene expressed only in serum-resistant variants of *Trypanosoma brucei rhodesiense*. Mol Biochem Parasitol 36:169–176

De Greef C, Chimfwembe E et al (1992) Only the serum-resistant bloodstream forms of *Trypanosoma brucei* rhodesiense express the serum resistance associated (SRA) protein. Ann Soc Belg Med Trop 72:13–21

Donelson JE (2003) Antigenic variation and the African trypanosome genome. Acta Trop 85:391–404

Engstler M, Weise F et al (2005) The membrane-bound histidine acid phosphatase TbMBAP1 is essential for endocytosis and membrane recycling in *Trypanosoma brucei*. J Cell Sci 118:2105–2118

Engstler M, Pfohl T et al (2007) Hydrodynamic flow-mediated protein sorting on the cell surface of trypanosomes. Cell 131:505–515

Ferguson MA (1999) The structure, biosynthesis and functions of glycosylphosphatidylinositol anchors, and the contributions of trypanosome research. J Cell Sci 112:2799–2809

Field MC, Carrington M (2004) Intracellular membrane transport systems in *Trypanosoma brucei*. Traffic 5:905–913

Field MC, Carrington M (2009) The trypanosome flagellar pocket. Nat Rev Microbiol 7:775–786

Gadelha C, Rothery S et al (2009) Membrane domains and boundaries: cytoskeletal influences on the flagellar pocket of bloodstream trypanosomes. Proc Natl Acad Sci USA 106:17425–17430

Garcia-Salcedo JA, Perez-Morga D et al (2004) A differential role for actin during the life cycle of *Trypanosoma brucei*. EMBO J 23:780–789

Gerrits H, Mussmann R et al (2002) The physiological significance of transferrin receptor variations in *Trypanosoma brucei*. Mol Biochem Parasitol 119:237–247

Ginger ML, Portman N et al (2008) Swimming with protists: perception, motility and flagellum assembly. Nat Rev Microbiol 6:838–850

Gull K (1999) The cytoskeleton of trypanosomatid parasites. Annu Rev Microbiol 53:629–655

Gull K (2009) The parasite point of view: Insights and questions on the cell biology of *Trypanosoma* and *Leishmania* parasite-phagocyte interactions. In: Russell D, Gordon S (eds) Phagocyte-pathogen interactions: macrophages and the host response to infection. ASM Press, Washington, pp 453–462

Hall B, Allen CL et al (2004) Both of the Rab5 subfamily small GTPases of *Trypanosoma brucei* are essential and required for endocytosis. Mol Biochem Parasitol 138:67–77

Hall BS, Smith E et al (2005) Developmental variation in Rab11-dependent trafficking in *Trypanosoma brucei*. Eukaryot Cell 4:971–980

Harrington J, Howell S et al (2009) Membrane permeabilization by trypanosome lytic factor, a cytolytic human high-density lipoprotein. J Biol Chem 284:13505–13512

Hertz-Fowler C, Ersfeld K et al (2001) CAP5.5, a life-cycle-regulated, cytoskeleton-associated protein is a member of a novel family of calpain-related proteins in *Trypanosoma brucei*. Mol Biochem Parasitol 116:25–34

Hoare C, Wallace F (1966) Developmental stages of trypanosomatid flagellates: a new terminology. Nature 212:1385–1386

Horn D (2004) The molecular control of antigenic variation in *Trypanosoma brucei*. Curr Mol Med 4:563–576

Hung CH, Qiao X et al (2004) Clathrin-dependent targeting of receptors to the flagellar pocket of procyclic-form *Trypanosoma brucei*. Eukaryot Cell 3:1004–1014

Ilg T (2000) Proteophosphoglycans of *Leishmania*. Parasitol Today 16:489–497

Jeffries TR, Morgan GW et al (2001) A developmentally regulated rab11 homologue in *Trypanosoma brucei* is involved in recycling processes. J Cell Sci 114:2617–2626

Lacomble S, Vaughan S et al (2009) Three-dimensional cellular architecture of the flagellar pocket and associated cytoskeleton in trypanosomes revealed by electron microscope tomography. J Cell Sci 122:1081–1090

LaCount DJ, Barrett B et al (2002) *Trypanosoma brucei* FLA1 is required for flagellum attachment and cytokinesis. J Biol Chem 277:17580–17588

Lee MG, Bihain BE et al (1990) Characterization of a cDNA encoding a cysteine-rich cell surface protein located in the flagellar pocket of the protozoan *Trypanosoma brucei*. Mol Cell Biol 10:4506–4517

Ligtenberg MJ, Bitter W et al (1994) Reconstitution of a surface transferrin binding complex in insect form *Trypanosoma brucei*. EMBO J 13:2565–2573

Linder JC, Staehelin LA (1977) Plasma membrane specializations in a trypanosomatid flagellate. J Ultrastruct Res 60:246–262

Liu J, Qiao X et al (2000) Receptor-mediated endocytosis in the procyclic form of *Trypanosoma brucei*. J Biol Chem 275:12032–12040

Maier A, Steverding D (1996) Low affinity of *Trypanosoma brucei* transferrin receptor to apotransferrin at pH 5 explains the fate of the ligand during endocytosis. FEBS Lett 396:87–89

McLintock LM, Turner CM et al (1993) Comparison of the effects of immune killing mechanisms on *Trypanosoma brucei* parasites of slender and stumpy morphology. Parasite Immunol 15:475–480

Moreira-Leite FF, Sherwin T et al (2001) A trypanosome structure involved in transmitting cytoplasmic information during cell division. Science 294:610–612

Morgan GW, Allen CL et al (2001) Developmental and morphological regulation of clathrin-mediated endocytosis in *Trypanosoma brucei*. J Cell Sci 114:2605–2615

Morgan GW, Hall BS et al (2002a) The kinetoplastida endocytic apparatus. Part I: a dynamic system for nutrition and evasion of host defences. Trends Parasitol 18:491–496

Morgan GW, Hall BS et al (2002b) The endocytic apparatus of the kinetoplastida. Part II: machinery and components of the system. Trends Parasitol 18:540–546

Morgan GW, Goulding D et al (2004) The single dynamin-like protein of *Trypanosoma brucei* regulates mitochondrial division and is not required for endocytosis. J Biol Chem 279:10692–10701
Morriswood B, He CY et al (2009) The bilobe structure of *Trypanosoma brucei* contains a MORN-repeat protein. Mol Biochem Parasitol 167:95–103
Mussmann R, Janssen H et al (2003) The expression level determines the surface distribution of the transferrin receptor in *Trypanosoma brucei*. Mol Microbiol 47:23–35
Natesan SK, Peacock L et al (2007) Activation of endocytosis as an adaptation to the mammalian host by trypanosomes. Eukaryot Cell 6:2029–2037
Navarro M, Penate X et al (2007) Nuclear architecture underlying gene expression in *Trypanosoma brucei*. Trends Microbiol 15:263–270
Nolan DP, Geuskens M et al (1999) N-linked glycans containing linear poly-N-acetyllactosamine as sorting signals in endocytosis in *Trypanosoma brucei*. Curr Biol 9:1169–1172
O'Beirne C, Lowry C et al (1998) Both IgM and IgG anti-VSG antibodies initiate a cycle of aggregation-disaggregation of bloodstream forms of *Trypanosoma brucei* without damage to the parasite. Mol Biochem Parasitol 91:165–193
Oli MW, Cotlin LF et al (2006) Serum resistance-associated protein blocks lysosomal targeting of trypanosome lytic factor in *Trypanosoma brucei*. Eukaryot Cell 5:132–139
Overath P, Engstler M (2004) Endocytosis, membrane recycling and sorting of GPI-anchored proteins: *Trypanosoma brucei* as a model system. Mol Microbiol 53:735–744
Overath P, Stierhof YD et al (1997) Endocytosis and secretion in trypanosomatid parasites – Tumultuous traffic in a pocket. Trends Cell Biol 7:27–33
Pal A, Hall BS et al (2003) Rab5 and Rab11 mediate transferrin and anti-variant surface glycoprotein antibody recycling in *Trypanosoma brucei*. Biochem J 374:443–451
Pays E (2005) Regulation of antigen gene expression in *Trypanosoma brucei*. Trends Parasitol 21:517–520
Pays E, Vanhollebeke B (2009) Human innate immunity against African trypanosomes. Curr Opin Immunol 21:493–498
Pedersen LB, Rosenbaum JL (2008) Intraflagellar transport (IFT) role in ciliary assembly, resorption and signalling. Curr Top Dev Biol 85:23–61
Perez-Morga D, Vanhollebeke B et al (2005) Apolipoprotein L-I promotes trypanosome lysis by forming pores in lysosomal membranes. Science 309:469–472
Plattner H, Kissmehl R (2003) Molecular aspects of membrane trafficking in *Paramecium*. Int Rev Cytol 232:185–216
Qiao X, Chuang BF et al (2006) Sorting signals required for trafficking of the cysteine-rich acidic repetitive transmembrane protein in *Trypanosoma brucei*. Eukaryot Cell 5:1229–1242
Radwanska M, Chamekh M et al (2002) The serum resistance-associated gene as a diagnostic tool for the detection of *Trypanosoma brucei* rhodesiense. Am J Trop Med Hyg 67:684–690
Robinson DR, Gull K (1991) Basal body movements as a mechanism for mitochondrial genome segregation in the trypanosome cell cycle. Nature 352:731–733
Rogalski AA, Bouck GB (1982) Flagellar surface antigens in *Euglena*: immunological evidence for an external glycoprotein pool and its transfer to the regenerating flagellum. J Cell Biol 93:758–766
Rogers ME, Ilg T et al (2004) Transmission of cutaneous leishmaniasis by sand flies is enhanced by regurgitation of fPPG. Nature 430:463–467
Rogers M, Kropf P et al (2009) Proteophosophoglycans regurgitated by *Leishmania*-infected sand flies target the L-arginine metabolism of host macrophages to promote parasite survival. PLoS Pathog 5:e1000555
Rosenbaum JL, Witman GB (2002) Intraflagellar transport. Nat Rev Mol Cell Biol 3:813–825
Salmon D, Geuskens M et al (1994) A novel heterodimeric transferrin receptor encoded by a pair of VSG expression site-associated genes in *T. brucei*. Cell 78:75–86
Salmon D, Paturiaux-Hanocq F et al (2005) *Trypanosoma brucei*: growth differences in different mammalian sera are not due to the species-specificity of transferrin. Exp Parasitol 109:188–194

Schwartz KJ, Peck RF et al (2005) GPI valence and the fate of secretory membrane proteins in African trypanosomes. J Cell Sci 118:5499–5511

Shiflett A, Bishop J et al (2005) Human high density lipoproteins are platforms for the assembly of multi-component innate immune complexes. J Biol Chem 280:32578–32585

Stephan A, Vaughan S et al (2007) An essential quality control mechanism at the eukaryotic basal body prior to intraflagellar transport. Traffic 8:1323–1330

Stierhof YD, Ilg T et al (1994) Characterization of polymer release from the flagellar pocket of *Leishmania mexicana* promastigotes. J Cell Biol 125:321–331

Stockdale C, Swiderski MR et al (2008) Antigenic variation in *Trypanosoma brucei*: joining the DOTs. PLoS Biol 6:e185

Taylor JE, Rudenko G (2006) Switching trypanosome coats: what's in the wardrobe? Trends Genet 22:614–620

Tyler KM, Fridberg A et al (2009) Flagellar membrane localization via association with lipid rafts. J Cell Sci 122:859–866

van Meirvenne N, Maginus E et al (1976) The effect of normal human serum on trypanosomes of distinct antigenic type (ETat 1 to 12) isolated from a strain of *Trypanosoma brucei rhodesiense*. Ann Soc Belg Med Trop 56:55–63

Vanhollebeke B, Lecordier L et al (2007a) Human serum lyses *Trypanosoma brucei* by triggering uncontrolled swelling of the parasite lysosome. J Eukaryot Microbiol 54:448–451

Vanhollebeke B, Nielsen MJ et al (2007b) Distinct roles of haptoglobin-related protein and apolipoprotein L-I in trypanolysis by human serum. Proc Natl Acad Sci USA 104:4118–4123

Vanhollebeke B, De Muylder G et al (2008) A haptoglobin-hemoglobin receptor conveys innate immunity to *Trypanosoma brucei* in humans. Science 320:677–681

Vickerman K (1973) The mode of attachment of *Trypanosoma vivax* in the proboscis of the tsetse fly Glossina fuscipes: an ultrastructural study of the epimastigote stage of the trypanosome. J Protozool 20:394–404

Vickerman K, Tetley L (1990) Flagellar surfaces of parasitic protozoa and their role in attachment. In: Bloodgood R (ed) Ciliary and flagellar membranes. Plenum, New York, pp 267–304

Vieira OV, Gaus K et al (2006) FAPP2, cilium formation, and compartmentalization of the apical membrane in polarized Madin-Darby canine kidney (MDCK) cells. Proc Natl Acad Sci USA 103:18556–18561

Wickstead B, Gull K (2006) A "holistic" kinesin phylogeny reveals new kinesin families and predicts protein functions. Mol Biol Cell 17:1734–1743

Wickstead B, Gull K (2007) Dyneins across eukaryotes: a comparative genomic analysis. Traffic 8:1708–1721

Widener J, Nielsen M et al (2007) Hemoglobin is a co-factor of human trypanosome lytic factor. PLoS Pathog 3:1250–1261

Yang H, Russell DG et al (2000) Sequence requirements for trafficking of the CRAM transmembrane protein to the flagellar pocket of African trypanosomes. Mol Cell Biol 20:5149–5163

Yoshikawa H, Furuki J et al (1990) Freeze-fracture study of the bloodstream form of *Trypanosoma brucei gambiense*. J Protozool 37:27–32

Zamze SE, Ashford DA et al (1991) Structural characterization of the asparagine-linked oligosaccharides from *Trypanosoma brucei* type II and type III variant surface glycoproteins. J Biol Chem 266:20244–20261

Reservosomes of *Trypanosoma cruzi*

Narcisa Leal Cunha-e-Silva, Celso Sant'Anna, Miria G. Pereira, and Wanderley de Souza

Contents

Abstract Reservosomes are lysosome-related organelles (LROs) of *Trypanosoma cruzi* with the special capacity of nutrient storage and hydrolase accumulation. They represent the final compartment of epimastigote endocytic pathway and the site of activity of cruzipain, the major *T. cruzi* protease. They are essential for epimastigote differentiation into trypomastigote infective forms. Trypomastigotes, as well as amastigotes, present related LROs that do not have storage capacity. Typical epimastigote reservosomes have an electrondense protein matrix in which planar membrane units, rare vesicles, and lipid inclusions are immersed. More than

N.L. Cunha-e-Silva (✉) and M.G. Pereira
Laboratório de Ultraestrutura Celular Hertha Meyer, Instituto de Biofísica Carlos Chagas Filho, Universidade Federal do Rio de Janeiro, Av. Carlos Chagas Filho, 373, bloco G subsolo, Cidade Universitária, Ilha do Fundão, Rio de Janeiro 21941-902, Brazil
e-mail: narcisa@biof.ufrj.br

C. Sant'Anna and W. de Souza
Laboratório de Ultraestrutura Celular Hertha Meyer, Instituto de Biofísica Carlos Chagas Filho, Universidade Federal do Rio de Janeiro, Av. Carlos Chagas Filho, 373, bloco G subsolo, Cidade Universitária, Ilha do Fundão, Rio de Janeiro 21941-902, Brazil
Diretoria de Programas, Instituto Nacional de Metrologia, Normalização e Qualidade Industrial-INMETRO, Av. Nossa Senhora das Graças, 50, Xerém, Duque de Caxias 25250-020, Rio de Janeiro, Brazil

W. de Souza (ed.), *Structures and Organelles in Pathogenic Protists*,
Microbiology Monographs 17, DOI 10.1007/978-3-642-12863-9_5,
© Springer-Verlag Berlin Heidelberg 2010

700 proteins are identified in reservosome proteome, confirming the known organelle functions and indicating its participation in other still unexplored pathways.

1 Introduction: Endocytosis in Higher Eukaryotic Cells

Eukaryotic cells possess a remarkably integrated and strongly regulated system that is essential for the internalization of nutrients. In addition to supplying the energy needs of a cell, endocytosis plays key roles in a variety of processes, including membrane homeostasis, cell surface receptor recycling, and cell signaling. In most cells, there are many routes of macromolecule internalization. Although there is considerable heterogeneity in these different routes, particularly in terms of the molecular machinery responsible for governing the entire process, they coexist in the same cell type. Depending on the cargo size, endocytosis can be subdivided into phagocytosis and pinocytosis.

Phagocytosis has been mainly characterized in professional phagocytes, such as macrophages and neutrophils and in the protozoan parasite amoeba (Stuart and Ezekowitz 2005). In this mechanism, large molecules (>250 nm), including whole cells, are ingested in an actin-dependent manner through a pseudopod protrusion of the plasma membrane that subsequently fuses to form a closed cargo-containing vacuole. A substantial role for phagocytosis in the defense against pathogen infection and the elimination of apoptotic bodies has been uncovered by programmed cell death studies. Macropinocytosis is characterized by the mobilization of actin microfilaments, with subsequent formation of membrane ruffles and fusion with the plasma membrane; this process results in the formation of a macropinosome. In response to external stimuli, such as platelet-derived growth factor (PDGF) that is involved in cell migration or bacterial toxins that lead to Rho-GTPase activation, internalization of bacteria, and subsequent replication, a substantial volume of extracellular fluid is ingested (Ridley 2001).

Small solutes are internalized by pinocytosis. The uptake of molecules or nutrients by pinocytosis involves a set of proteins and lipids that coordinate vesicle budding and fusion with target compartments. In higher eukaryotes, clathrin-dependent endocytosis is the best understood of these processes. It requires the assembly of protein 2 complex (AP-2), which is responsible for accessory proteins, adaptors, and clathrin coat recruitment, and phosphatidylinositol (4,5)-bisphosphate (PIP_2), which mediates cargo clustering and interactions between dynamin and its adaptors. Clathrin-independent pathways have also been reported; of these, the best characterized is the endocytosis that is mediated by highly dynamic structures called caveolae. These are stable flask-shaped membrane domains that are found in a wide variety of cellular types, such as endothelial cells, muscles tissues, or macrophages. They are formed by oligomerized caveolin that is associated with cholesterol and the latter is responsible for maintaining the shape of the structure. The presence of dynamin aids in the scission of cargo-containing vesicles. To gain access to the intracellular environment, toxins, bacteria, and viruses may either fuse first with caveosomes and then with the plasma membrane (transcytosis) or fuse directly with early endosomes (Nichols and Lippincott-Schwartz 2001).

The constitutive formation of small vesicles via caveolin- and clathrin-independent routes occurs in all cell types, and the cellular controls of their formation are unclear. Furthermore, these vesicles are not surrounded by a protein coat. The heterogeneity of these vesicles primarily concerns the variable participation of dynamin in each of these routes; it is present in some vesicle budding, but not in others. Regardless of the participation of clathrin or caveolin, the vesicles express Rab5 and fuse with early endosomes. Endosomes are complex tubular networks that are acidified via a V-type-H^+-ATPase, and in which most receptor–ligand complexes uncouple. The fusion is dependent on the presence of Rab5 effectors on endosomes, such as EEA-1 and PI-(3)-P (phosphatidylinositol (3)-phosphate). Subsequently, Rab11-positive vesicles carrying cargo are directed towards recycling endosomes, while those that present Rab7 follow a degradative pathway in late endosomes/lysosomes (Wieffer et al. 2009).

After gaining entrance to the plasma membrane, macromolecules follow a highly ordered flow of vesicles and pass through a series of biochemically, functionally, and morphologically different organelles that comprise the complex and integrated cell endo-lysosomal system. A general endocytic pathway comprises early and late endosomes (degradative route), recycling endosomes (recycling route), and lysosomes (the degradative organelle). This system is closely integrated with the secretory pathway, via the Golgi complex. For a more detailed review of endocytic processes, see Doherty and McMahon (2009).

2 Endocytosis in Trypanosomatids

Studies of endocytic pathways have mainly considered mammalian cells and yeast and a restricted number of protists, including trypanosomatid family members such as *Leishmania*, *Trypanosoma brucei*, and *Trypanosoma cruzi* (De Souza et al. 2009). In the first two cases, endocytosis has an additional function besides obtaining substrates for energy: it is also responsible for evading the host immune system. Although simpler than in mammalian cells, the endocytic process in trypanosomes is governed by several conserved pathways, organelles, and molecular machinery. It requires the presence of surface receptors, the assembly of clathrin coated vesicles, and the fusion of endocytic vesicles, endosomes, and lysosomes. Indeed, the structures and organelles that are related to endocytosis in this family are both unique and interesting. Therefore, they constitute an attractive model for understanding the evolution of the endocytic pathway, for elucidating the role of these processes in cell survival, and also as a possible chemotherapeutic target (see De Souza et al. 2009 for recent review).

Trypanosome's endocytic pathway is a highly polarized system that is formed by compartments that are located at conserved positions within the cytoplasm.

Although similar in many respects, there are also certain peculiarities in the structural organization of each member. In *T. brucei* bloodstream trypomastigotes, all of the endocytic organelles are confined to the posterior region of the parasite, between the nucleus and the kinetoplast (Morgan et al. 2002). In *Leishmania* promastigotes, the main structure of the endosomal/lysosomal system is a long multivesicular tubule that extends from the anterior to the posterior region (Weise et al. 2000). In *T. cruzi* epimastigotes, nutrients are mainly obtained via an additional entry site – the cytosome – that is absent in both *Leismania* and *T. brucei* (Soares and De Souza 1991; Porto-Carreiro et al. 2000). In addition to these structures, there are also organelles that store nutrients and concentrate lysosomal hydrolases called reservosomes (Soares and De Souza 1988); these will be reviewed in this chapter.

3 Endocytic Pathway in *Trypanosoma cruzi*

In addition to trypomastigotes, the infective forms that are found in the feces of invertebrate vectors and in the host blood of vertebrates, *T. cruzi* includes two distinct replicative forms: amastigotes, which are found inside vertebrate host cells, and epimastigotes, which are present in the lumen of the invertebrate host's midgut. Despite the high energy requirements that are an inherent feature of cell proliferation, endocytosis has only been demonstrated in epimastigotes (Soares and De Souza 1991; Porto-Carreiro et al. 2000).

Endocytosis in epimastigotes occurs via both flagellar pocket and cytostome (Soares and De Souza 1991). A flagellar pocket is a well-defined structure that has been described as a lateral cell membrane depression that is continuous with the flagellar membrane and that forms a confined domain from which a flagellum emerges, and where endocytic and exocytic events take place (Landfear and Ignatushchenko 2001). The cytostome (a round opening at the plasma membrane near the flagellar pocket) and its deep invagination (the cytopharynx) form a stable complex (Milder and Deane 1969) that can extend to the nuclear region. It is observed only in epimastigote and amastigote forms of *Schizotrypanum* subgenus, such as *T. cruzi*, *Trypanosoma vespertilionis*, and *Trypanosoma dionisii*. Vesicles originate from both entry sites, and they fuse with early endosomes (the branched-tubular compartments that extend from the anterior region of epimastigotes to the posterior end). The accumulation of acridine orange has demonstrated that early endosomes are acidic structures (Porto-Carreiro et al. 2000). Although Rab5, an early endosome molecular marker in mammalian cells, is expressed in all *T. cruzi* developmental stages, its precise location has yet to be determined (Araripe et al. 2005). Cargo-containing vesicles bud off from early endosomes and fuse with the reservosomes; these are special types of compartment that have been classified as organelle-related to lysosomes (Sant'Anna et al. 2008a).

4 Reservosomes: A Historical View

The first descriptions of reservosomes were based on ultrastructural observations (Boiso et al. 1977; De Souza et al. 1978). Endocytosis assays, using exogenous horseradish peroxidase (HRP) and visualized by ultrastructural cytochemistry, showed that epimastigotes take in high amounts of HRP via the flagellar pocket and cytostome. Close to the cytostome, pinocytic vesicles with an average diameter of 80 nm could be visualized by a positive reaction for HRP. In these studies, the authors proposed that homotypic fusion between vesicles gives rise to large endocytic organelles, called multivesicular structures, that are located in the posterior region of the epimastigotes (Fig. 1b) (De Souza et al. 1978). Freeze-fracture analysis provided further evidence for the existence of inner vesicles, ascribing the compartments as multivesicular structures (Fig. 1c). Subsequently, cytochemical studies to detect basic protein (using ethanolic phosphotungstic acid) and lipids (using imidazole-buffered osmium for enhancing lipid staining) showed that the multivesicular structures are composed of electron-lucent lipid inclusions that are dispersed in an electron-dense protein matrix (Fig. 1a, d). This indicated that the multivesicular structures are sites of lipid and protein accumulation that could represent nutrient storage sites in *T. cruzi* (Soares and De Souza 1988). In contrast to previous reports (De Souza et al. 1978), Soares and De Souza (1988) were unable to observe internal membranes using routine examination with transmission electron microscopy of ultrathin sections. In the absence of membrane units that contained these inclusions, they proposed that the name "multivesicular structure" was inaccurate; thus, the name "reservosome" (based on the unusual reserve

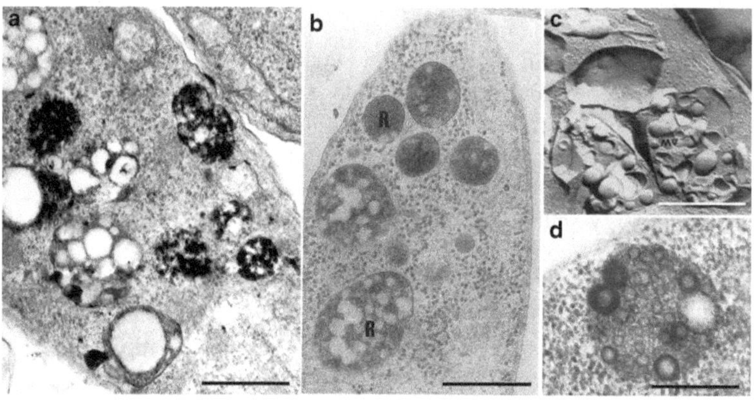

Fig. 1 *Reservosomes cytochemical characterization:* (**a**) After endocytosis, HRP accumulated in reservosomes and were revealed by DAB- osmium cytochemistry; big electron-lucent inclusions were also observed. (**b**) Cytochemistry for basic proteins showed positive reaction at reservosome matrix. (**c**) Freeze fracture of reservosomes indicated that they are multivesicular organelles. (**d**) Osmium-imidazole cytochemistry demonstrated that reservosome electron-lucent inclusions contain neutral lipids. *Bars*: (**a**) 0.5 μm; (**b**) 0.4 μm; (**c**) 0.5 μm; (**d**) 0.4 μm

characteristic) was launched. Later, while studying endocytosis in epimastigote forms at the ultrastructural level using gold-labeled proteins (albumin, transferrin, and peroxidase) and low-density lipoproteins (LDL), Soares and De Souza (1991) traced the endocytic pathway in epimastigotes and confirmed the reserve trait of reservosomes.

5 Reservosome Morphological and Biochemical Characterization

Reservosomes are large membrane-bound organelles, with a mean diameter of 600 nm, that are located in the posterior region of these parasites (Fig. 2a) (Cunha-e-Silva et al. 2006). Although they are typically round in shape, reservosomes with asymmetric morphology are also found. Each epimastigote has several reservosomes (Soares and De Souza 1988). During epimastigote division, the reservosomes are distributed between the new daughter cells (Figueiredo et al. 1994). Whether or not the partition process is symmetric has not thus far been determined. Ultrastructural analysis combined with cytochemistry revealed that the reservosome lumen is composed of an electron-dense protein matrix that contains lipid droplets. A stereological study demonstrated that reservosomes occupy about 5% of the entire cell volume (Soares and De Souza 1988). The evaluation of the reservosome volume during increasing periods of cultivation in LIT medium (3, 5, 7, 9, and 12 days) showed that it is smaller in early cultures, increases up to a period of 9 days, and decreases again at later stages (Figueiredo et al. 1994). This suggests that the content of the reservosome is consumed to provide energy for survival of the parasite during periods of starvation. Transmission electron microscopy analysis showed that whereas the lumen of reservosomes in young cultures is composed of an electron-dense protein matrix and lipid inclusions, the lumen in older cultures does not have a dense appearance. Stereological studies carried out during the differentiation of epimastigotes to trypomastigotes (metacyclogenesis) showed a gradual involution of the reservosomes resulting from a reduction of the organelle volume (Soares et al. 1989), emphasizing the importance of reservosome stored content as an energy source during metacyclogenesis. Although the existence of inner membranes was considered to be controversial in earlier studies (De Souza et al. 1978; Soares and De Souza 1988), recent ultrastructural analyses of reservosomes in situ and of isolated organelles showed that these compartments do indeed contain internal membranes; thus, the age-old dilemma has been resolved (Sant'Anna et al. 2008b). Internal membranes could also be seen either as planar structures that crossed the organelle (Fig. 2c) or similar to vesicles (Fig. 2b, d). As the vesicles were always present in low numbers and were not accessible to the endocytic tracer, their origin needs to be clarified. Exogenous cargo reaches reservosomes via fusion of endocytic vesicles with the organelle's boundary membrane (Fig. 2f). The vesicle contents are then delivered directly into the reservosome lumen (Fig. 2g). Moreover, the reservosomes also display rectangular (or rod-shaped) lipid bodies that are

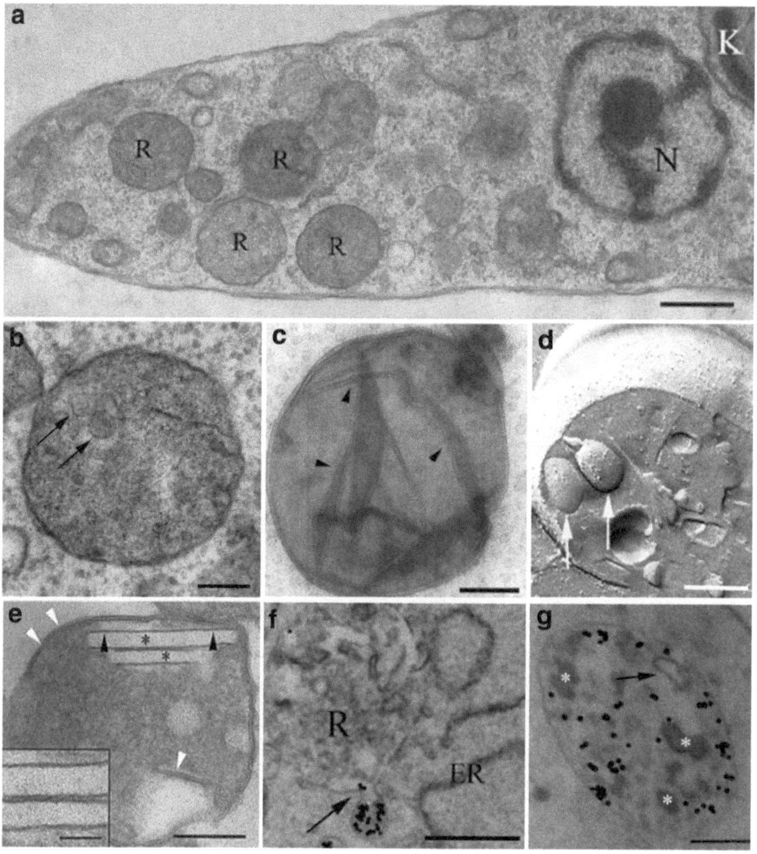

Fig. 2 *Ultrastructural analysis of in situ or isolated reservosomes:* (**a**) Longitudinal section showing general *T. cruzi* epimastigote morphology, where it is possible to observe reservosomes at their typical position (posterior region of the parasite). (**b**) Thin section of an in situ reservosome containing inner vesicles (*arrows*). (**c**) Electron microscopy of uranyl acetate stained isolated reservosome showing long inner membrane profiles spread along the lumen of the organelle (*arrowheads*). (**d**) Image of freeze fractured reservosome showing the inner vesicles containing IMPs (*arrows*). (**e**) Ultrastructure of isolated reservosomes showing large electron-lucent rod shaped bodies (*asterisks*). Note that these structures are surrounded by a monolayer of membrane (*black arrowheads*), different from the boundary membrane (*white arrowheads*). The *inset* is a high magnification of the rod edge area, detailing the surrounding monolayer. (**f**) Osmium-imidazole treated parasite showing an endocytic vesicle containing TF–Au particles fusing with a reservosome (*arrow*). (**g**) In situ reservosome containing endocytosed TF–Au in an osmium-imidazole treated parasite; notice that the gold-particles are spread in the lumen of reservosome and not inside inner vesicles (*arrow*). *Asterisks* indicate lipid inclusions positive to the neutral lipid reaction. *N* nucleus; *R* reservosome; *K* kinetoplast. *Bars*: (**a**) 0.5 μm; (**b**) 0.2 μm; (**c**) 0.09 μm; (**d**) 0.2 μm; (**e**) 0.2 μm; (*inset*) 0.1 μm; (**f**) 0.5 μm; (**g**) 0.3 μm. After Sant'Anna et al. (2008a)

surrounded by a phospholipid monolayer (Fig. 2e), besides spherical lipid inclusions. Imidazole buffered osmium revealed a negative core in the rod-shape inclusions, in contrast to the positive reaction in spherical ones, probably due to a differentiated saturated lipid composition in the former. The lipid inclusions of reservosomes are morphologically similar to the cholesterol crystals that have been observed in foam cell-differentiated macrophages during the development of atheromatous plaques. Although *T. cruzi* does not synthesize cholesterol, epimastigotes take up and store high concentrations of cholesterol from the medium (Soares and De Souza 1991; Cunha-e-Silva et al. 2002). Indeed, it is possible that under conditions of low pH, high concentrations of cholesterol reproduce the same conditions under which there is nucleation of cholesterol crystals inside macrophage lysosomes (Kellner-Weibel et al. 1999). Biochemical characterization of isolated reservosomes demonstrated that these organelles present twice as many lipids as proteins; however, in whole epimastigotes the ratio is 1:1. Thin layer chromatography revealed a high concentration of cholesteryl ester in the organelle that comprised 55% of the neutral lipid composition (Cunha-e-Silva et al. 2002).

Furthermore, Torres et al. (2004) characterized an ABCA1 transporter in *T. cruzi* epimastigotes. It was suggested that TcABCA1 is localized in the plasma membrane, flagellar pocket, and reservosomes. Proteomic analysis of isolated reservosomes (Sant'Anna et al. 2009) revealed the presence of an ABC transporter. ABC (*A*TP *B*inding *C*assette) transporters belong to a class of small proteins that use the energy from ATP hydrolysis to pump substrates across biological membranes (Davidson and Maloney 2007). In mammalian cells, this process is involved in lipid metabolism. These transporters are also present in many organisms such as prokaryotes, in which they are involved in the control of nutrient acquisition (Weiner et al. 1971) or the export of virulence factors. In humans, they are important in various processes, including the exclusion of xenobiotics (MDR-multi drug resistance) (Davidson and Maloney 2007), in cholesterol efflux, and in HDL (high-density lipoprotein) formation (Oram and Yokoyama 1996; Takahashi et al. 2005); the latter constitutes the initial step in the pathway of reverse cholesterol transport. Cholesterol efflux back to the plasma membrane occurs in both early and late endosomes, and is dependent on ABCA1 transporter with coparticipation of ApoA-I (Neufeld et al. 2004). Similar to higher eukaryotes, TcABCA1 can traffic from the plasma membrane to reservosomes and regulate the internal pool of cholesterol by releasing excess cholesterol via an unknown mechanism.

Interestingly, proteomic analysis has revealed that reservosomes also contain a homologue to Rab18 (TcRab18) (Sant'Anna et al. 2009). Rab18 is a small GTPase that is associated with lipid droplets. Lipid droplets or lipid bodies are a class of organelles that originate from the cytoplasmic leaflet of the endoplasmic reticulum and comprise phospholipid molecules, cholesterol, and proteins organized in a monolayer. They are particularly abundant in adipocytes and are associated with the provision of neutral lipids for use in many biological processes, such as β-oxidation or phospholipid and lipoprotein synthesis. Neutral lipids could conceivably reach reservosomes through LDL endocytosis, and thereafter be distributed to either other organelles or directly to the cytosol, according to cellular demand.

Curiously, proteomic analysis of reservosomes has also identified a P-glycoprotein transporter (Sant'Anna et al. 2009). In a previous study, which used specific inhibitors of this transporter, P-glycoprotein was suggested to be responsible for heme transport in *T. cruzi* epimastigotes (Lara et al. 2007). As heme is stored in reservosomes, it is possible that an additional transporter, such as a P-glycoprotein, could operate in reservosomes.

Reservosomes were considered to be both acidic compartments that are involved in controlling the cellular digestion of ingested material and key organelles for providing the metabolite substrates that are required for metacyclogenesis (Soares 1999). Due to the absence of lysosomal markers such as LAMP1 or LAMP2 and lgp120, the reservosomes were categorized as prelysosomal compartments or late endosomes (Soares et al. 1992). However, the description of reservosomes as late endosomes was somewhat controversial. As Rab 11 is commonly found in recycling endosomes while Rab 7 is a molecular marker of late endosomes in mammalian cells, the localization of the monomeric GTPase TcRab7 to the Golgi complex (Araripe et al. 2004), rather than to reservosomes, and the presence of TcRab11 in these organelles exemplify these inconsistencies (Mendonca et al. 2000). The low pH of reservosomes was inferred from studies that used acridine orange (Soares and De Souza 1991). Subsequently, the use of DAMP (a weak base that is concentrated in acidic compartments and can be detected by immunocytochemistry under the electron microscope) suggested that the pH of the organelle was 6.0 (Soares et al. 1992). The acidification process in the endocytic compartments of other eukaryotes is mediated by a V-type-H^+-ATPase (Beyenbach and Wieczorek 2006). Intriguingly, in *T. cruzi* epimastigotes, endocytic compartments are acidified by the action of a P-type H^+-ATPase, which has only been found in the plasma membranes of yeast, plants (Serrano et al. 1992), and recently in *T. brucei* (Luo et al. 2006). Three isoforms of the P-type proton ATPase were found in reservosomes: TcHA1, TcHA2 (Vieira et al. 2005), and TcHA3 (Sant'Anna et al. 2009). While TcHA1 was identified in the plasma membrane, cytostome, endocytic vesicles, and reservosomes, TcHA2 was located only in reservosomes. TcHA3 was identified by proteomic analysis of isolated reservosomes.

Reservosomes are important sites of lysosomal hydrolase action. Cazzulo et al. (1990) characterized a cathepsin L-like cysteine protease called cruzipain (CZP) as the major protease in epimastigote forms. It was also demonstrated that substantial quantities of CZP are accumulated in reservosomes (Souto-Padron et al. 1990; Soares et al. 1992). The isolation of these organelles enabled verification that cruzipain is active in reservosomes (Cunha-e-Silva et al. 2002). Although the pathway that targets cruzipain to reservosomes is not fully understood, previous work has suggested that this enzyme is transported by secretory vesicles from the Golgi complex to the reservosomes (Soares et al. 1992) via a mannose 6-phosphate-independent pathway (Cazzulo et al. 1990). Using an immunocytochemical approach, CZP was also localized to the parasite's plasma membrane and flagellar pocket (Souto-Padron et al. 1990). Therefore, it is possible that at least a fraction of CZP is first delivered to the parasite surface and then internalized as part of the endocytic pathway before being delivered to reservosomes. During its journey from the Golgi complex to the reservosomes, the prodomain of CZP undergoes cleavage; this maturation process has been demonstrated to be necessary for the correct delivery of the enzyme to its accumulation site (Engel et al. 1998, 2000). Even after cleavage, the propeptide is a potent inhibitor of the mature enzyme (Reis et al. 2007); this suggests a role in activity modulation that is in addition to its targeting function. The presence of a CZP natural inhibitor, a 12-kDa protein referred to as chagasin (Monteiro et al. 2001), has already been reported. This has been shown to modulate the action of the proteases inside the reservosomes and Golgi apparatus (Santos et al. 2005).

Another important protease, serine carboxypeptidase (a hydrolase that catalyses the hydrolysis of the protein carboxyterminal bonds), was characterized in *T. cruzi* epimastigotes and was suggested to lie in reservosomes (Parussini et al. 2003). Carboxypeptidases are identified by their lysosomal location in mammalian cells and in the endosomal/vacuolar compartments of yeast and plants. After using a proteomic approach, new reservosomal hydrolases were identified; these included α-mannosidase, calpain cysteine peptidase, and membrane bound acid phosphatase (Sant'Anna et al. 2009). Although the presence of several proteases has been described in reservosomes, at least one characteristic lysosome enzyme, aryl sulfatase, was detected by cytochemistry in small vesicles but not in reservosomes (Adade et al. 2007).

Reservosomes are also associated with autophagic events, and are particularly important for parasite survival during period of starvation or differentiation (Soares et al. 1989). Recently, it was demonstrated that components similar to those of the yeast Atg8 pathway are expressed in all developmental forms of *T. cruzi*. At least one of them, a homologue of yeast Atg8 and human LC-3 protein (TcAtg8.1), is translocated to reservosomes after prolonged starvation in phosphate-buffered saline (PBS) (Alvarez et al.). The components of this autophagic pathway are also upregulated during metacyclogenesis (Alvarez et al. 2008a, b).

Although a preliminary biochemical characterization of reservosomes was accomplished by Cunha-e-Silva et al. (2002), the multifactorial function of reservosomes was evaluated using only proteomic analysis of isolated reservosomes and their membranes (Fig. 3a, b); the latter revealed the presence of a set of proteins that were implicated in multiple diverse functions (Sant'Anna et al. 2009). For example, enzymes such as farsenylated protein tyrosine phosphatase (TcPRL-1) (Cuevas

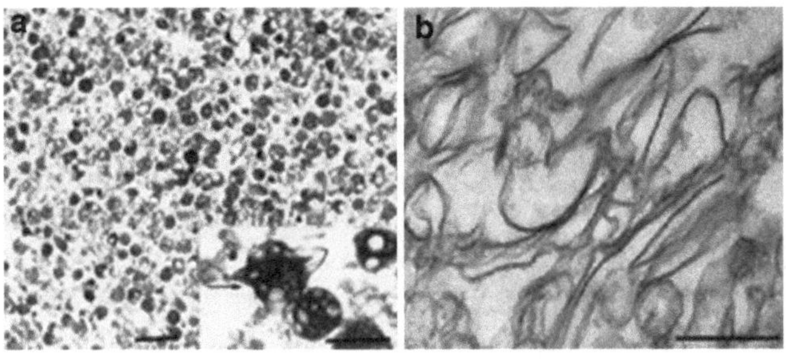

Fig. 3 *Subcellular fractionation of reservosomes:* (**a**) Ultrathin section showing highly purified reservosomal fraction; inset, higher magnification showing organelles after fractionation procedure. (**b**) Purified total reservosome membrane fraction. *Bars*: (**a**) 5 μm; (**b**) 1 μm. After Cunha-e-Silva et al. (2002) (**a**) and Sant'Anna et al. (2009) (**b**)

et al. 2005), serine/threonine protein kinases, and phosphatases that are related to signaling events were detected.

The totality of these findings reinforce two points about reservosomes (1) the organelles are formed as a result of the fusion of Golgi and endocytic vesicles (Sant'Anna et al. 2004), and this accounts for their biochemical composition; (2) an intense traffic is established between the secretory route and the endosomal compartments, in which cellular events could include uptake, storage, and macromolecule digestion; the expression of chagasin and the recycling of other cellular components suggest that these processes probably occur in a modulating fashion. In this regard, the reservosomes could be interpreted as lysosome-related organelles (LROs) (Sant'Anna et al. 2008a); these can be defined as a class of heterogeneous organelles that share lysosomal/late endosomal features (such as low internal pH, some specific proteins, and a common pathway formation), but are morphologically, functionally, and biochemically distinct (Huizing et al. 2008) as suggested by their display of a set of specific proteins that are required for their individual functions (Huizing et al. 2008). Melanosomes (melanocytes), azurophil granules (neutrophils), lamellar bodies (lung type II epithelial cells), and neuramin granules are all examples of LROs (Tribl et al. 2006; Raposo et al. 2007; Huizing et al. 2008). Abnormalities in LROs or lysosomes are associated with human genetic diseases, such as Hermansky-Pudlak syndrome (HRS) or Chediak-Higashi syndrome (CHS); this emphasizes the importance of these compartments in cell homeostasis.

In *T. cruzi,* reservosomes are a key component of the intricate endosomal pathway. The precise mechanisms that the parasites employ for signaling the degradative pathway or for initiating cargo storage in reservosomes require further clarification. Epimastigotes are proliferative forms that develop in the midgut vector. At the posterior end of the vector midgut, they differentiate to metacyclic trypomastigotes and lose their ability to divide. As trypomastigotes and amastigotes lack the type of endocytic apparatus that is a feature of epimastigotes, determining the mechanism of nutrient acquisition in trypomastigotes is of considerable interest. Recently, Sant'Anna et al. (2008a) demonstrated the presence of LROs in trypomastigote and amastigote forms. These compartments, which were also localized at the posterior region of the parasites, share many features with epimastigote reservosomes (Figs. 4a–d and 5a–c); these include the presence of a P-type H^+-ATPase, the accumulation of CZP and serine carboxypeptidase, and the electron-lucent rods that have been observed in epimastigote reservosomes (Sant'Anna et al. 2008b).

6 Concluding Remarks

It is interesting to observe that *T. cruzi* reservosomes are key components of an intricate intracellular network. At the crossroads between endocytic and secretory apparatus, reservosomes constitute an instructive model (Fig. 6) for understanding the mechanisms by which epimastigotes coordinate the acquisition of nutrients, their degradation, and subsequent distribution, as well as the accumulation and modulation

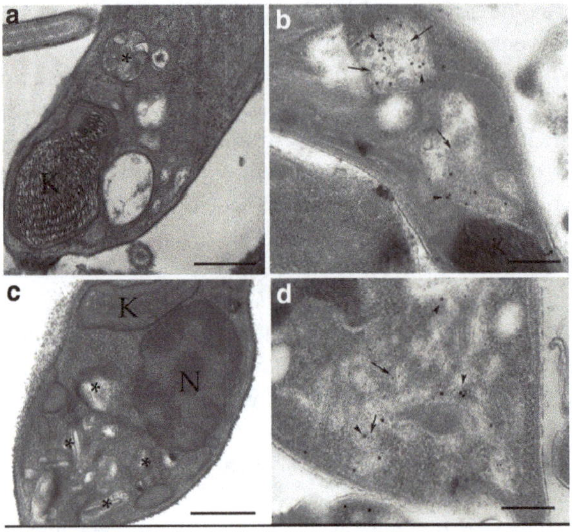

Fig. 4 *Ultrastructure of trypomastigote and amastigote lysosome related organelles* (LROs)*:* Ultrathin sections of the posterior region of a trypomastigote (**a**) and an amastigote (**c**), showing LROs (*asterisks*). These compartments were shown to be positive for serine carboxypeptidase (5 nm gold particles, *arrowheads*) and cruzipain (15 nm gold particles, *arrows*) in trypomastigotes (**b**) and amastigotes (**d**). *Bars*: (**a**) 0.5 μm; (**b**) 0.1 μm; (**c**) 0.5 μm; (**d**) 0.2 μm

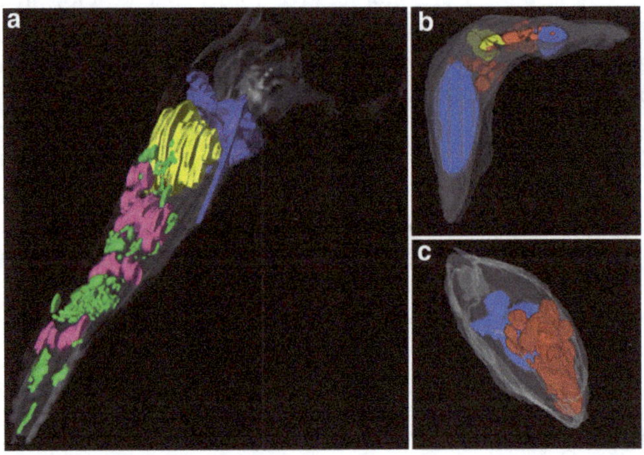

Fig. 5 *3D reconstruction of TcLROs in T. cruzi stages*: (**a**) Spatial distribution of reservosomes in epimastigotes. Epimastigote reservosomes are concentrated in the posterior region of the parasite. (**b**) In trypomastigotes, TcLROs are confined between nucleus and kinetoplast. (**c**) In amastigotes, TcLROs have polyphormic shape and size and are distributed from perinuclear to posterior region. (**a**) *Blue*, kinetoplast; *Yellow*, nucleus; *Green*, early endosomes; *Purple*, reservosomes; *Gray*, plasma membrane; (**b**, **c**) *Yellow*, Golgi complex; *Blue*, nucleus and kinetoplast; *Red*, TcLROs; *Gray*, plasma membrane. After Porto-Carreiro et al. (2000) (**a**) and Sant'Anna et al. (2008b) (**b**, **c**)

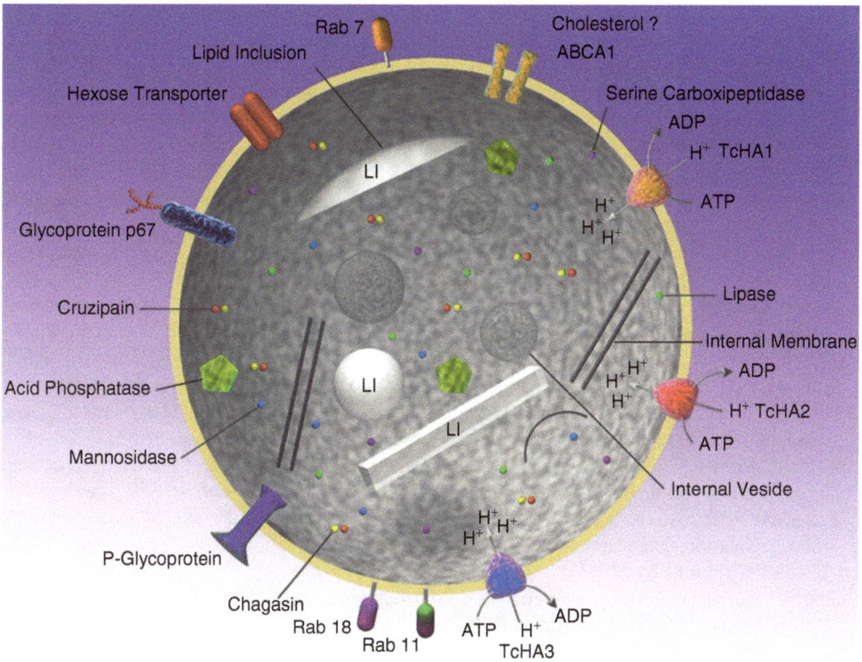

Fig. 6 *Schematic model of a T. cruzi reservosome*: A typical reservosome, with its eletrondense matrix, surrounded by a unit membrane (*yellow*), internal vesicles, and planar membranes and lipid inclusions (LI). The LIs could be visualized in *spherical* or *rectangular* shapes surrounded by a phospholipid monolayer. The main proteins identified by proteomic analysis are represented as serine carboxipeptidase (*purple*), mannosidase (*blue*), lipases (*green*), and cruzipain (*red*), which are modulated by chagasin (*yellow*). Three P-type-H^+-ATPases are responsible for organelle acidification, represented by TcHA1, TcHA2, and TcHA3. The GTPases TcRab7, TcRab11, and TcRab18 have been identified although their exact localization has not been addressed, as well as the ABCA1 transporter, P-glycoprotein or hexose transporter

of hydrolases. In addition, identification of the protein and/or lipid arsenals in *T. cruzi* LROs could suggest alternative targets for chemotherapeutic drugs.

Acknowledgments The authors received financial support from Conselho Nacional de Desenvolvimento Científico e Tecnológico (CNPq), Fundação Carlos Chagas Filho de Amparo à Pesquisa no Estado do Rio de Janeiro (FAPERJ), and Coordenação de Aperfeiçoamento de Pessoal de Nivel Superior (CAPES).

References

Adade CM, de Castro SL, Soares MJ (2007) Ultrastructural localization of *Trypanosoma cruzi* lysosomes by aryl sulphatase cytochemistry. Micron 38:252–256

Alvarez VE, Kosec G, Sant Anna C, Turk V, Cazzulo JJ, Turk B (2008a) Blocking autophagy to prevent parasite differentiation: a possible new strategy for fighting parasitic infections? Autophagy 4:361–363

Alvarez VE, Kosec G, Sant'Anna C, Turk V, Cazzulo JJ, Turk B (2008b) Autophagy is involved in nutritional stress response and differentiation in *Trypanosoma cruzi*. J Biol Chem 283: 3454–3464

Araripe JR, Cunha-e-Silva NL, Leal ST, de Souza W, Rondinelli E (2004) *Trypanosoma cruzi*: TcRAB7 protein is localized at the Golgi apparatus in epimastigotes. Biochem Biophys Res Commun 321:397–402

Araripe JR, Ramos FP, Cunha-e-Silva NL, Urmenyi TP, Silva R, Leite Fontes CF, da Silveira JF, Rondinelli E (2005) Characterization of a RAB5 homologue in *Trypanosoma cruzi*. Biochem Biophys Res Commun 329:638–645

Beyenbach KW, Wieczorek H (2006) The V-type H+ ATPase: molecular structure and function, physiological roles and regulation. J Exp Biol 209:577–589.

Boiso JF, Docampo R, Stoppani AOM (1977) Investigacion citoquimica por microscopia electronica de la presencia de microperoxisomas em *Trypanosoma cruzi*. Medicina (Buenos Aires) 37:365–370

Cazzulo JJ, Hellman U, Couso R, Parodi AJ (1990) Amino acid and carbohydrate composition of a lysosomal cysteine proteinase from *Trypanosoma cruzi*. Absence of phosphorylated mannose residues. Mol Biochem Parasitol 38:41–48

Cuevas IC, Rohloff P, Sanchez DO, Docampo R (2005) Characterization of farnesylated protein tyrosine phosphatase TcPRL-1 from Trypanosoma cruzi. Eukaryot Cell 4:1550–1561.

Cunha-e-Silva NL, Atella GC, Porto-Carreiro IA, Morgado-Diaz JA, Pereira MG, De Souza W (2002) Isolation and characterization of a reservosome fraction from *Trypanosoma cruzi*. FEMS Microbiol Lett 214:7–12

Cunha-e-Silva N, Sant'Anna C, Pereira MG, Porto-Carreiro I, Jeovanio AL, de Souza W (2006) Reservosomes: multipurpose organelles? Parasitol Res 99:325–327

Davidson AL, Maloney PC (2007) ABC transporters: how small machines do a big job. Trends Microbiol 15:448–455

De Souza, de Carvalho TU, Benchimol M, Chiari E (1978) Trypanosoma cruzi: ultrastructural, cytochemical and freeze-fracture studies of protein uptake. Exp Parasitol 45(1):101–115.

De Souza W, Sant'Anna C, Cunha-e-Silva NL (2009) Electron microscopy and cytochemistry analysis of the endocytic pathway of pathogenic protozoa. Prog Histochem Cytochem 44:67–124

Doherty GJ, McMahon HT (2009) Mechanisms of endocytosis. Annu Rev Biochem 78:857–902

Engel JC, Doyle PS, Palmer J, Hsieh I, Bainton DF, McKerrow JH (1998) Cysteine protease inhibitors alter Golgi complex ultrastructure and function in *Trypanosoma cruzi*. J Cell Sci 111:597–606

Engel JC, Torres C, Hsieh I, Doyle PS, McKerrow JH (2000) Upregulation of the secretory pathway in cysteine protease inhibitor-resistant *Trypanosoma cruzi*. J Cell Sci 113: 1345–1354

Figueiredo RC, Steindel M, Soares MJ (1994) The reservosomes of epimastigote forms of *Trypanosoma cruzi*: occurrence during in vitro cultivation. Parasitol Res 80:517–522

Huizing M, Helip-Wooley A, Westbroek W, Gunay-Aygun M, Gahl WA (2008) Disorders of lysosome-related organelle biogenesis: clinical and molecular genetics. Annu Rev Genomics Hum Genet 9:359–386

Kellner-Weibel G, Yancey PG, Jerome WG, Walser T, Mason RP, Phillips MC, Rothblat GH (1999) Crystallization of free cholesterol in model macrophage foam cells. Arterioscler Thromb Vasc Biol 19:1891–1898

Landfear SM, Ignatushchenko M (2001) The flagellum and flagellar pocket of trypanosomatids. Mol Biochem Parasitol 115:1–17

Lara FA, Sant'anna C, Lemos D, Laranja GA, Coelho MG, Reis Salles I, Michel A, Oliveira PL, Cunha-e-Silva N, Salmon D, Paes MC (2007) Heme requirement and intracellular trafficking in *Trypanosoma cruzi* epimastigotes. Biochem Biophys Res Commun 355:16–22

Luo S, Fang J, Docampo R (2006) Molecular characterization of *Trypanosoma brucei* P-type H+-ATPases. J Biol Chem 281:21963–21973

Mendonca SM, Nepomuceno da Silva JL, Cunha e-Silva N, de Souza W, Gazos Lopes U (2000) Characterization of a Rab11 homologue in *Trypanosoma cruzi*. Gene 243:179–185

Milder R, Deane MP (1969) The cytostome of *Trypanosoma cruzi* and *T. conorhini*. J Protozool 16:730–737

Monteiro AC, Abrahamson M, Lima AP, Vannier-Santos MA, Scharfstein J (2001) Identification, characterization and localization of chagasin, a tight-binding cysteine protease inhibitor in *Trypanosoma cruzi*. J Cell Sci 114:3933–3942

Morgan GW, Hall BS, Denny PW, Field MC, Carrington M (2002) The endocytic apparatus of the kinetoplastida. Part II: machinery and components of the system. Trends Parasitol 18:540–546

Neufeld EB, Stonik JA, Demosky SJ Jr, Knapper CL, Combs CA, Cooney A, Comly M, Dwyer N, Blanchette-Mackie J, Remaley AT, Santamarina-Fojo S, Brewer HB Jr (2004) The ABCA1 transporter modulates late endocytic trafficking: insights from the correction of the genetic defect in Tangier disease. J Biol Chem 279:15571–15578

Nichols BJ, Lippincott-Schwartz J (2001) Endocytosis without clathrin coats. Trends Cell Biol 11:406–412

Oram JF, Yokoyama S (1996) Apolipoprotein-mediated removal of cellular cholesterol and phospholipids. J Lipid Res 37:2473–2491

Parussini F, Garcia M, Mucci J, Aguero F, Sanchez D, Hellman U, Aslund L, Cazzulo JJ (2003) Characterization of a lysosomal serine carboxypeptidase from *Trypanosoma cruzi*. Mol Biochem Parasitol 131:11–23

Porto-Carreiro I, Attias M, Miranda K, De Souza W, Cunha-e-Silva N (2000) *Trypanosoma cruzi* epimastigote endocytic pathway: cargo enters the cytostome and passes through an early endosomal network before storage in reservosomes. Eur J Cell Biol 79:858–869

Reis FC, Costa TF, Sulea T, Mezzetti A, Scharfstein J, Brömme D, Ménard R, Lima AP (2007) The propeptide of cruzipain–a potent selective inhibitor of the trypanosomal enzymes cruzipain and brucipain, and of the human enzyme cathepsin F. FEBS J 274:1224–1234.

Raposo G, Marks MS, Cutler DF (2007) Lysosome-related organelles: driving post-Golgi compartments into specialisation. Curr Opin Cell Biol 19:394–401

Ridley AJ (2001) Rho proteins: linking signaling with membrane trafficking. Traffic 2:303–310

Sant'Anna C, de Souza W, Cunha-e-Silva N (2004) Biogenesis of the reservosomes of *Trypanosoma cruzi*. Microsc Microanal 10:637–646

Sant'Anna C, Parussini F, Lourenco D, de Souza W, Cazzulo JJ, Cunha-e-Silva NL (2008a) All *Trypanosoma cruzi* developmental forms present lysosome-related organelles. Histochem Cell Biol 130:1187–1192

Sant'Anna C, Pereira MG, Lemgruber L, de Souza W, Cunha-e-Silva NL (2008b) New insights into the morphology of *Trypanosoma cruzi* reservosome. Microsc Res Tech 71:599–605

Sant'Anna C, Nakayasu ES, Pereira MG, Lourenco D, de Souza W, Almeida IC, Cunha-e-Silva NL (2009) Subcellular proteomics of *Trypanosoma cruzi* reservosomes. Proteomics 9:1782–1794

Santos CC, Sant'Anna C, Terres A, Cunha-e-Silva NL, Scharfstein J, de Ana Paula Lima C (2005) Chagasin, the endogenous cysteine-protease inhibitor of *Trypanosoma cruzi*, modulates parasite differentiation and invasion of mammalian cells. J Cell Sci 118:901–915

Serrano R, Villalba JM, Palmgren MG, Portillo F, Parets-Soler A, Roldan M, Ferguson C, Montesinos C (1992) Studies of the plasma membrane H(+)-ATPase of yeast and plants. Biochem Soc Trans 20:562–566

Soares MJ (1999) The reservosome of *Trypanosoma cruzi* epimastigotes: an organelle of the endocytic pathway with a role on metacyclogenesis. Mem Inst Oswaldo Cruz 94(Suppl 1):139–141

Soares MJ, De Souza W (1988) Cytoplasmic organelles of trypanosomatids: a cytochemical and stereological study. J Submicrosc Cytol Pathol 20:349–361

Soares MJ, de Souza W (1991) Endocytosis of gold-labeled proteins and LDL by *Trypanosoma cruzi*. Parasitol Res 77:461–468

Soares MJ, Souto-Padron T, Bonaldo MC, Goldenberg S, de Souza W (1989) A stereological study of the differentiation process in *Trypanosoma cruzi*. Parasitol Res 75:522–527

Soares MJ, Souto-Padron T, De Souza W (1992) Identification of a large pre-lysosomal compartment in the pathogenic protozoon *Trypanosoma cruzi*. J Cell Sci 102:157–167

Souto-Padron T, Campetella OE, Cazzulo JJ, de Souza W (1990) Cysteine proteinase in *Trypanosoma cruzi*: immunocytochemical localization and involvement in parasite-host cell interaction. J Cell Sci 96:485–490

Stuart LM, Ezekowitz RA (2005) Phagocytosis: elegant complexity. Immunity 22:539–550

Takahashi K, Kimura Y, Nagata K, Yamamoto A, Matsuo M, Ueda K (2005) ABC proteins: key molecules for lipid homeostasis. Med Mol Morphol 38:2–12

Torres C, Perez-Victoria FJ, Parodi-Talice A, Castanys S, Gamarro F (2004) Characterization of an ABCA-like transporter involved in vesicular trafficking in the protozoan parasite *Trypanosoma cruzi*. Mol Microbiol 54:632–646

Tribl F, Marcus K, Meyer HE, Bringmann G, Gerlach M, Riederer P (2006) Subcellular proteomics reveals neuromelanin granules to be a lysosome-related organelle. J Neural Transm 113:741–749

Vieira M, Rohloff P, Luo S, Cunha-e-Silva NL, de Souza W, Docampo R (2005) Role for a P-type H+-ATPase in the acidification of the endocytic pathway of Trypanosoma cruzi. Biochem J 392:467–474

Weiner JH, Furlong CE, Heppel LA (1971) A binding protein for L-glutamine and its relation to active transport in E. coli. Arch Biochem Biophys 142:715–717

Weise F, Stierhof YD, Kuhn C, Wiese M, Overath P (2000) Distribution of GPI-anchored proteins in the protozoan parasite *Leishmania*, based on an improved ultrastructural description using high-pressure frozen cells. J Cell Sci 24:4587–4603

Wieffer M, Maritzen T, Haucke V (2009) SnapShot: endocytic trafficking. Cell 137:381–383

Megasomes in *Leishmania*

Diane McMahon-Pratt, Tania Ueda-Nakamura, and Yara M. Traub-Csekö

Contents

Abstract Leishmaniasis is a serious public health problem in the whole world. There are various forms of the disease that are caused by parasites of the genus *Leishmania*. Leishmaniasis is transmitted by phlebotomine sandflies and two forms

---

D. McMahon-Pratt
Yale University School of Public Health, New Haven, CT, USA

T. Ueda-Nakamura
Departamento de Análises Clínicas, Universidade Estadual de Maringá, Centro de Ciências da Saúde, Maringá, Paraná, Brazil

Y.M. Traub-Csekö
Laboratório de Parasitas e Vetores, Instituto Oswaldo Cruz, Fiocruz, Rio de Janeiro, RJ, Brazil
e-mail: ytraub@ioc.fiocruz.br

W. de Souza (ed.), *Structures and Organelles in Pathogenic Protists*,
Microbiology Monographs 17, DOI 10.1007/978-3-642-12863-9_6,
© Springer-Verlag Berlin Heidelberg 2010

of the parasite exist, the amastigote form that occurs in the mammalian host and the promastigote form found in the insect.

Leishmania protein trafficking is not fully understood, but there seems to exist a classic pathway. One striking aspect of this pathway is the existence of very large lysosome-like structures in the *Leishmania mexicana* complex, called megasomes. These organelles were found in both axenic and lesion amastigotes and contain large amounts of lysosomal hydrolases, including cysteine proteinases, that are frequently used as a megasomal markers. There is plenty of evidence for the importance of lysosomes for *Leishmania* virulence and survival, suggesting that the resident enzymes and cellular targeting mechanisms might be good targets for control.

1 Introduction

1.1 Leishmaniasis and Leishmania

Leishmaniasis is a serious public health problem worldwide, with a recent increase of cases and endemic regions. Urban migrations and environmental changes are partially responsible for this increase, as well as coinfection with HIV (Shaw 2007). Protozoans of the genus *Leishmania* are responsible for a variety of clinical manifestations, which range from cutaneous, muco-cutaneous, diffuse to visceral disease, depending on the parasite species and the host immunological status.

Leishmania are members of the *Trypanosomatidae* family, order *Kinetoplastidae* characterized by a unique mitochondrion, containing a network of mini- and maxicircles, denominated the kinetoplast. The main species that affect humans include the *Leishmania donovani* complex (*Leishmania donovani, Leishmania infantum,* and *Leishmania chagasi*), the *Leishmania mexicana* complex (*L. mexicana, Leishmania amazonensis, Leishmania pifanoi,* and *Leishmania venezuelensis*), *Leishmania tropica, Leishmania major, Leishmania aethiopica*, and the subgenus *Viannia* with four major species (*Leishmania braziliensis, Leishmania guyanensis, Leishmania panamensis,* and *Leishmania peruviana*) (http://www.dpd.cdc.gov/dpdx/HTML/Leishmaniasis.htm).

1.2 Leishmania *Life Cycle and Differentiation*

The *Leishmania* life cycle involves a phlebotomine vector that acquires the parasite by biting an infected host. In the insect, the *Leishmania* goes through various stages of development that will culminate in the infective metacyclic promastigote form, which will be transmitted to the host through the female insect bite.

There are marked differences between the noninfective promastigotes and the infective metacyclic forms. An important molecule during *Leishmania* metacyclogenesis is an abundant surface lipophosphoglycan (LPG), with an important role in vector competence (Sacks 2001). LPG is also one of the most important molecules used by the parasite to subvert macrophage responses during infection of the mammalian host, leading to distortion of specific intracellular signaling cascades (Olivier et al. 2005).

As mentioned before, during the developmental cycle of *Leishmania* there are two main forms, the long, flagellated insect promastigote form and the round amastigote form that contains a rudimentary internal flagellum, and infects the mammalian host. Morphological differences are associated with molecular changes that are responsible for the differentiation and survival of the parasite in different environments.

1.3 Molecular Aspects of Differentiation

Recent DNA microarray analysis was employed in studies of global gene expression of *Leishmania* promastigote and amastigote life stages. Surprisingly, low numbers of differentially expressed genes were detected between amastigotes and promastigotes (Dumas et al. 2006; Cohen-Freue et al. 2007; Subba Raju et al. 2008). Quantitative proteomic analyses revealed a weak correlation to gene expression, indicating a posttranscriptional regulation. This was shown in further work by Rosenzweig et al. (2008) using isobaric tagging for relative and absolute quantitative proteomics in *L. donovani* to detect protein phosphorylation, methylation, acetylation, and glycosylation throughout differentiation. Posttranslational modifications were observed during differentiation, suggesting a role in intracellular development. Phosphorylation in relation to differentiation was studied in *L. donovani* by Morales et al. (2008), who also found a correlation between this specific modification and differentiation.

Many molecules have been known for a long time to be modulated during parasite differentiation. Among these are the abundant surface glycoprotein GP63 (Kweider et al. 1987), which is differentially expressed and also has different subcellular locations when comparing amastigotes and promastigotes (Medina-Acosta et al. 1989). GP63 expression at the level of RNA and protein increases during in vitro metacyclogenesis and is drastically reduced in amastigote forms (Schneider et al. 1992).

It has been well-established that promastigotes and amastigotes differ in metabolic aspects, such as CO_2 fixation and phenolpyruvate metabolism. These metabolic differences were observed for various *Leishmania* species and impacted on responses of these parasites to drugs currently in use for leishmaniasis treatment (Mottram and Coombs 1985). If one considers how different the environments are that *Leishmania* encounter during development, promastigotes a glucose-rich, slightly alkaline environment in the insect gut, amastigotes an acidic environment in macrophages, sparse in glucose and abundant in amino acids, such differences in the parasites' metabolism are not unexpected. Proteome analysis through promastigote to amastigote differentiation (Schneider et al. 1992; Walker et al. 2006) showed the appearance of enzymes that evidenced parasite shifting from the use of sugars to that of fatty acids and amino acids as energy sources; these observations confirmed expectations, based upon the known environments inhabited by these two life cycle stages.

The production of monoclonal antibodies against amastigotes and promastigotes also demonstrated that protein and glycolipid components were selectively expressed during development (Handman and Hocking 1982; Jaffe and Rachamim 1989). The use of membrane-enriched fractions also showed the antigenic specificity of the two forms (Pan and McMahon-Pratt 1988; Eperon and McMahon-Pratt 1989). Antibodies prepared against either axenic or animal derived amastigotes were specific for this form, indicating that antigenically cultured amastigotes are closely related to macrophage-derived parasites. Promastigote derived antibodies were specific for this parasite form. These antibodies were very useful for further characterization of differentiation as well as taxonomic, epidemiologic, and other aspects of leishmaniasis.

1.4 Morphological Aspects and Markers of Differentiation

Although many early studies investigated morphological aspects of *Leishmania* development (Rudzinska et al. 1964; Bray 1974), not until the 1980s were specific differences between the two developmental forms investigated. Pan and Pan (1986) used electron microscopy to analyse structural aspects of *L. pifanoi*, a member of the *L. mexicana* complex; promastigotes and both culture and lesion amastigotes were examined. While axenic and animal derived amastigotes were very similar in structure, they differed from promastigotes in many aspects. Among these differences, they verified that amastigotes lacked the paraxial rod that in promastigotes originates at the axosome level within the flagellar pocket, which was much larger in amastigote forms. The subpellicullar microtubules extended fully to the posterior end, while in amastigotes the subpellicullar microtubules ended subterminally. Finally, electron dense and membranous vesicles were highly evident in the flagellar pocket area of the amastigotes; the electron dense material was considered to derive from the megasomal/electron dense organelles that could be found abutting the flagellar pocket.

The transformation of promastigotes into amastigotes in vivo was also investigated in *L. amazonensis* using some markers of differentiation: presence of megasomes, the large lysosomes seen in some *Leishmania* species, cysteine proteinases, and sensitivity to L-leucine-methyl ester (LeuOMe) (Galvao-Quintao et al. 1990). It had been previously demonstrated that both intracellular and isolated *L. amazonensis* amastigotes are killed by LeuOMe in a mechanism that involves the targeting of parasite cysteine proteinases, while promastigotes are much more resistant

(Rabinovitch et al. 1987; Ramazeilles and Rabinovitch 1989). Megasomes and high levels of cysteine proteinases (Pral et al. 1993) are characteristic of amastigotes of the *L. mexicana* complex, as we will see in detail below. The authors concluded that intracellular differentiation is rather slow, with full development of megasomes and consequent high levels of cysteine proteinases and sensitivity to LeuOMe only on the 7th day after promastigote infection of macrophages.

All these results indicate fundamental differences between *Leishmania* promastigote and amastigote forms, which were also found in relation to the secretory/endocytic pathway.

2 Secretory/Endocytic Pathway

In eukaryotic cells in general, the secretory pathway is complex, involving an endomembrane system that encompasses the endoplasmic reticulum (ER), Golgi apparatus, vesicles, and vacuoles. Molecules utilize this complex membrane system to travel through the cell and reach their various destinations. In opposition to the secretary pathway, also called the biosynthetic pathway, is the endocytic or retrograde pathway that originates at the plasma membrane and normally leads to the lysosome. These paths are interconnected, creating a complex and dynamic intracellular trafficking network, where vesicular transport appears to be the major vehicle.

In trypanosomatids, including *Leishmania*, protein trafficking is not fully understood. Nevertheless, there is evidence that secreted or surface molecules follow a classic trafficking pathway: ER to Golgi to surface, with the involvement of fusion of vesicles with the flagellar pocket membrane (Overath et al. 1997).

2.1 Differentiation and the Secretory/Endocytic Pathways in Leishmania

Among the many striking physiological and morphological aspects that differentiate the insect and mammalian *Leishmania* forms, organelles involved in the secretory /endocytic pathways are significantly affected by this differentiation. Both promastigotes and amastigotes are covered by a thick layer of subpellicular microtubules, leaving the flagellar pocket (Landfear and Ignatushchenko 2001), an invagination of the plasma membrane situated on the anterior part of the parasite, as the only endocytic connection of the parasite with the outside world. The flagellar pocket is highly specialized for uptake and secretion of molecules by the parasite. As a result of this spatial organization, many organelles related to the secretory /endocytic pathways are located in the proximity of the flagellar pocket (Waller and McConville 2002).

2.2 *Differentiation and the Secretory/Endocytic Pathways in* Leishmania *Amastigotes*

There are few studies on the secretory organelles of amastigotes; ultrastructural studies of *L. pifanoi* indicate that amastigotes have a defined Golgi (Pan and Pan 1986). Although the ER is relatively inconspicuous (Waller and McConville 2002), confocal studies using the ER marker BiP (Kar et al. 2000) suggest a dispersed cellular localization for the amastigote, as found for the promastigote (Knuepfer et al. 2001). On the other hand, the lysosomes of *Leishmania* amastigotes are expanded, which is most apparent in the species belonging to the *L. mexicana* complex that contain the very large lysosomal vacuoles termed megasomes (Alexander and Vickerman 1975; Coombs 1986). These large amastigote lysosomes (up to 15% of the cell volume) correlate with high levels of developmentally regulated cysteine proteinase (Duboise et al. 1994; Brooks et al. 2000; Ueda-Nakamura et al. 2002), which in many studies is used as a lysosomal or megasomal marker.

2.3 *Differentiation and the Secretory/Endocytic Pathways in* Leishmania *Promastigotes*

The endocytic pathway in *Leishmania* promastigotes has been more extensively studied and many interesting aspects were found (McConville et al. 2002a, b). Among these are the multivesicular bodies, which are found near the anterior end of promastigotes and, most interestingly, an unusual multivesicular tubule (MVT)-lysosome was found to run along the anterior–posterior axis of the parasite (Ghedin et al. 2001).

In order to investigate the intracellular localization of the glycosylphosphotidylinositol (GPI) biosynthetic enzymes, Ilgoutz et al. (1999a, b) constructed a molecule containing the enzyme dolichol-phosphate-mannose synthase (DPMS), used mainly by the GPI biosynthetic pathway fused to GFP. Surprisingly, when this construct was expressed in *L. mexicana* promastigotes, strong fluorescence was observed in a unique tubular structure extending along the anterior–posterior axis of the parasite (Ilgoutz et al. 1999b). Depolymerization of subpellicular microtubules caused the disruption of this tubular structure, suggesting a close relation between these two cellular structures. The authors suggested that the tubular structure is a stable transitional ER through which vesicles containing proteins and lipids are taken to the Golgi apparatus. This compartment was further characterized through the construction of chimeras containing GFP fused to the *Leishmania* surface protein 3′ nucleotidase/nuclease (which contains a transmembrane (TM) domain) or to a signal peptide and the GPI addition signal from GP63 (Ghedin et al. 2001). These constructs not only reached their expected destination (surface membrane) but also accumulated in the tubular structure. The molecules which contained GPI-anchors trafficked more efficiently to the cell surface, and, as a result, could only be seen accumulating in the tubular structure upon strong drug selection, which led to overproduction of GFP-tagged 3′-NTase. On the other hand, the GFP-construct containing the TM signal, probably due to its slower intracellular trafficking, appeared to preferentially accumulate in the tubules. The authors concluded that this unique compartment, similar to endosomes of higher eukaryotes, probably has an important role in acquisition and digestion of nutrients; this compartment was also responsible for the degradation of over-expressed proteins. This amplified transitional organelle in promastigotes seems to indicate a high secretory capacity, but is also related to the over-expression of proteins targeted to the lysosome. When *L. major* was transfected with a construct expressing GFP fused to a lysosomal targeting signal for an abundant cysteine proteinase from *L. pifanoi*, similar tubular structures were also seen in cells submitted to high doses of selective drugs (Costa-Pinto et al. 2001). It was suggested that this tubule corresponds to a stable transitional ER, where lipids and proteins are packaged into vesicles for transport to the Golgi apparatus (Ilgoutz et al. 1999b).

3 Lysosomal Targeting

Protein targeting in trypanosomatids, although similar in some ways to that of higher eukaryotes, has also revealed differences for most subcellular compartments investigated thus far (Clayton et al. 1995; Costa-Pinto et al. 2001). In mammalian cells, the primary lysosomal targeting mechanism for hydrolases involves mannose-6-phosphate receptors (MPR) (von Figura 1991), although the use of alternative mechanisms has been demonstrated (Ni et al. 2006; Luzio et al. 2007). In yeast, the transport of soluble hydrolases to the vacuole is mediated by various targeting mechanisms (reviewed in Luzio et al. 2007), which include the classic secretory pathway using a targeting signal (an amino acid domain) (Valls 1987), a direct pathway that avoids the multivesicular body (MVB) compartment going directly to the vacuole (Piper et al. 1997), or the cytoplasmatic-vacuolar transportation (Cvt) system (Martinez et al. 1999) and autophagy (Klionsky 2005), which target to the vacuole from the cytoplasm. Interestingly, a lysosomal targeting signal was recognized by yeast and not by mammalian cells (Marin-Villa et al. 2008a).

In *T. cruzi*, no phosphorylated mannose residues were found in the lysosomal cysteine proteinase, indicating a targeting mechanism different from that of MPR. Also, *N*-acetylglucosamine-1-phosphotransferase, the first enzyme responsible for the formation of the mannose-6-phosphate ligands, has not been detected in *T. cruzi* epimastigotes or *L. amazonensis* promastigotes (Cazzullo et al. 1990). Nevertheless, it seems that in trypanosomatids, most lysosomal integral membrane proteins or hydrolases are targeted through the secretory pathway. This assertion is based on the observations that (1) many lysosomal proteins contain a signal peptide on their N-terminus, which permits the entrance of the nascent peptide into the ER (Kelly et al. 1999; Costa-Pinto et al. 2001); (2) some of these proteins undergo glycosylation in the ER and Golgi (Parodi 1995); (3) nonprocessed lysosomal proteases accumulate in the Golgi and flagellar pocket (Brooks et al. 2000); and (4) exposure of parasites to protease inhibitors leads to severe morphological alterations of the Golgi (Engel et al. 1998). Two possible lysosomal targeting mechanisms are known in trypanosomatids: the classical secretory pathway, where proteins are targeted to the lysosome from the Golgi, passing through some intermediate endosomal organelles, or indirectly, from the parasite surface, from where some proteins are taken through the flagellar pocket to be endocytosed and transported to the lysosome via endosomes (Selzer et al. 1999; McConville et al. 2002a).

3.1 Lysosomal Targeting in Leishmania

Morphological and biochemical changes in the lysosome have been described for different stages of *Leishmania*. In the promastigote, an MVT was described, extending from the flagellar pocket to the posterior end of the cell (Costa-Pinto et al. 2001; Ghedin et al. 2001; Mullin et al. 2001). In stationary stage promastigotes of *L. amazonensis* and *L. mexicana*, an MVT structure was described that matures into multiple electron dense vacuoles, correlating with the biogenesis of megasomes in amastigotes (Coombs 1986; Weise et al. 2000; Pimenta et al. 1991; Ueda-Nakamura et al. 2001, 2002). Also, transformation of procyclic form into infective metacyclic *Leishmania* was shown to be dependent on the late endosome formation and to rely directly on autophagy (Besteiro et al. 2006).

Lysosomal targeting of only a few proteins has been studied in *Leishmania*, with some classical examples for different pathways. The ER resident TM protein glycosyltransferase DPMS, when fused to GFP, was found to be constitutively directed to the MVT of *L. mexicana* (Ilgoutz et al. 1999b; Ghedin et al. 2001). Another example is the soluble, abundant, megasome specific cysteine proteinase Lpcys2 of *L. pifanoi*, a member of the *L. mexicana* complex (Traub-Cseko et al. 1993; Duboise et al. 1994; Boukai et al. 2000a; Ueda-Nakamura et al. 2002) which is related to the cysteine proteinase B of *L. mexicana, L. donovani,* and *L. major* (Mottram et al. 1997; Selzer et al. 1997; Hide and Banuls 2008). In the case of Lpcys2 of *L. pifanoi*, it was demonstrated that the classical mannose-6-phosphate receptor mechanism is not used for lysosome targeting, as the elimination of potential N-glycosylation sites through site directed mutagenesis (Boukai et al. 2000b) failed to effect lysosomal targeting. Further, a lysosomal targeting signal was found in the pro domain of Lpcys2, by fusing the signal peptide and this region to GFP and observing fluorescence in lysosomes (Costa-Pinto et al. 2001). It was possible to visualize the MVT when both the pre-pro domain and the C-terminal domain of the cysteine proteinase Lpcys2 were fused to GFP (Costa-Pinto et al. 2001). The same construct marked the megasome of amastigotes of *L. pifanoi* (unpublished results). In *T. cruzi,* there is also a targeting signal in the pro region of a lysosomal cysteine proteinase that was recognized by *Leishmania* (Huete-Perez et al. 1999). Also, it was found that the GPI anchor of GP63 and the TM domain of

3′nucleosidase/nuclease were able to direct a reporter protein to the *Leishmania* MVT (Ghedin et al. 2001).

4 Megasomes in *Leishmania*

The term megasome was first used in *Leishmania* by Alexander and Vickerman in 1975. The name reflected the size of the large lysosomal organelles first observed in amastigote parasites of the *L. mexicana* complex. These amastigotes were found to contain much higher activities than cultured promastigotes of typical lysosomal hydrolases as cysteine proteinase, arylsulfatase, and β-glucuronidase (Pupkis et al. 1986), and these were shown by immunolocalization studies to be localized in large electron dense lysosomal organelles. In these initial studies, *L. amazonensis* was compared to *L. donovani* and *L. major* revealing that amastigotes of the two *L. mexicana* complex species contained high cysteine proteinase activity and presence of megasomes, while the other two species lacked both these features. This suggested the occurrence of megasomes to be characteristic of the *L. mexicana* complex (Pupkis et al. 1986). Studies were performed in *L. mexicana* to compare proteinase activities and megasomes in axenic amastigote-like forms, lesion amastigotes, and promastigotes. Megasomes were not seen in promastigotes but were present in both amastigote stages, albeit homogeneous and were of a smaller size in cultured amastigotes than in amastigotes isolated from lesions. We also observed the presence of megasomes in both axenic and lesion amastigotes of *L. mexicana* (Fig. 1). The same pattern of proteinase activities was observed in axenic and lesion amastigotes, although a more representative pool of inactive enzyme precursors was seen in axenic parasites (Pral et al. 1993). Similar results of the development of megasomes and enzyme expression were observed in later work by the same group (Pral et al. 2003) using amastigotes of *L. amazonensis*. Interestingly,

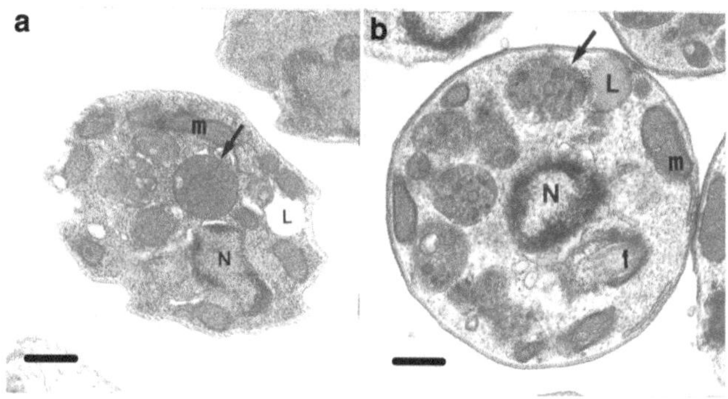

Fig. 1 Transmission electron microscopy of axenic (**a**) and lesion-derived amastigote (**b**) of *Leishmania mexicana*. The *arrows* indicate the megasomes. *f* flagellum; *L* Lipid inclusion; *m* mitochondrion; *N* nucleus. *Bars* = 1 μm

stationary growth phase cultured amastigotes had more typical megasomes, and increased expression of amastigote-specific proteinases, although only about a third of the parasites had typical megasomes, as compared with at least 80% of lesion amastigotes.

Later work by Alberio et al. (2004) showed, by transmission electron microscopy analysis and three-dimensional reconstruction, large structures compatible with megasomes in amastigote forms of *L. chagasi*. Further characterization of the megasome in *L. chagasi* amastigotes was carried out by immunolabeling of cysteine proteinase, whereas the lysosomal content of amastigotes and promastigotes was confirmed by arylsulfatase cytochemistry. Nevertheless, these megasomes only occupied about 5% of the cell volume in amastigotes of *L. chagasi*. In contrast, in previous three-dimensional reconstruction of the subcellular organelles/structures of amastigotes of *L. mexicana*, the lysosome-like "megasomes" comprised as much as 15% of the total cell volume (Coombs 1986).

4.1 Megasome Biogenesis

An investigation of megasome biogenesis was carried out by transmission electron microscopy and cysteine proteinase immunolocalization during *L. amazonensis* promastigote into amastigote transformation (Ueda-Nakamura et al. 2001). As expected, primarily small lysosomal organelles were observed in the promastigote, with increases in both volume and density of these structures occurring during transformation into amastigote. Cysteine proteinase, used as a lysosomal marker, was seen to localize in this structure by immunocytochemistry (Fig. 2).

Further studies on megasome biogenesis during *L. amazonensis* differentiation were performed, which included the investigation of cysteine proteinase expression both by northern and western blots, owing to the potential importance of these proteinases on host–parasite interaction, differentiation process, and intracellular survival (Ueda-Nakamura et al. 2002). Transformation of promastigotes to amastigotes brought an increase in Lpcys2 production, activity, and processing rate. More recent electron microscopic studies (Ueda-Nakamura et al. 2007) of various species of the *L. mexicana* complex revealed variations in size and number of megasomes according to the species, and also between amastigotes obtained from axenic cultures and from infected animals. Using three-dimensional reconstruction, stereology and immunocytochemical localization of cysteine proteinase revealed significant differences between *L. amazonensis*, *L. mexicana*, and *L. pifanoi*. The relative volume of megasomes of lesion-derived *L. mexicana* amastigotes was 2–3 times bigger than in *L. amazonensis*, while it was smaller in axenic amastigotes of *L. pifanoi*. The presence of high cysteine proteinase activity in some *Leishmania* species has been linked to virulence (Mottram et al. 2004). Also, the production of knockout mutants for some cysteine proteinases of *L. mexicana* produced parasites with reduced virulence (Mottram et al. 1996). Consistent with these findings, a smaller relative megasome volume and lower cysteine proteinase activity were also

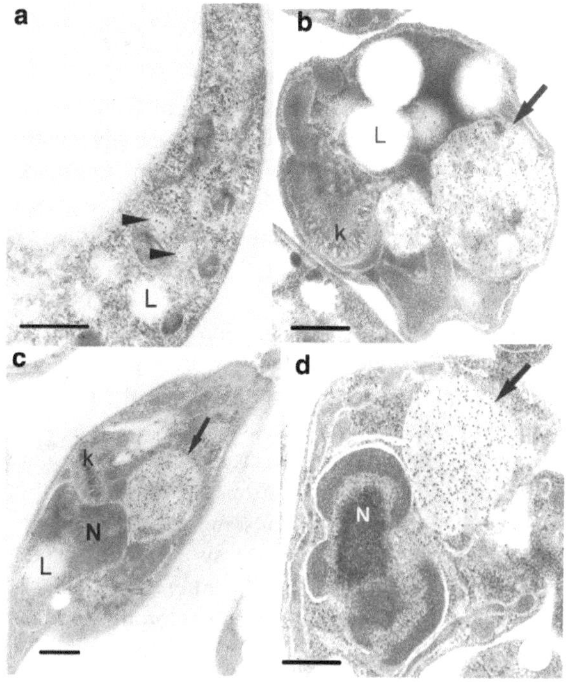

Fig. 2 Immunolocalization of cysteine proteinase Lpcys2 by transmission electron microscopy: (**a**) lysosomes (*arrowheads*) in promastigote forms of *L. mexicana;* megasomes (*arrows*) are shown in axenic amastigotes of *L. mexicana* (**b**), *L. pifanoi* (**c**), and *L. amazonensis* (**d**); *L* lipid inclusion; *N* nucleus; *k* kinetoplast; *Bars* = 1 μm

found to correlate with virulence and the development of lesions in animals (Ueda-Nakamura et al. 2007). These results support the idea that megasomes and their constituents may be involved in infectivity and virulence of *Leishmania* species.

5 Lysosomes as Targets for Kinetoplastid Control

The importance of lysosomes and lysosomal enzymes for *Leishmania* virulence and survival has been evident from genetic and biological studies (as indicated above), suggesting that the enzymes and cellular targeting mechanisms might represent approaches for control. Consequently, biochemical studies have focused on structure and function with the objective of developing drugs that target this organelle (Rabinovitch et al. 1986; Antoine et al. 1989; McKerrow 1999). Moreover, the lysosome and its associated molecules have been shown to be potential therapeutic targets among the kinetoplastids (McKerrow 1999; Siles et al. 2006; Mallari et al. 2008).

Similar to *Leishmania*, in the African trypanosomes as well as in the American trypanosome, *T. cruzi*, lysosomal morphology and hydrolytic activities vary throughout the life cycle (Pamer et al. 1989; Cazzulo 2002; Sant'Anna et al. 2008) with the highest level of expression associated with the stages found in the mammalian host. In the case of *T. brucei*, the lysosomes are essential for nutrition acquisition, the destruction of potentially lytic antibody (Balber et al. 1979; Mackey et al. 2004), as well as recycling of the variant surface glycoproteins (VSG) (Pal et al. 2003). Furthermore, African trypanosomes export the by-products of lysosomal digestion to the external milieu and have among the highest rates of endocytosis (Coppens et al. 1987; Engstler et al. 2004). Moreover, this endocytic rate has been shown to be developmentally regulated, with the highest rate in the bloodstream forms (Natesan et al. 2007); a rapid downregulation of endocytic markers (Rab11, clathrin) take place upon transformation into the insect stage (Hall et al. 2005). This is consistent with the fact that remodeling of the parasite surface is essential within the mammalian host but not for survival within the insect vector.

Notably, the lysosomes of African trypanosomes are also the target of the host innate immunity through the action of a serum-derived high density lipoprotein called trypanolytic factor (TLF) (Rifkin 1978; Hager et al. 1994; Harrington et al. 2009) which results in the killing of trypanosomes. Although the precise composition of active TLF and mechanism of trypanolysis are somewhat controversial, it is generally considered that after receptor-mediated endocytosis and delivery to the lysosomes, TLF initiates a lytic cascade that compromises lysosomal integrity, leading to parasite destruction (Hager et al. 1994; Raper et al. 1999; Shimamura et al. 2001; Pays et al. 2006; Vanhollebeke et al. 2007). However, drugs targeting the African trypanosome cathepsins (Nkemgu et al. 2003; Mallari et al. 2008), which are found in the lysosomes, have also proven effective in parasite destruction, presumably through blocking the essential functions of the organelle – nutrient acquisition, detoxification, and/or immune evasion.

In the case of *Leishmania* parasites immune evasion may also be a key biological function of the lysosome/megasome. Evidence indicates that major hitocompatibility complex (MHC) class II antigen presentation is significantly down regulated for the amastigote stage of the parasite (Fruth et al. 1993; Kima et al. 1996). As MHC class II presentation is essential for the activation of CD4+ T cells and control of infection, the modulation of MHC class II is critical for parasite survival. Insightful cell biological studies have demonstrated that *L. amazonensis* and *L. mexicana* amastigotes within the macrophage parasitophorous vacuole (PV; a late-endosomal compartment within the macrophage) sequester MHC class II molecules to discrete areas of the PV membrane and then subsequently degrade both MHC class II as well as their chaperone molecules, H-2M (Antoine et al. 1999, 2004). The molecules of the MHC class II pathway are degraded within the parasite's lysosomal compartment – the megasome. Although the cellular events required for this process are still not well understood, these cellular processes and the megasome therefore represent potential targets for immunomodulation and disease control. Interestingly, *Leishmania* lysosomal cysteine proteinases themselves have been shown to be vaccine candidate molecules capable of eliciting protection against infection (Rafati et al. 2000; Ferreira et al. 2008; Khoshgoo et al. 2008; Marin-Villa et al. 2008b).

Therefore, in *Leishmania* as well as in the related kinetoplastids, *T. brucei* and *T. cruzi,* the lysosomes are potential immunologic and pharmacologic targets for disease control, which are sorely needed. Recent promising results show that metal complexes targeted at parasite cysteine proteases might be employed in the treatment of Chagas' disease and leishmaniasis (Fricker et al. 2008). Further understanding of the mechanisms involved in subcellular targeting and/or enzyme action is warranted. The unique and exaggerated structure of "megasome" in the *L. mexicana* complex should provide an excellent model for such studies.

References

Alberio SO, Dias SS, Faria FP, Mortara RA, Barbiéri CL, Freymüller Haapalainen E (2004) Ultrastructural and cytochemical identification of megasome in *Leishmania (Leishmania) chagasi*. Parasitol Res 92:246–254

Alexander J, Vickerman K (1975) Fusion of host cell secondary lysosomes with the parasitophorous vacuole of *Leishmania mexicana*-infected macrophages. J Protozool 22:502–508

Antoine JC, Jouanne C, Ryter A (1989) Megasomes as the targets of leucine methyl ester in *Leishmania amazonensis* amastigotes. Parasitology 99(Pt 1):1–9

Antoine JC, Lang T, Prina E, Courret N, Hellio R (1999) H-2M molecules, like MHC class II molecules, are targeted to parasitophorous vacuoles of Leishmania-infected macrophages and internalized by amastigotes of *L. amazonensis* and *L. mexicana*. J Cell Sci 112:2559–2570

Antoine JC, Prina E, Courret N, Lang T (2004) *Leishmania* spp.: on the interactions they establish with antigen-presenting cells of their mammalian hosts. Adv Parasitol 58:1–68

Balber AE, Bangs JD, Jones SM, Proia RL (1979) Inactivation or elimination of potentially trypanolytic, complement-activating immune complexes by pathogenic trypanosomes. Infect Immun 24:617–627

Besteiro S, Williams RA, Morrison LS, Coombs GH, Mottram JC (2006) Endosome sorting and autophagy are essential for differentiation and virulence of *Leishmania major*. J Biol Chem 281:11384–11396

Boukai LK, McMahon-Pratt D, Traub-Cseko YM (2000a) Evidence for a recent mutation giving rise to a truncated copy of a cysteine proteinase gene in *Leishmania pifanoi*. Parasitol Int 49:301–307

Boukai LK, Costa-Pinto D, Soares MJ, McMahon-Pratt D, Traub-Cseko YM (2000b) Trafficking of cysteine proteinases to *Leishmania* lysosomes: lack of involvement of glycosylation. Mol Biochem Parasitol 107:321–325

Bray RS (1974) *Leishmania*. Annu Rev Microbiol 28:189–217

Brooks DR, Tetley L, Coombs GH, Mottram JC (2000) Processing and trafficking of cysteine proteases in *Leishmania mexicana*. J Cell Sci 113:4035–4041

Cazzullo JJ, Hellman U, Couso R, Parodi AJ (1990) Amino acid and carbohydrate composition of a lysosomal cysteine proteinase from *Trypanosoma cruzi*. Absence of phosphorylated mannose residues. Mol Biochem Parasitol 38:41–48

Cazzulo JJ (2002) Proteinases of *Trypanosoma cruzi*: potential targets for the chemotherapy of Chagas disease. Curr Top Med Chem 2:1261–1271

Clayton C, Hausler T, Blattner J (1995) Protein trafficking in kinetoplastid protozoa. Microbiol Rev 59:325–344

Cohen-Freue G, Holzer TR, Forney JD, McMaster WR (2007) Global gene expression in *Leishmania*. Int J Parasitol 37:1077–1086

Coombs GH (1986) Three-dimentional structure of *Leishmania* amastigotes as revealed by computer-aided reconstruction from serial section. Parasitology 92:13–23

Coppens I, Opperdoes FR, Courtoy PJ, Baudhuin P (1987) Receptor-mediated endocytosis in the bloodstream form of *Trypanosoma brucei*. J Protozool 34:465–473

Costa-Pinto D, Trindade LS, McMahon-Pratt D, Traub-Cseko YM (2001) Cellular traficking in trypanosomatids: a new target for therapies? Int J Parasitol 31:537–544

Duboise SM, Vannier-Santos MA, Costa-Pinto D, Rivas L, Pan AA, Traub-Cseko Y, de Souza W, McMahon-Pratt D (1994) The biosynthesis, processing, and immunolocalization of *Leishmania pifanoi* amastigote cysteine proteinase. Mol Biochem Parasitol 68:119–132

Dumas C, Chow C, Muller M, Papadopoulou B (2006) A novel class of developmentally regulated noncoding RNAs in Leishmania. Eukaryot Cell 5:2033–2046

Engel JC, Doyle PS, Palmer J, Hsieh I, Bainton DF, McKerrow JH (1998) Cysteine protease inhibitors alter Golgi complex ultrastructure and function in *Trypanosoma cruzi*. J Cell Sci 111:597–606

Engstler M, Thilo L, Weise F, Grunfelder CG, Schwarz H, Boshart M, Overath P (2004) Kinetics of endocytosis and recycling of the GPI-anchored variant surface glycoprotein in *Trypanosoma brucei*. J Cell Sci 117:1105–1115

Eperon S, McMahon-Pratt D (1989) Extracellular amastigote-like forms of *Leishmania panamensis* and *L. braziliensis*. II. Stage- and species-specific monoclonal antibodies. J Protozool 36:510–518

Ferreira JH, Gentil LG, Dias SS, Fedeli CE, Katz S, Barbieri CL (2008) Immunization with the cysteine proteinase Ldccys1 gene from *Leishmania (Leishmania) chagasi* and the recombinant Ldccys1 protein elicits protective immune responses in a murine model of visceral leishmaniasis. Vaccine 26:677–685

Fricker SP, Mosi RM, Cameron BR, Baird I, Zhu Y, Anastassov V, Cox J, Doyle PS, Hansell E, Lau G, Langille J, Olsen M, Qin L, Skerlj R, Wong RS, Santucci Z, McKerrow JH (2008) Metal compounds for the treatment of parasitic diseases. J Inorg Biochem 102:1839–1845

Fruth U, Solioz N, Louis JA (1993) *Leishmania major* interferes with antigen presentation by infected macrophages. J Immunol 150:1857–1864

Galvao-Quintao L, Alfieri SC, Ryter A, Rabinovitch M (1990) Intracellular differentiation of *Leishmania amazonensis* promastigotes to amastigotes: presence of megasomes, cysteine proteinase activity and susceptibility to leucine-methil ester. Parasitology 101:7–13

Ghedin E, Debrabant A, Engel JC, Dwyer DM (2001) Secretory and endocytic pathways converge in a dynamic endosomal system in a primitive protozoan. Traffic 2:175–188

Hager KM, Pierce MA, Moore DR, Tytler EM, Esko JD, Hajduk SL (1994) Endocytosis of a cytotoxic human high density lipoprotein results in disruption of acidic intracellular vesicles and subsequent killing of African trypanosomes. J Cell Biol 126:155–167

Hall BS, Smith E, Langer W, Jacobs LA, Goulding D, Field MC (2005) Developmental variation in Rab11-dependent trafficking in *Trypanosoma brucei*. Eukaryot Cell 4:971–980

Handman E, Hocking RE (1982) Stage-specific, strain-specific, and cross-reactive antigens of *Leishmania* species identified by monoclonal antibodies. Infect Immun 37:28–33

Harrington JM, Howell S, Hajduk SL (2009) Membrane permeabilization by trypanosome lytic factor, a cytolytic human high density lipoprotein. J Biol Chem 284:13505–13512

Hide M, Banuls AL (2008) Polymorphisms of cpb multicopy genes in the *Leishmania (Leishmania) donovani* complex. Trans R Soc Trop Med Hyg 102:105–106

Huete-Perez JA, Engel JC, Brinen LS, Mottram JC, McKerrow JH (1999) Protease trafficking in two primitive eukaryotes is mediated by a prodomain protein motif. J Biol Chem 274: 16249–16256

Ilgoutz SC, Zawadski JL, Ralton JE, McConville MJ (1999a) Evidence that free GPI glycolipids are essential from growth of *L. mexicana*. EMBO J 18:2746–2755

Ilgoutz SC, Mullin KA, Southwell BR, McConville MJ (1999b) Glycosylphosphatidylinositol biosynthetic enzymes are localized to a stable tubular subcompartment of the endoplasmic reticulum in *Leishmania mexicana*. EMBO J 18:3643–3654

Jaffe CL, Rachamim N (1989) Amastigote stage-specific monoclonal antibodies against *Leishmania major*. Infect Immun 57:3770–3777

Kar S, Soong L, Colmenares M, Goldsmith-Pestana K, McMahon-Pratt D (2000) The immunologically protective P-4 antigen of *Leishmania* amastigotes. A developmentally regulated single strand-specific nuclease associated with the endoplasmic reticulum. J Biol Chem 275: 37789–37797

Kelly RJ, Alexander DL, Cowan C, Balber AE, Bangs JD (1999) Molecular cloning of p67, a lysosomal membrane glycoprotein from *Trypanosoma brucei*. Mol Biochem Parasitol 98: 17–28

Khoshgoo N, Zahedifard F, Azizi H, Taslimi Y, Alonso MJ, Rafati S (2008) Cysteine proteinase type III is protective against *Leishmania infantum* infection in BALB/c mice and highly antigenic in visceral leishmaniasis individuals. Vaccine 26:5822–5829

Kima PE, Soong L, Chicharro C, Ruddle NH, McMahon-Pratt D (1996) *Leishmania*-infected macrophages sequester endogenously synthesized parasite antigens from presentation to CD4+ T cells. Eur J Immunol 26:3163–3169

Klionsky DJ (2005) The molecular machinery of autophagy: unanswered questions. J Cell Sci 118:7–18

Knuepfer E, Stierhof YD, McKean PG, Smith DF (2001) Characterization of a differentially expressed protein that shows an unusual localization to intracellular membranes in *Leishmania major*. Biochem J 356:335–344

Kweider M, Lemesre JL, Darcy F, Kusnierz JP, Santoro F (1987) Infectivity of Leishmania braziliensis promastigotes is dependent on the increasing expression of a 65, 000 dalton surface antigen. J Immunol 138:299–305

Landfear SM, Ignatushchenko M (2001) The flagellum and flagellar pocket of trypanosomatids. Mol Biochem Parasitol 115:1–17

Luzio JP, Pryor PR, Bright NA (2007) Lysosomes: fusion and function. Nat Rev Mol Cell Biol 8:622–632

Mackey ZB, O'Brien TC, Greenbaum DC, Blank RB, McKerrow JH (2004) A cathepsin B-like protease is required for host protein degradation in *Trypanosoma brucei*. J Biol Chem 279:48426–48433

Mallari JP, Shelat AA, Obrien T, Caffrey CR, Kosinski A, Connelly M, Harbut M, Greenbaum D, McKerrow JH, Guy RK (2008) Development of potent purine-derived nitrile inhibitors of the trypanosomal protease TbcatB. J Med Chem 51:545–552

Marin-Villa M, Sampaio-Morgado G, Roy D, Traub-Cseko YM (2008a) *Leishmania* lysosomal targeting signal is recognized by yeast and not by mammalian cells. Parasitol Res 103:983–988

Marin-Villa M, Vargas-Inchaustegui D, Chaves S, Tempone A, Dutra J, Soares M, Ueda-Nakamura T, Mendonça S, Rossi-Bergmann B, Soong L, Traub-Cseko YMT (2008b) The C-terminal extension of *Leishmania pifanoi* amastigote-specific cysteine proteinase Lpcys2: a putative function in macrophage infection. Mol Biochem Parasitol 162:52–59

Martinez E, Seguí-Real B, Silles E, Mazón MJ, Sandoval IV (1999) The propeptide of vacuolar aminopeptidase I is necessary and sufficient to target the fluorescent protein GFP to the vacuole of yeast by the Ccvt pathway. Mol Microbiol 33:52–62

McConville MJ, Mullin KA, Ilgoutz SC, Teasdale RD (2002a) Secretory pathway of trypanosomatid parasites. Microbiol Mol Biol Rev 66:122–154

McConville MJ, Ilgoutz SC, Teasdale RD, Foth BJ, Matthews A, Mullin KA, Gleeson PA (2002b) Targeting of the GRIP domain to the trans-Golgi network is conserved from protists to animals. Eur J Cell Biol 81:485–495

McKerrow JH (1999) Development of cysteine protease inhibitors as chemotherapy for parasitic diseases: insights on safety, target validation, and mechanism of action. Int J Parasitol 29:833–837

Medina-Acosta E, Karess RE, Schwartz H, Russell DG (1989) The promastigote surface protease (gp63) of *Leishmania* is expressed but differentially processed and localized in the amastigote stage. Mol Biochem Parasitol 37:263–273

Morales MA, Watanabe R, Laurent C, Lenormand P, Rousselle JC, Namane A, Späth GF (2008) Phosphoproteomic analysis of *Leishmania donovani* pro- and amastigote stages. Proteomics 8:350–363

Mottram JC, Coombs GH (1985) *Leishmania mexicana*: enzyme activities of amastigotes and promastigotes and their inhibition by antimonials and arsenicals. Exp Parasitol 59:151–160

Mottram JC, Souza AE, Hutchison JE, Carter R, Frame MJ, Coombs GH (1996) Evidence from disruption of the *lmcpb* gene array of *Leishmania mexicana* that cysteine proteinases are virulence factors. Proc Natl Acad Sci U S A 93:6008–6013

Mottram JC, Frame MJ, Brooks DR, Tetley L, Hutchison JE, Souza AE, Coombs GH (1997) The multiple cpb cysteine proteinase genes of *Leishmania mexicana* encode isoenzymes that differ in their stage regulation and substrate preferences. J Biol Chem 272:14285–14293

Mottram JC, Coombs GH, Alexander J (2004) Cysteine peptidases as virulence factors of *Leishmania*. Curr Opin Microbiol 7:375–381

Mullin KA, Foth B, Ilgoutz SM, Callaghan J, McFadden GM, McConville MJ (2001) Regulated degradation of ER membrane proteins in a novel tubular lysosome in *Leishmania mexicana*. Mol Biol Cell 12:2364–2377

Natesan SK, Peacock L, Matthews K, Gibson W, Field MC (2007) Activation of endocytosis as an adaptation to the mammalian host by trypanosomes. Eukaryot Cell 6:2029–2037

Ni X, Canuel M, Morales CR (2006) The sorting and trafficking of lysosomal proteins. Histol Histopathol 21:899–913

Nkemgu NJ, Grande R, Hansell E, McKerrow JH, Caffrey CR, Steverding D (2003) Improved trypanocidal activities of cathepsin L inhibitors. Int J Antimicrob Agents 22:155–159

Olivier M, Gregory DJ, Forget G (2005) Subversion mechanisms by which *Leishmania* parasites can escape the host immune response: a signaling point of view. Clin Microbiol Rev 18:293–305

Overath P, Stierhof YD, Wiese M (1997) Endocytosis and secretion in trypanosomatid parasites – tumultuous traffic in a pocket. Trends Cell Biol 7:27–33

Pal A, Hall BS, Jeffries TR, Field MC (2003) Rab5 and Rab11 mediate transferrin and anti-variant surface glycoprotein antibody recycling in *Trypanosoma brucei*. Biochem J 374:443–451

Pamer EG, So M, Davis CE (1989) Identification of a developmentally regulated cysteine protease of *Trypanosoma brucei*. Mol Biochem Parasitol 33:27–32

Pan AA, McMahon-Pratt D (1988) Monoclonal antibodies specific for the amastigote stage of *Leishmania pifanoi*. I. Characterization of antigens associated with stage- and species-specific determinants. J Immunol 140:2406–2414

Pan AA, Pan SC (1986) *Leishmania mexicana*: comparative fine structure of amastigotes and promastigotes in vitro and in vivo. Exp Parasitol 62:254–265

Parodi AJ (1995) The presence of complex-type oligosaccharides at the C-terminal domain glycosylation site of some molecules of cruzipain. Mol Biochem Parasitol 69:247–255

Pays E, Vanhollebeke B, Vanhamme L, Paturiaux-Hanocq F, Nolan DP, Perez-Morga D (2006) The trypanolytic factor of human serum. Nat Rev Microbiol 4:477–486

Pimenta PF, Saraiva EM, Sacks DL (1991) The comparative fine structure and surface glycoconjugate expression of three life stage of *Leishmania major*. Exp Parasitol 72:191–204

Piper RC, Bryant NJ, Stevens TH (1997) The membrane protein alkaline phosphatase is delivered to the vacuole by a route that is distinct from the *VPS*-dependent pathway. J Cell Biol 138:531–545

Pral EM, Bijovsky AT, Balanco JM, Alfieri SC (1993) *Leishmania mexicana*: proteinase activities and megasomes in axenically cultivated amastigote-like forms. Exp Parasitol 77:62–73

Pral EM, da Moitinho ML, Balanco JM, Teixeira VR, Milder RV, Alfieri SC (2003) Growth phase and medium pH modulate the expression of proteinase activities and the development of megasomes in axenically cultivated *Leishmania (Leishmania) amazonensis* amastigote-like organisms. J Parasitol 89:35–43

Pupkis MF, Tetley L, Coombs GH (1986) *Leishmania mexicana*: amastigote hydrolases in unusual lysosomes. Exp Parasitol 62:29–39

Rabinovitch M, Zilberfarb V, Ramazeilles C (1986) Destruction of *Leishmania mexicana amazonensis* amastigotes within macrophages by lysosomotropic amino acid esters. J Exp Med 163:520–535

Rabinovitch M, Zilberfarb V, Pouchelet M (1987) *Leishania mexicana*: destruction of isolated amastigotes by aminoacid esters. Am Trop Med Hyg 36:288–293

Rafati S, Baba AA, Bakhshayesh M, Vafa M (2000) Vaccination of BALB/c mice with *Leishmania major* amastigote-specific cysteine proteinase. Clin Exp Immunol 120:134–138

Ramazeilles C, Rabinovitch M (1989) *Leishmania amazonensis*: uptake and hydrolysis of 3H-amino acid methyl esters by isolated amastigotes. Exp Parasitol 68:135–143

Raper J, Fung R, Ghiso J, Nussenzweig V, Tomlinson S (1999) Characterization of a novel trypanosome lytic factor from human serum. Infect Immun 67:1910–1916

Rifkin MR (1978) Identification of the trypanocidal factor in normal human serum: high density lipoprotein. Proc Natl Acad Sci U S A 75:3450–3454

Rosenzweig D, Smith D, Myler PJ, Olafson RW, Zilberstein D (2008) Post-translational modification of cellular proteins during *Leishmania donovani* differentiation. Proteomics 8:1843–1850

Rudzinska MA, D'Alesandro PA, Trager W (1964) The fine structure of *Leishmania* to leptomonad transformation. J Protozool 11:166–191

Sacks DL (2001) *Leishmania*-sand fly interactions controlling species-specific vector competence. Cell Microbiol 3:189–196

Sant'Anna C, Pereira MG, Lemgruber L, de Souza W, Cunha e Silva NL (2008) New insights into the morphology of *Trypanosoma cruzi* reservosome. Microsc Res Tech 71:599–605

Schneider P, Rosat JP, Bouvier J, Louis J, Bordier C (1992) *Leishmania major*: differential regulation of the surface metalloprotease in amastigote and promastigote stages. Exp Parasitol 75:196–206

Selzer PM, Chen X, Chan VJ, Cheng M, Kenyon GL, Kuntz ID, Sakanari JA, Cohen FE, McKerrow JH (1997) *Leishmania major*: molecular modeling of cysteine proteases and prediction of new nonpeptide inhibitors. Exp Parasitol 87:212–221

Selzer PM, Pinel S, Hsieh I, Ugele B, Chan VJ, Engel JC, Bogyo M, Russell DG, Sakanari JA, McKerrow JH (1999) Cysteine protease inhibitors as chemotherapy: lessons from a parasite target. Proc Natl Acad Sci U S A 96:11015–11022

Shaw J (2007) The leishmaniases-survival and expansion in a changing world. A mini-review. Mem Inst Oswaldo Cruz 102:541–547

Shimamura M, Hager KM, Hajduk SL (2001) The lysosomal targeting and intracellular metabolism of trypanosome lytic factor by *Trypanosoma brucei brucei*. Mol Biochem Parasitol 115:227–237

Siles R, Chen SE, Zhou M, Pinney KG, Trawick ML (2006) Design, synthesis, and biochemical evaluation of novel cruzain inhibitors with potential application in the treatment of Chagas' disease. Bioorg Med Chem Lett 16:4405–4409

Subba Raju BV, Singh R, Sreenivas G, Singh S, Salotra P (2008) Genetic fingerprinting and identification of differentially expressed genes in isolates of *Leishmania donovani* from Indian patients of post-kala-azar dermal leishmaniasis. Parasitology 135:23–32

Traub-Cseko YM, Duboise SM, McMahon-Pratt D (1993) Identification of two cysteine proteinase genes of *Leishmania pifanoi* axenic amastigotes using PCR. Mol Biochem Parasitol 57:101–116

Ueda-Nakamura T, Attias M, de Souza W (2001) Megasome biogenesis in *Leishmania amazonensis*: a morphometric and cytochemical study. Parasitol Res 87:89–97

Ueda-Nakamura T, Sampaio MCR, Cunha-e-Silva NL, Traub-Cseko YM, de Souza W (2002) Expression and processing of megasome cysteine proteinases during *Leishmania amazonensis* differentiation. Parasitol Res 88:332–337

Ueda-Nakamura T, Attias M, de Souza W (2007) Comparative analysis of megasomes in members of the *Leishmania mexicana* complex. Res Microbiol 158:456–462

Valls LA (1987) Protein sorting in yeast: the localization determinant of yeast vacuolar carboxypeptidase Y resides in the propeptide. Cell 48:887–897

Vanhollebeke B, Nielsen MJ, Watanabe Y, Truc P, Vanhamme L, Nakajima K, Moestrup SK, Pays E (2007) Distinct roles of haptoglobin-related protein and apolipoprotein L-I in trypanolysis by human serum. Proc Natl Acad Sci U S A 104:4118–4123 von Figura K (1991) Molecular recognition and targeting of lysosomal proteins. Curr Opin Cell Biol 3:642–646

Walker J, Vasquez JJ, Gomez MA, Drummelsmith J, Burchmore R, Girard I, Ouellette M (2006) Identification of developmentally-regulated proteins in *Leishmania panamensis* by proteome profiling of promastigotes and axenic amastigotes. Mol Biochem Parasitol 147:64–73

Waller RF, McConville MJ (2002) Developmental changes in lysosome morphology and function *Leishmania* parasites. Int J Parasitol 32:1435–1445

Weise F, Stierhof YD, Kuhn C, Overath P (2000) Distribution of GPI-anchored proteins in the protozoan parasite leishmania, based on an improved ultrastructural description using high-pressure frozen cells. J Cell Sci 113:4587–4603

Organelles and Trafficking in *Entamoeba histolytica*

Sherri S. Smith and Nancy Guillen

Contents

Abstract Publication of the *Entamoeba histolytica* genome has provided strong support for cell biology studies on this lower eukaryote, in which clear cellular compartments involving vesicular trafficking have not yet been well established. Protein trafficking through the endoplasmic reticulum and the Golgi apparatus is an important issue linked to pathogenesis due to the important number of factors secreted by this parasite to accomplish cell killing, tissue invasion, and the onset of inflammation in humans. The advancements in the cellular and molecular tools available have allowed previously unidentifiable organelles to be identified based on gene homologies and video microscopy. The aim of this chapter is to provide a review of the current literature as it pertains to *Entamoeba histolytica* organelles, protein transport, and nutrient uptake, specifically by examining the Golgi, ER, mitosome, phagocytosis, and endocytosis.

S.S. Smith and N. Guillen (✉)
Institut Pasteur, Cell Biology of Parasitism Unit, 25–28 rue du Docteur Roux, 75015 Paris, France
Inserm U786, 25–28 rue du Docteur Roux, 75015 Paris, France
e-mail: nguillen@pasteur.fr

W. de Souza (ed.), *Structures and Organelles in Pathogenic Protists*,
Microbiology Monographs 17, DOI 10.1007/978-3-642-12863-9_7,
© Springer-Verlag Berlin Heidelberg 2010

1 Introduction

Entamoeba histolytica is a human intestinal parasite and the causative agent of amoebiasis. *E. histolytica* is the major *Entamoeba* species known to cause disease in humans; although the species *E. gingivalis* has been shown to cause periodontal disease in HIV infected individuals (Lucht et al. 1998; Ali et al. 2008). Amoebiasis is prevalent worldwide, but is most commonly found in developing countries. Infection occurs through ingestion of food or water containing *E. histolytica* cysts. Excystation into trophozoites occurs within the terminal ileum or colon (Stanley 2003). There are four steps of interactions between trophozoites and target cells (1) adherence, (2) extracellular cytolysis, (3) contact-dependent cytolysis, and (4)phagocytosis (Ravdin 1988). The extremely motile trophozoites feed on erythrocytes, leukocytes, epithelial cells, and bacteria in the large intestine. Proteolytic enzymes (i.e., cysteine proteases) are secreted by the trophozoites, which allow the organism to invade the submucosal tissue leading to amoebic colitis. Most infections are asymptomatic (~90%) and usually clear without any signs of the disease (Stanley 2003), however, symptomatic infection characteristics may include bloody diarrhea, abdominal pain, and tenderness. In some instances, *E. histolytica* invades the liver and forms liver abscesses (ALA) that can occur over the course of months to years after travel or residency in an endemic area. ALA is the most common extraintestinal manifestation of the disease; causing fever, right upper quadrant pain and substantial hepatic tenderness. Acute symptoms usually last 10 days or less, but can become chronic.

Amoebic virulence is defined by the ability to produce ALA in susceptible experimental animals (i.e., gerbils and hamsters). *E. histolytica* virulence is attenuated by prolonged axenic cultivation (Bos and Van de Griend 1977; Mattern et al. 1982; Phillips 1973), however, virulence can be regained by passaging the amoebas through the hamster liver (Das and Ghoshal 1976; Diamond et al. 1974; Lushbaugh et al. 1978; Olivos-García et al. 2009) or by coculturing with bacteria (Padilla-Vaca et al. 1999). Molecules have been identified that are important for amoeba virulence. Attachment to host cells is aided by surface components such as the galactose/*N*-acetylgalactosamine-inhibitable lectin (Gal/GalNac) (reviewed by Petri et al. 2002), the serine-rich *E. histolytica* protein (SREHP) (Teixeira and Huston 2008), and the lysine-rich protein KERP1 (Seigneur et al. 2005). Cytoskeleton related proteins such as Myosin IB and PAK kinase play a role in phagocytosis in *E. histolytica* (Voigt et al. 1999; Voigt and Guillén 1999; Labruyère et al. 2003). Cysteine proteinases (Bruchhaus et al. 1996; Que and Reed 2000) and pore-forming peptides (i.e., amoebapores) (Leippe et al. 1994; Leippe 1999) allow the amoeba to penetrate the epithelial cell layer in the intestine.

E. histolytica trophozoites are between 20 and 40 μm in size and their cytoplasm often contains ingested red blood cells. An interesting aspect of *E. histolytica* is the lack of visible classical structures/organelles, such as peroxisomes, rough endoplasmic reticulum (rER), and Golgi bodies (Lohia 2003). Although homologs of α, β, and γ tubulin have been identified, the amoebas are missing a typical microtubular spindle structure (Sanchez et al. 1994; Katiyar and Edlind 1996; Ray et al. 1997). The amino acid sequences of α, β, and γ tubulin differ significantly from other eukaryotic tubulins, sharing only 50%, between 54% and 58% and 46% identity, respectively (Sanchez et al. 1994; Ray et al. 1997; Vayssié et al. 2004; Meza et al. 2006). Unlike other organisms, *E. histolytica* tubulin is not a major component of the cytoskeletal network; instead it is localized to the nucleus and forms an atypical spindle structure (Vayssié et al. 2004; Roy and Lohia 2004). Vayssié et al. (2004) demonstrated that γ-tubulin localizes to the central region of the nucleus where the microtubules radiate. In fact, *E. histolytica* trophozoites often contain multiple nuclei (Das and Lohia 2002; Lohia et al. 2007; Mukherjee et al. 2008), which results from the formation of multiple microtubule organizing centers (MTOCs) and multipolar spindles (Mukherjee et al. 2009). Cytokinesis in *E. histolytica* is not coupled with nuclear division, therefore, cell division can be asymmetrical leading to daughter cells with zero, one or multiple nuclei (Mukherjee et al. 2009).

2 Organelles and Protein Transport in *E. histolytica*

E. histolytica lack visible classical structures/organelles, however, the amoebas still have Golgi/ER associated functions and contain remnant mitochondrial organelles. The focus of this chapter is to review the currently known information regarding the organelles and protein transport in *E. histolytica*.

2.1 Mitosomes

Mitosomes are defined as remnant mitochondrial organelles that have lost their capacity to synthesize ATP and are commonly found in organisms that reside in an oxygen poor environment (Aguilera et al. 2008). They were identified in supposedly "amitochondrial" parasitic protozoa such as *Trachipleistophora hominis* (microsporidian), *Giardia intestinalis* (diplomonad), *E. histolytica*, and *Cryptosporidium parvum* (alveolate) (Williams et al. 2002; Riordan et al. 2003; Tovar et al. 2003; Putignani et al. 2004). Mitosomes contain mitochondrial enzymes typically found in mitochondria from higher eukaryotes [i.e., mitochondrial type Hsp70 (mtHsp70)] (Matouschek et al. 2000; Williams et al. 2002). The mitosomes are surrounded by a double membrane and contain only a small number of mitochondrial marker proteins (Aguilera et al. 2008). While the morphology of the mitosomes has not been completely resolved, transmission electron microscopy revealed the presence of small organelles approximately 0.5 μm in diameter surrounded by a double membrane in *E. histolytica* (Ghosh et al. 2000; León-Avila and Tovar 2004). *E. histolytica* mitosomes are more than double the size of mitosomes in other organisms (Williams et al. 2002; Tovar et al. 2003; Putignani et al. 2004; Slapeta and Keithly 2004). No evidence of extranuclear DNA was identified in *E. histolytica* mitosomes indicating that there was a loss of the entire mitosome genome during evolution (Tovar et al. 1999; León-Avila and Tovar 2004).

In 1995 Clark and Roger identified and cloned two *E. histolytica* genes encoding the mitochondrial marker proteins, chaperonin 60 (Cpn60) and pyridine nucleotide transhydrogenase (PNT) (Clark and Roger 1995; Aguilera et al. 2008). Cpn60 and PNT both have N-terminal domains that contain many hydroxylated and basic amino acids similar to known targeting motifs of mitochondrial and hydrogenosomal organelles. EhCpn60 is involved in ATP-dependent folding of organellar proteins after import into the mitosome (Martin 1997; Chan et al. 2005). EhCpn10, another chaperonin, was identified as a functional partner of EhCpn60. Neither EhCpn60 nor EhCpn10 are upregulated in response to heat shock (van der Giezen et al. 2005). Chaperones (such as heat shock proteins) are upregulated during heat shock, protect cells from stresses, prevent protein aggregation, and participate in the refolding and/or destruction of misfolded proteins (Voellmy and Boellmann 2007). EhCpn10 lacks a signal targeting sequence and is potentially imported into the mitosome by a mechanism that does not require amino-terminal extensions (van der Giezen et al. 2005). In fact, EhCpn10 is similar to other Cpn10 proteins, which also lack the amino-terminal extensions (Hartman et al. 1992). Additionally, the indentified *E. histolytica* heat shock protein 70 (Hsp70) was cloned and found to localize to the mitosomal fraction (Bakatselou et al. 2000; Tovar et al. 2007).

Characteristically, mitochondria make use of mitochondrial-type ADP/ATP carriers (mitochondrial carrier family proteins (MCF)) to mediate the exchange of substrates, including ATP, within the cytosol of the cell. It was determined that *Encephalitozoon cuniculi*, a microsporidian, had lost all genes encoding for MCF proteins and instead utilizes four unique nucleotide transporters to transport ATP from the host into the cytosol. One nucleotide transporter colocalizes with Hsp70 at the mitosome, indicating that the cytoplasm of *E. cuniculi* provides the ATP necessary for mitosome function (Tsaousis et al. 2008). *E. histolytica* also utilizes an alternative member of the MCF to transport ATP across the mitosomal membrane (Chan et al. 2005; Hackstein et al. 2006). The *E. histolytica* mitosomal carrier does not respond to the classically used inhibitors, carboxyatractyloside, or bongkrekic acid (Brandolin et al. 1993) because of the lack of conservation in the residues typically responsible for binding these inhibitors (Chan et al. 2005). In addition, *E. histolytica* does not use an electrogenic transport mechanism for adenine nucleotide exchange (which results in transport of one negative charge across the membrane), but instead uses an electroneutral exchange, which does not rely on membrane potential (Chan et al. 2005).

Currently, it is unclear what function the mitosomes play in *E. histolytica*; however, a large number are present indicating that a major function is associated with these organelles. Biosynthesis of iron–sulfur (FeS) clusters is a newly discovered function for the mitochondria. FeS clusters are catalysts for chemical reactions, electron carriers in redox reactions, regulatory sensors, and stabilizers of protein structure (Lill and Mühlenhoff 2006, 2008). FeS cluster metabolism is considered a most likely candidate for mitosome function for two reasons (1) FeS cluster assembly is essential for cell survival under aerobic and anaerobic growth conditions and (2) FeS cluster biosynthesis has been shown to be a major function of *Giardia* mitosomes and trichomonad hydrogenosomes (Tovar et al. 2003; Sutak et al. 2004). Recently, it was shown that iron (frataxin) and sulfur (NfsI) donors along with scaffolding proteins (IsuI) colocalize with mitochondrial Hsp70 at the mitosomes in *E. cuniculi* (Goldberg et al. 2008), indicating that the mitosome may be a site for FeS cluster biosynthesis. However, *Trachipleistophora hominis* mitochondrial Hsp70 and NfsI are located inside the mitosome whereas IsuI and frataxin are cytosolic making it difficult to understand how *T. hominis* generates FeS clusters (Goldberg et al. 2008). *E. histolytica* is also unique in that it is the only eukaryote that has a bacterial type FeS cluster assembly system instead of a mitochondrial type (Ali et al. 2004; van der Giezen et al. 2004). A large proteomic analysis is still pending to decipher whether FeS cluster biogenesis enzymes are located in the mitosomes.

2.2 Golgi/ER

The endomembrane system is defined as membranes in the cytoplasm of a eukaryotic cell that divide the cell into functional and structural compartments or organelles, and is composed of the nuclear envelope, endoplasmic reticulum (ER), Golgi apparatus, lysosomes, and endosomes. Protein transport between the organelles of the endomembrane system is carried out by vesicle transport (Bonificino and Glick 2004; Mellman and Warren 2000). No typical ER and Golgi structures have been identified in *E. histolytica*, however, recent evidence suggests the presence of primitive structures. Figure 1 shows the localization of BiP, a molecular marker of the ER in *E. histolystica*. The endocytic vacuoles are also localized using fluorescent dextran. Typically, the ER is organized into branching tubules and flattened sacs that are interconnected and is responsible for protein folding and packaging of proteins into coatomer-coated vesicles (COPII) for transport to the Golgi. COPII vesicles fuse with the *cis*-Golgi and empty their contents into the Golgi lumen; where they undergo modification, sorting and packaging for subsequent transport and/or secretion. The Golgi is composed of stacks of four to six flattened, membrane-enclosed cisternae. Molecules pass from the *cis*-Golgi to the *trans*-Golgi and are modified as they pass through the stacks in order to be correctly transported to other regions of the cell or secreted. Escaped ER resident proteins are subsequently returned to the ER via COPI vesicles.

Resident luminal ER proteins containing a KDEL/HDEL (KDEL for *E. histolytica* – see below) sequence motif are returned to the ER via the ER retention receptor, ERD2. ERD2 is a *cis*-Golgi-associated transmembrane protein that binds to the C-terminal KDEL motif. An ERD2 homolog was identified in the *E. histolytica* genome. EhERD2 amino acid sequence (accession number AJ002138) shares a high homology with other mammalian ERD2 genes and is one of the few intron containing genes in *E. histolytica* (Sanchez-Lopez et al. 1998). The amoebic ERD2 has been cloned and expressed (Sanchez-Lopez et al. 1998). Two proteins,

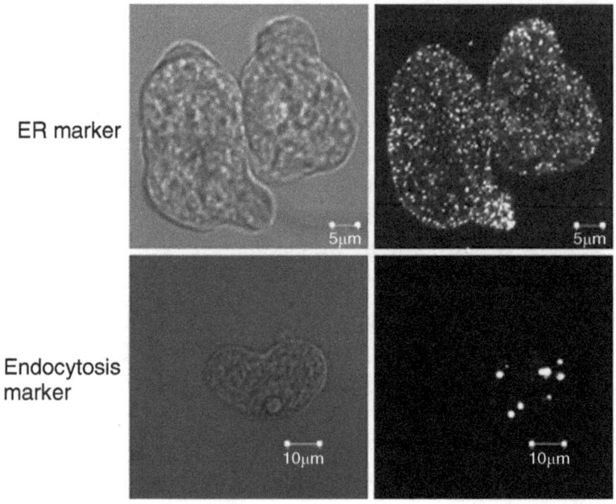

Fig. 1 Cellular localization of the endoplasmic reticulum and endocytic vesicles in *Entamoeba histolytica*. An antibody against the ER marker, BiP, was used to label *Entamoeba histolytica* ER vesicles (*top panel*) by immunoflurescence. ER vesicles are localized throughout the cytoplasm of *E. histolytica*. Endocytic vesicles were stained by incubating *E. histolytica* with fluorescent dextran (*bottom panel*). Phase contrast microscope images of the amoebas are noted in the *left column* and fluorescent images in the *right column* were acquired by confocal microscopy

ERD2 and protein disulfide isomerase (PDI), exhibit redistribution after incubating amoebas with Brefeldin A (BFA) (Manning-Cela et al. 2003). BFA is a fungal metabolite that causes the redistribution of Golgi proteins to the ER and the subsequent collapse of the Golgi stacks. Although there is a redistribution of EhERD2 and EhPDI with BFA, it does not cause the same type of redistribuition as observed in other mammalian cells (Burdett 2002; Tamaki and Yamashina 2002). In addition to the effects on these two proteins, BFA also stopped the trafficking of the heavy subunit of the Gal/GalNAc lectin to the parasites cell surface.

E. histolytica lacks a visible Golgi, however, genes encoding Golgi homologs proteins were found in the genome suggesting the presence of a primitive or reduced Golgi apparatus (Dacks et al. 2003; Dacks and Doolittle 2004). Amoebas lack a Golgi with tight lamellae; however vesicles from *E. histolytica* were reported to exhibit activity of glucose-6-phosphatase and thiamine pyrophosphatase, two Golgi enzymes. These vesicles also labeled positive for NBD-ceramide, a Golgi label, and a fluorescent substrate used to study the activity of UDP-glucose: ceramide glucosyltransferase (Paul et al. 1996; Mazzuco et al. 1997). The basic vesicular trafficking pathway appears to be present as data are available regarding (1) N-linked protein glycosylation, (2) N-terminal signal sequences, and (3) C-terminal ER retention peptides. BiP, an ER chaperone protein with a KDEL sequence motif, was identified in *E. histolytica* and localized to numerous vesicles of varying size near the nucleus (Ghosh et al. 1999). A FLAG–GFP–KDEL construct was generated and used to determine KDEL localization within *E. histolytica*.

FLAG–GFP–KDEL was found to (1) colocalize with BiP, (2) localize within a membrane-enclosed compartment, (3) be excluded from the psuedopod, (4) distribute extensively in the reticular network in visible three-way junctions, and (5) be present in a single, continuous compartment (Teixeira and Huston 2008). Chitinase, a secretory protein expressed by encysting amoeba, is retained in the putative ER after adding a C-terminal KDEL tag (de la Vega et al. 1997; Ghosh et al. 1999). Expression of chitinase during encystation produces many chitinase-associated secretory vesicles that do not overlap with pinosomes or phagosomes but are consistent in size with lysosomes (although experiments attempting to colocalize the chitinase-associated secretory vesicles were never carried out) (Ghosh et al. 1999).

An important component of membrane trafficking is the soluble *N*-ethylmaleimide-sensitive fusion protein attachment receptors (SNAREs) because they are involved in a variety of processes including vesicle tethering (Ungermann et al. 2000), docking (Ungermann et al. 1998), fusion (Nickel et al. 1999), and vesicular transport specificity within eukaryotic cells (McNew et al. 2000). Transport vesicles contain specific v(esicle)-SNARE (renamed as Q-SNARE) proteins on their membranes. Q-SNAREs contain either a glutamine or aspartate in the central "SNARE" domain and are classified into three groups: Qa, Qb, and Qc (Bock et al. 2001; Ungermann and Langosch 2005). The v-SNAREs bind to specific t(arget)-SNARE (renamed as R-SNARE) proteins on the membrane where the vesicles are supposed to fuse (Mayer 2001). R-SNAREs contain an arginine in the "SNARE" domain.

Syntaxins are members of the Q class of SNAREs. Genome mining has identified the presence of two syntaxins (syntaxin 5 and syntaxin PM) in *E. histolytica* (Dacks and Doolittle 2004); however, not much else is known about the localization and interactions of the identified syntaxins. In other organisms syntaxin 5 is localized to the Golgi and is involved in both anterograde and retrograde transport between the Golgi and ER (Banfield et al. 1995); whereas syntaxin PM is localized to the plasma membrane and is involved in anterograde transport (Bennett et al. 1992; Aalto et al. 1993; Sanderfoot et al. 2000).

In other eukaryotes, the main function of *N*-ethylaleimide-sensitive factor (NSF) is to dissociate SNARE complexes after their fusion with target membranes by using the energy from ATP hydrolysis. An NSF was identified in the *E. histolytica* genome, is present as a single copy gene, and shares 60% similarity with NSF homologs from a wide range of organisms (Libros-Ziv et al. 2005). EhNSF exhibited ATPase activity and intra-Golgi transport activity in vitro; displaying expected properties for an NSF protein. In addition to the EhNSF gene homolog, other genes homologs have been identified including SNAP, tSNARE, and v-SNARE, which indicate that the necessary machinery for vesicle trafficking and membrane fusion is present in *E. histolytica* (Libros-Ziv et al. 2005).

The glycosylphosphatidylinositol (GPI) moiety is one way many cell surface proteins, such as the Gal/GalNAc lectin and proteophosphoglycans attach to the surface of *E. histolytica*. It is believed that these GPI-anchored molecules are involved in parasite adhesion to cells, to mucus, and to the extracellular matrix.

Biosynthesis of GPI begins in the ER with the transfer of *N*-acetyl glucosamine from UDP-*N*-acetyl glucosamine to the phosphatidylinositol (PI) residing at the ER membrane. This step is catalyzed by an *N*-acetylglucosamine transferase located in the ER membrane. Then the intermediate is deacetylated to form GlcN–PI by the ER enzyme GPI-deacetylase (PIG-L). It is thought that GlcN–PI is then flipped into the ER lumen by a set of flippases. Next, a set of mannosyl transferases act on GlcN–PI to add three mannose moieties successively to form $(Man)_3$–GlcN–PI (Almeida et al. 2000). The analysis of the *E. histolytica* genome identified genes involved in the GPI biosynthetic pathway (Vats et al. 2005) including all catalytic subunits of the enzymatic complexes sustaining GPI biosynthesis. Parasites were highly sensitive to the complement immune response after blocking GPI biosynthesis (Weber et al. 2008).

Nucleotide-sugar transporters (NSTs) transport nucleotide diphosphate sugars from the cytosol into the lumen of the ER or Golgi apparatus (Hirschberg et al. 1998; Caffaro and Hirschberg 2006). Multiple NSTs are found in most fungi and metazoans allowing for the transport of multiple nucleotide sugars. Using several approaches Bredeston et al. (2005) provided the first evidence for Golgi-like functions such as glycosylation in *E. histolytica* and also showed that *E. histolytica* has Golgi transporters for UDP-galactose and UDP-glucose. Genes encoding three putative nucleotide sugar transporters (EhNst1, EhNst2, EhNst3) (Bredeston et al. 2005) were identified. The transporters are similar to other eukaryotes in that they are membrane bound and are part of vesicles containing luminal ectonucleoside triphosphate diphosphohydrolase, a soluble enzyme that hydrolyzes purine nucleoside diphosphates (Hirschberg et al. 1998). The function for the three NSTs has been identified: EhNst1 transports UDP-Gal, EhNst2 transports UDP-Glc and EhNst3 (which contains the ER retrieval motif KKXX) transports UDP-*N*-acetylglucosamine (UDP-GlcNAc) (Bredeston et al. 2005; Banerjee et al. 2008).

Glycoprotein folding is mediated by a set of proteins known as *N*-glycan-dependent quality control proteins, which include the proteins UDP-glucose: glycoprotein glucosyltransferase (UGGT), calreticulin (CRT) or calnexin (CNX), glucosidase 2 (Gls2) and ERGIC-53 (Banerjee et al. 2007). UGGT glucosylates misfolded proteins, CRT/CNX binds to and refolds the glucosylated proteins, Gls2 removes the Glc, and finally ERGIC-53 moves properly folded proteins to the Golgi (Hauri et al. 2000; Trombetta and Parodi 2003; Appenzeller-Herzog et al. 2004; Helenius and Aebi 2004; Moremen and Molinari 2006; Banerjee et al. 2007). *E. histolytica* contains UGGT (activity located in the ER and not the Golgi fraction), CRT, Gls2, and ERGIC-53 (Zamarripa-Morales et al. 1999; Bravo-Torres et al. 2004; Bredeston et al. 2005; Banerjee et al. 2007; Girard-Misguich et al. 2008).

A unique vacuole is present in *E. histolytica* termed the prephagosomal vacuole (PPV). The PPV is involved in processing, activation, and storage of digestive proteins before their transport to the phagosomes (Saito-Nakano et al. 2004). EhRab5 and EhRab7A are two GTPases involved in the formation of the PPV (Saito-Nakano et al. 2005). EhRab7A is one of nine Rab7 isotypes and is the closest homolog to mammalian Rab7 even though it does not share a similar function

(Saito-Nakano et al. 2004). In addition to aiding in the formation of the PPV, EhRab7A interacts with a retromer-like complex in *E. histolytica* and is dependent on being in the GTP bound state (Nakada-Tsukui et al. 2005).

In mammalian systems, the retromer complex recycles proteins from the endosome back to the TGN and is typically composed of five subunits (Vps5p, Vps17p, Vps26p, Vps29p, and Vps35p) (Nothwehr et al. 1999; Reddy and Seaman 2001). Using both biochemical and phenotypic characteristics the retromer complex can be broken down into two subcomplexes. The first subcomplex is responsible for cargo selection and consists of Vps35p, Vps29p, and Vps26p (i.e., Vps35p interacts with the cargo protein Vps10p (Nothwehr et al. 1999)); while the second subcomplex plays a necessary structural role and consists of Vps5p and Vps17p (Seaman 2004, 2005). Mammalian retromer localizes to the endosomes and multivesicular bodies (MVB) and is required for retrieval of the cation-independent mannose-6-phosphate receptor (CI-MPR) from the endosomes to the Golgi (Arighi et al. 2004; Seaman 2005). The subunits Vps26p and Vps35p have been implicated in retrograde transport of the hydrolase receptor from the late endosomes to the TGN (Seaman 2005).

E. histolytica retromer-like complex is composed of Vps26, Vps29, and Vps35. Homologs for the other two subunits are absent or have not yet been identified (Nothwehr et al. 1999; Seaman 2005; Nakada-Tsukui et al. 2005). EhRab7 binds to the C-terminal region of the Vps26 subunit. Both EhRab7 and Vps26 colocalize at the PPV membrane during erythrocyte attachment to *E. histolytica*. Thirty minutes after *E. histolytica* have ingested several erythrocytes, EhRab7 and Vps26 colocalize with both the PPV and the phagosome (Nakada-Tsukui et al. 2005), indicating an important role for the retromer-like complex during erythrophagocytosis.

2.3 Phagocytosis

Endocytosis and phagocytosis play important roles in the virulence of *E. histolytica*. Phagocytosis is the process where cells can engulf particles that are greater than 0.5 μm in size (Groves et al. 2008). This is a process that involves particles or cells binding to surface receptors, cytoskeletal reorganization to form a phagocytic cup which closes around the ligand to form a phagosome, trafficking of the phagosome along the endocytic pathway, and fusion of the phagosome with the lysosome for digestion of the engulfed particle or cell (Silverstein et al. 1977; Rabinowitz et al. 1992; reviewed in Voigt and Guillén 1999).

Endocytosis and phagocytosis is very dynamic in the amoebas and plays an important role in the pathogenesis of *E. histolytica*. As the parasite colonizes the human it engulfs foreign cells, which includes microorganisms and host cells (i.e., erythrocytes and immune cells). Parasites that are phagocytosis-deficient have been shown to be avirulent (Orozco et al. 1983; Rodriguez and Orozco 1986). Amoebas show astounding heterogeneity with respect to their cellular functions; in fact most amoebas conduct macropinocytosis at the same time as phagocytosis (Meza and

Clarke 2004). *E. histolytica* digests bacteria and host cells within the phagolysosome using oxygen-independent mechanisms. Digestive proteins used include cysteine proteases, lysozymes, and amoebapores (Bruchhaus et al. 1996, 1995; Leippe 1997). Amoebapores are activated through acidification of the phagolysosome (Bruhn et al. 2003).

In a proteomic analysis, Okada et al. (2005) identified several proteins found in phagosomes including the Gal/GalNAc lectin, several small GTPases (i.e., Rab1, Rab7 A–E, Rac A, C, and G), hydrolytic enzymes, and degradative proteins (i.e., cysteine proteases (1, 2, 4, and 5), lysozyme, dipeptidylaminopeptidase, and phospholipase A_2) and a calcium-transporting ATPase. Proteins not identified in their study were those previously shown to be involved in red blood cell phagocytosis including cysteine protease 3 and myosin IB (Que et al. 2002; Marion et al. 2004). Myosin IB localizes to the phagocytic cup and phagosomes during ingestion of human red blood cells (Marion et al. 2004). Amoebapores and cysteine protease (CP) 2 and 3 are recruited to phagosomes and are involved in the uptake and degradation of ingested bacteria (Andrä et al. 2003; Que et al. 2002). CP activity in *E. histolytica* is 10–1,000-fold higher than in *E. dispar* (Bruchhaus et al. 2003). Calreticulin, a receptor for the collagenous tail of C1q, was also identified as an abundant protein in *E. histolytica* phagosomes (Boettner et al. 2008; Marion et al. 2005). C1q is structurally related to the collectin protein family. Collectins are pattern recognition molecules of the innate immune system and are ligands presented by apoptotic cells and bacteria, which are recognized by macrophages. C1q binds to apoptotic cells to aid in macrophage phagocytosis (Ogden et al. 2001) and *E. histolytica* trophozoites migrate towards human C1q and collectins, which then stimulates *E. histolytica* trophozoite phagocytosis of apoptotic lymphocytes (Teixeira et al. 2008).

E. histolytica induces host cell apoptosis before engulfing the host cell; preferentially "eating" apoptotic cells as opposed to necrotic cells (Huston et al. 2003). Amoebas induce host cell apoptosis using a caspase 3-dependent mechanism, which is independent of caspase 8 and caspase 9 as demonstrated in studies between Jurkat T cells and *E. histolytica* (Huston et al. 2000, 2003). Caspase 3-like activity was observed within a few minutes of host cell contact with *E. histolytica* and inhibition of caspase 3 using the inhibitor Ac–DEVD–CHO completely blocked Jurkat cell DNA fragmentation (Huston et al. 2000). Apoptosis of host cells requires the Gal/GalNAc lectin but does not require cysteine proteases (Huston et al. 2000). Exposure of phosphatidylserine at the outer leaflet of the cell membrane is one of the hallmarks of apoptosis (Lang et al. 2004). During amoebic induced apoptosis phosphatidylserine is exposed on the surface of Jurkat cells, enhancing amoebic phagocytosis (Huston et al. 2003). The immunodominant surface antigen, SREHP, was identified as an *E. histolytica* phagocytosis receptor with a role in adhering amoeba to the apoptotic cells (Teixeira and Huston 2008).

Phagocytosis can be classified as either classical phagocytosis or microphagocytosis. Classical phagocytosis is characterized by the extension of the plasma membrane to form an actin-rich phagocytic cup around the particle prior to engulfment. Microphagocytosis, on the other hand, is characterized by suctioning the particle into a cellular vesicle in the absence of membrane extensions (Lejeune and Gicquaud 1987; Lejeune and Gicquaud 1992). Phagocytosis in *E. histolytica* has been measured using gerbil red blood cells (Saito-Nakano et al. 2004), human red blood cells (hRBCs) (Voigt et al. 1999; Welter et al. 2005; Powell et al. 2006), bacteria (Pimenta et al. 2002) and magnetic or latex beads (Okada et al. 2005; Marion et al. 2005). *E. histolytica* appears to uptake hRBCs via microphagocytosis and are very efficient because each amoeba can ingest as many as nine hRBCs within 15 min (Welter et al. 2006). Erythrocytes pretreated with a Ca^{2+} ionophore, which induces surface changes similar to that of apoptotic cells, are preferentially phagocytosed by *E. histolytica* (Boettner et al. 2005).

The role of the unconventional myosin IB in the cytoskeleton remodeling process (inducing phagocytic cup formation and internalization) has been demonstrated in *E. histolytica*. Myosin IB is recruited to both the phagocytic cup and around internalized phagosomes (Voigt et al. 1999; Marion et al. 2004) in the presence of RBCs. Overexpression of the myosin IB heavy chain leads to an increase in cytoplasm viscosity through changes in the density of the F-actin network, indicating that myosin IB cross-links actin filaments through the two actin-binding sites within the heavy chain. The excess of myosin IB in over-expressing cells exaggerated the cross-linking activity. Myosin IB can regulate the dynamics of the F-actin network at the cell cortex; a crucial parameter for investigating fundamental cellular processes such as phagocytosis.

Phosphatidylinsitol 3-phosphate (PI3P) localizes to endosomes and recruits proteins with PI3P binding domains that aid in the regulation of endocytosis in mammalian cells. Proteins containing a FYVE-finger domain bind with high affinity and specificity to PI3P (Stenmark and Aasland 1999; Stenmark et al. 2002). The PI3-kinase inhibitors wortmannin and LY 294002 can inhibit endocytosis in *E. histolytica* (Ghosh and Samuelson 1997; Meza and Clarke 2004) as well as phagocytosis (Powell et al. 2006). After contact with hRBCs for 1 min, a GST-2xFYVE fusion construct (a GST construct containing two tandem FYVE domains) was enriched in the cups of developing erythrophagosomes and hRBC containing ring-like structures near the plasma membrane (Powell et al. 2006). No FYVE staining was observed on the ring structures after transport to the center of the cell indicating that PI3P is utilized during the formation of the phagosomes, but is not necessary for phagosome maturation. Treatment of *E. histolytica* with wortmannin blocked internalization of hRBCs but not the parasites ability to bind the hRBCs, which indicates that PI3P is not involved in binding to hRBCs, but rather is important for internalization (Powell et al. 2006). GST-PI3P accumulates on newly forming phagosomes containing hRBCs or bacteria but is not found on fluid phase pinosomes in *E. histolytica* (Powell et al. 2006).

Phosphatidylinositol (PI) is thought to be involved in the regulation of phagocytosis (Fratti et al. 2001; Botelho et al. 2004; Scott et al. 2005; Yeung et al. 2006). PI-3-phosphate (PI3P) is normally found on the surface of phagosomes and recruits early endosome antigen-1 (EEA1) and hepatocyte growth factor related tyrosine kinase substrate (Hrs) (Gillooly et al. 2001; Raiborg et al. 2001; Vieira et al. 2004). Although the *E. histolytica* genome is missing these effectors (i.e., EEA1 and Hrs)

(Loftus et al. 2005; Clark et al. 2007), PI3P was shown to be involved in *E. histolytica* phagocytosis and endocytosis (Ghosh and Samuelson 1997; Powell et al. 2006). Nakada-Tsukui et al. (2009) recently utilized a PI3P biomarker Hrs-FYVE to visualize the initiation of phagocytosis in *E. histolytica*. The biomarker was localized on the phagocytic cup, the phagosome and to structures connecting the plasma membrane with the phagosomes. Twelve *E. histolytica* proteins contain an FYVE domain (EhFP1–12) and all but one contains a Rho guanine nucleotide exchange factor (GEF) domain (Nozaki and Nakada-Tsukui 2006). The FYVE domain of EhFP4 provides binding specificity between EhFP4 and PI-P. The current work provides evidence that PI3P, PI(4)P, and EhFP4 regulate phagocytosis and phagosome maturation (Nakada-Tsukui et al. 2009).

Several research groups have undergone studies on ameoba phagocytosis to determine the differences between pathogenic (*E. histolytica*) and nonpathogenic (*E. dispar*) amoebae and between virulent and nonvirulent *E. histolytica*. *E. histolytica* loses virulence when it is axenically cultured, but it is unclear what factors change. In phagocytosis, acidification is important for the degradation of engulfed particles or cells. A study comparing the acidification of phagosomes between the pathogenic amoeba, *E. histolytica*, and the nonpathogenic amoeba, *E. dispar*, was conducted. In studies using bacteria it was demonstrated that *E. histolytica* phagosomes contain bacteria that were degraded extensively, whereas bacteria in phagosomes from *E. dispar* often retain their cell morphology (Mitra et al. 2005). Phagosomes from *E. dispar* were less acidic than those in *E. histolytica*, but both species of *Entamoeba* showed prolonged acidification of up to 12 h. Acidification of the phagosomes in *E. histolytica* occurs within 2 min, which is in contrast to the digestive vacuole in mouse macrophages and hemocytes of *Mytilus edulis, Amoeba proteus,* and *Dictyostelium discoideum* where acidification occurs between 7 and 15 min (Jensen and Bainton 1973; Geisow et al. 1981; Bassoe and Bjerknes 1985; Rupper et al. 2001). Additionally, the prolonged acidification is highly unusual as it is energetically unfavorable. In other organisms such as *D. discoideum* the vacuole is neutralized between 30 and 60 min (Rupper et al. 2001) and neutralization begins within 8 min with *Paramecium caudatum* (Fok et al. 1982). In a different study researchers compared virulent liver-passaged *E. histolytica* with the nonvirulent axenically cultured *E. histolytica*. This study revealed striking and/or surprising differences in phagosome acidification and degradation. Phagosomes from the virulent amoebas were acidified more slowly and degraded materials more slowly and less efficiently than phagosomes from the nonvirulent amoebas (Mitra et al. 2006).

Phagosome acidification is also dependent on functional myosin but not microtubules or actin. Inhibition of actin or microtubule polymerization showed only slight effects on the phagosome acidification whereas inhibition of myosin produced a strong effect. The initial drop of phagosome pH within 2 min of engulfment was most notably affected with myosin inhibition (Mitra et al. 2005). Inhibition of cysteine proteinases via the potent inhibitor E64 blocked degradation of engulfed GFP-labeled *Leishmania* parasites in *E. histolytica* only and had no effect on *E. dispar*.

Vacuolar H^+-ATPases (V-ATPase), a key enzyme that usually regulates vacuolar pH, is a multisubunit complex that moves protons across membranes against their electrochemical potential using ATP hydrolysis. In eukaryotic cells the V_1 complex is composed of nine subunits (A_3, B_3, C, D, E, G, G_2, H_{1-2}) and the V_0 complex is composed of 6 subunits (a, c_4, c', c'', d, e) (Nishi and Forgac 2002; Beyenback and Wieczorek 2006; Kane 2006). Homologs for all of the V_1 subunits and all subunits except the c' and e subunits of the V_0 complex have been identified in the *E. histolytica* genome (Meléndez-Hernández et al. 2008). In 1990 a vacuolar ATPase was identified in *E. histolytica* and three vacuolar ATPase subunits have been cloned: Ehvma1, Ehvma2, Ehvma3 (Yi and Samuelson 1994; Descoteaux et al. 1994; Meléndez-Hernández et al. 2008). Ehvma1 encodes subunit A of the V_1 complex (Yi and Samuelson 1994), Ehvma2 encodes subunit B of the V_1 complex (Meléndez-Hernández et al. 2008) and Ehvma3 encodes subunit C of the V_0 complex (Descoteaux et al. 1994). EhV-ATPase B localizes to many small cytoplasmic vesicles, organelles called EhKO's, but are not present in the nucleus. During erythrocyte phagocytosis EhV-ATPase B was localized to the membranes throughout the cell with the exception of the nucleus, but a concentration of EhV-ATPase B was noted in the phagocytic vesicles surrounding RBCs.

2.4 Endocytosis

Recently, studies characterizing the vesicular system in *E. histolytica* have identified lysosomal proteins, acidic pH and stage-specific markers in many of the vesicles, which follow variable maturation pathways (Ghosh et al. 1999; Temesvari et al. 1999; Batista et al. 2000; Welter et al. 2002). Fluid uptake by *E. histolytica* was demonstrated to occur through macropinocytosis, which had not previously been observed for this parasite (Meza and Clarke 2004). Macropinocytosis is the process where the cell membrane forms an invaginated pocket, which pinches off into the cell to form a vesicle filled with extracellular fluid that fuses with endosomes and subsequently with the lysosomes.

Endocytosis is a process whereby the cell takes up macromolecules from the extracellular milieu. Early endosomes mature into late endosomes and then fuse with lysosomes, a terminal degradative compartment of the endocytic pathway (Luzio et al. 2007). Common markers for early endosomes are the mannose-6-phosphate receptor (MPR), early endosome autoantigen 1 (EEA1) and the small GTPase, Rab5; whereas late endosomes lose Rab5 and gain Rab7. Late endosomes contain more luminal vesicles (often referred to as MVBs) than the early endosomes. Lysosomes are distinct from endosomes because they lack the MRP and can fuse with different cellular membranes including endosomes, autophagosomes, phagosomes and the plasma membrane (for membrane repair) (Luzio et al. 2007). *E. histolytica* early endosomes are rapidly acidified and the contents are neutralized before their release; a process that requires 2 h (Meza and Clarke 2004). In addition to *E. histolytica* endosomes being able to fuse together, fusion can also occur between endosomes and phagosomes containing phagocytosed bacteria (Meza and Clarke 2004) demonstrating the ability for *E. histolytica* endosomes to fuse in a homeotypic and heterotypic fashion. *E. histolytica* endosomes (both early and late) are enriched in acid phosphatase activity, cysteine proteases and colocalize with both EhRab7A and EhRab11A (Temesvari et al. 1999).

E. histolytica requires a large supply of iron for metabolism and reproduction in axenic culture (Serrano-Luna et al. 1998). Potential sources for iron include hemoglobin, holo-transferrin, holo-lactoferrrin and ferrintin (Serrano-Luna et al. 1998; Reyes-López et al. 2001; León-Sicairos et al. 2005; López-Soto et al. 2009). Iron-bound transferrin (holo-Tf) is taken up by higher eukaryotes via clathrin-dependent receptor-mediated endocytosis or otherwise known as micropinocytosis (Dautry-Varsat 1986). The involvement of clathrin in endocytosis of holo-Tf was investigated; however, it was determined that *E. histolytica* uptakes holo-Tf in a receptor-independent fashion (Welter et al. 2006), potentially through an identified Tf-binding protein (Reyes-López et al. 2001). Lactoferrin has been observed in tubular invaginations in *E. histolytica* that were independent of clathrin (Batista et al. 2000). Holo-lactoferrin is able to support the growth of *E. histolytica* and a specific binding protein for holo-lactoferrin was identified in *E. histolytica* (León-Sicairos et al. 2005). Holo-lactoferrin is endocytosed via filipin-sensitive vesicles that are recognized by caveolin antibodies, indicating that lactoferrin is taken up in a caveolin dependent method (León-Sicairos et al. 2005). An *E. histolytica* hemoglobin binding protein, Ehhmpb45, was previously identified and cloned, where it was shown that Ehhmbp45 was upregulated during iron starvation (León-Sicairos et al. 2005). In 2009, another hemoglobin binding protein was identified (Ehhmbp26) from the genome. Ehhmbp26 binds hemoglobin, but is negatively regulated by iron as the amoebas only express Ehhmbp26 during iron restrictive conditions (Cruz-Castañeda et al. 2009). The acquisition of ferritin on the other hand occurs through clathrin-coated vesicles and within 2 min after incubation with the amoebas. By 30 min after endocytosis, ferritin was localized to a compartment containing the lysosomal marker, LAMP1. Neither hemoglobin nor holo-transferrin competes with ferritin for binding to the amoebas indicating that the amoebas utilize several methods to acquire iron (López-Soto et al. 2009) as a way to survive in different host tissues.

2.5 Rabs and Trafficking

Rab proteins belong to the Ras superfamily of small GTPases that are essential for regulating vesicular trafficking in both the endocytic and exocytic/secretory pathways in eukaryotic cells (Zerial and McBride 2001; Mitra et al. 2007). Rab GTPases are localized to membrane-bound compartments where they have specific functions based on the specific Rab and/or location (Novick and Zerial 1997; Chavrier and Goud 1999; Mitra et al. 2007). Rab proteins are either ubiquitous between all cell types (i.e., Rab5, Rab7, and Rab11) or unique to specific cells and/or tissues (i.e., Rab3 and Rab27A) (Stenmark and Olkkonen 2001; Schulter et al. 2002; Hume et al. 2001; Stinchcombe et al. 2001). In the active GTP bound state, Rab effectors recognize and bind to Rab GTPases to aid in protein recruitment and vesicle mobility, while also bringing the correct membranes into contact (Picazarri et al. 2005).

E. histolytica has 91–105 Rab genes (Lal et al. 2005; Saito-Nakano et al. 2005); however only a select few of these genes have been characterized. The role Rab proteins play in phagocytosis and endocytosis has been investigated. Research has identified a rab-like gene, EhRabB. Studies have determined that EhRabB contains the four domains characteristic of Ras-related proteins involved in guanine nucleotide binding (Olkkonen and Stenmark 1997) and specifically binds both GTP and GDP (Rodriguez et al. 2000). During the initial steps of phagocytosis, EhRabB is located on the plasma membrane and phagocytic mouths. Ten minutes after phagocytosis EhRabB is no longer detected on the phagosomes and is typically localized to small vesicles throughout the cytoplasm (Rodriguez et al. 2000). Furthermore, an *E. histolytica* G protein-coupled receptor, EhGPCR-1, binds to EhRabB based on yeast two-hybrid experiments but it is unknown at this time whether EhGPCR-1 binds to the GTP-bound or GDP-bound EhRabB, if they colocalize together or where the receptor localizes (Picazarri et al. 2005).

Five Rab7 isotypes have been found in latex bead-containing phagosomes, which hint at the involvement of Rab7 in phagosome biogenesis (Okada et al. 2005). EhRab7A plays a role in the biogenesis of the prephagosomal vacuole and is concentrated in endosome-enriched fractions (Temesvari et al. 1999) containing lysosomal proteins such as amoebapore-A and cysteine proteases (Saito-Nakano et al. 2007). EhRab7A aids in the transport of CPs and amoebapore to phagosomes via the PPV and also regulates recycling of a CP receptor from the phagosomes to the *trans*-Golgi network via the retromer (Saito-Nakano et al. 2004; Nakada-Tsukui et al. 2005; Nozaki and Nakada-Tsukui 2006). EhRab7B localizes to a nonacidic compartment, partially colocalizes with lysosomal proteins and regulates the transport of cysteine proteases to lysosomes (Saito-Nakano et al. 2007).

E. histolytica possesses four Rab11 homologs (Rab11A–D) (Temesvari et al. 1999; McGugan and Temesvari 2003; Saito-Nakano et al. 2001, 2005). Rab11A has been postulated to be involved in transport during conditions of starvation and encystation (McGugan and Temesvari 2003). Rab11B does not localize with the lysosomes nor does it localize with the ER (Mitra et al. 2007), however, it partially overlaps with cysteine protease 5 (CP5). Rab11B was shown to be important for the transport and secretion of cysteine proteases and is cytolytic towards mammalian cells (Mitra et al. 2007).

Rab5 on the other hand is typically associated with early endosomes in higher eukaryotes; however, studies have shown that it is excluded from the endocytic pathway in *E. histolytica* (Saito-Nakano et al. 2004). EhRab5 is associated with phagocytosis, but not directly; instead EhRab5 colocalizes with EhRab7A to generate the PPV (Saito-Nakano et al. 2004), which then fuses with the phagosomes. Unlike other organisms where several Rab5 isotypes are present, EhRab5 is present with only one gene based on genomic database mining (Singer-Krüger et al.

1994; Bucci et al. 1995; Field et al. 1998; Ueda and Nakano 2002; Saito-Nakano et al. 2004). EhRab5 is required for the efficient uptake of red blood cells and has an important role in amoebapore transport. When red blood cells are not present both EhRab5 and EhRab7A are localized throughout the cytoplasm in small vesicles, however, large vacuoles appear that contain both EhRab5 and EhRab7A within 5 min after addition of red blood cells. Ten minutes after addition of red blood cells EhRab5 starts to dissociate from the large vacuoles and by 30 min EhRab5 was completely dissociated from the vacuoles (Saito-Nakano et al. 2004). Using EhRab5 constructs that are constitutively GDP-bound or GTP-bound both showed defects in erythrophagocytosis and PPV formation, which is in contrast to other mammalian systems where wild type and GTP-bound Rab5 exhibit similar behaviors (Stenmark et al. 1994; Saito-Nakano et al. 2004).

3 Conclusions

Protein transport and secretion are vital to the pathogenicity of *E. histolytica*. Advances in technology and the completion of the *E. histolytica* genome continue to provide more information regarding protein transport within the parasite. These advancements have allowed the identification of previously unidentified organelles, such as the Golgi apparatus and ER. It has also furthered our understanding of the phagocytosis and endocytosis processes within the parasite.

Acknowledgments The work of the Unité Biologie Cellulaire du Parasitisme (BCP unit) is sponsored in part by the Pasteur–Weizmann Research Council and the French granting agency, ANR. Sherri Smith is supported by a grant from the Pasteur Foundation of New York.

References

Aalto MK, Ronne H, Keranen S (1993) Yeast syntaxins Sso1p and Sso2p belong to a family of related membrane proteins that function in vesicular transport. EMBO J 12:4095–4104

Aguilera P, Barry T, Tovar J (2008) *Entamoeba histolytica* microsomes: organelles in search of a function. Exp Parasitol 118:10–16

Ali V, Shigeta Y, Tokumoto U, Takahashi Y, Nozaki T (2004) An intestinal parasitic protist, *Entamoeba histolytica*, possesses a non-redundant nitrogen fixation-like system for iron-sulphur cluster assembly under anaerobic conditions. J Biol Chem 279:16863–16874

Ali IK, Clark CG, Petri WA Jr (2008) Molecular epidemiology of amebiasis. Infect Genet Evol 8:698–707

Almeida IC, Camargo MM, Procópio DO, Silva LS, Mehlert A, Travassos LR, Gazzinelli RT, Ferguson MA (2000) Highly purified glycosylphosphatidylinositols from *Trypansoma cruzi* are potent proinflammatory agents. EMBO J 19:1476–1485

Andrä J, Herbst R, Leippe M (2003) Amoebapores, archaic effector peptides of protozoan origin, are discharged into phagosomes and kill bacteria by permeabilizing their membranes. Dev Comp Immunol 27:291–304

Appenzeller-Herzog C, Roch A, Nufer O, Hauri H (2004) pH-induced conversion of the transport Lectin ERGIC-53 triggers glycoprotein release. J Biol Chem 279:12943–12950

Arighi CN, Hartnell LM, Aguilar RC, Haft CR, Bonifacino JS (2004) Role of the mammalian retromer in sorting of the cation-independent mannose 6-phosphate receptor. J Cell Biol 165:123–133

Bakatselou C, Kidgell C, Clark CG (2000) A mitochondrial-type hsp70 gene of *E. histolytica*. Mol Biochem Parasitol 110:177–182

Banerjee S, Vishwanath P, Cui J, Kelleher DJ, Gilmore R, Robbins PW, Samuelson J (2007) The evolution of N-glycan-dependent endoplasmic reticulum quality control factors for glycoprotein folding and degradation. Proc Natl Acad Sci USA 104:11676–11681

Banerjee S, Cui J, Robbins PW, Samuelson J (2008) Use of *Giardia*, which appears to have a single nucleotide-sugar transporter for UDP-GlcNAc, to identify the UDP-Glc transporter of *Entamoeba*. Mol Biochem Parasitol 159:44–53

Banfield DK, Lewis MJ, Pelham HR (1995) A SNARE-like protein required for traffic through the Golgi complex. Nature 375:806–809

Bassoe CF, Bjerknes R (1985) Phagocytosis by human leukocytes, phagosomal pH and degradation of seven species of bacteria measured by flow cytometry. J Med Microbiol 19:115–125

Batista EJ, de Menezes Feitosa LF, de Souza W (2000) The endocytic pathway in *Entamoeba histolytica*. Parasitol Res 86:881–890

Bennett MK, Calakos N, Scheller RH (1992) Syntaxin: a synaptic protein implicated in docking of synaptic vesicles at presynaptic active zones. Science 257:255–259

Beyenback KW, Wieczorek H (2006) The V-type H^+ATPase: molecular structure and function, physiological roles and regulation. J Exp Biol 209:577–589

Bock JB, Matern HT, Peden AA, Scheller RH (2001) A genomic perspective on membrane compartment organization. Nature 409:839–841

Boettner DR, Huston CD, Sullivan JA, Petri WA Jr (2005) *Entamoeba histolytica* and *Entamoeba dispar* utilize externalized phosphatidylserine for recognition and phagocytosis of erythrocytes. Infect Immun 73:3422–3430

Boettner DR, Huston CD, Linford AS, Buss SN, Houpt E, Sherman NE, Petri WA Jr (2008) *Entamoeba histolytica* phagocytosis of human erythrocytes involves PATMK, a member of the transmembrane kinase family. PLoS Pathog 4:0122–0133

Bonificino JS, Glick BS (2004) The mechanisms of vesicle budding and fusion. Cell 116:153–166

Bos HJ, Van de Griend RJ (1977) Virulence and toxicity of axenic *Entamoeba histolytic*. Nature 265:341–343

Botelho RJ, Scott CC, Grinstein S (2004) Phosphoinositide involvement in phagocytosis and phagosome maturation. Curr Top Microbiol Immunol 282:1–30

Brandolin G, Le Saux A, Trezeguet V, Lauquin GJ, Vignais PV (1993) Chemical immunological, enzymatic, and genetic approaches to studying the arrangement of the peptide chain of the ADP/ATP carrier in the mitochondrial membrane. J Bioenerg Biomembr 25:459–472

Bravo-Torres JC, Villagómez-Castro JC, Calvo-Méndez C, Flores-Carreón A, López-Romero E (2004) Purification and biochemical characterization of a membrane-bound α-glucosidase from the parasite *Entamoeba histolytica*. Int J Parasitol 34:455–462

Bredeston LM, Caffaro CE, Samuelson J, Hirschberg CB (2005) Golgi and endoplasmic reticulum functions take place in different subcellular compartments of *Entamoeba histolytica*. J Biol Chem 280:32168–32176

Bruchhaus I, Jacobs T, Leippe M, Tannich E (1996) *Entamoeba histolytica* and *Entamoeba dispar*: differences in numbers and expression of cysteine proteinase genes. Mol Microbiol 22:255–263

Bruchhaus I, Loftus BJ, Hall N, Tannich E (2003) The intestinal protozoan parasite *Entamoeba histolytica* contains 20 cysteine protease genes, of which only a small subset is expressed during in vitro cultivation. Eukaryot Cell 2:501–509

Bruhn H, Riekens B, Berninghausen O, Leippe M (2003) Amoebapores and NK-lysin, members of a class of structurally distinct antimicrobial and cytolytic peptides from protozoa and mammals: a comparative functional analysis. Biochem J 375:737–744

Bucci C, Lütcke A, Steele-Mortimer O, Olkkonen VM, Dupree P, Chiariello M, Bruni CB, Simons K, Zerial M (1995) Co-operative regulation of endocytosis by three Rab5 isoforms. FEBS Lett 366:65–71

Burdett ID (2002) Effects of Brefeldin A on disassembly of the Golgi body in MDCK cells subjected to a Ca^{2+} shift at low temperature. Eur J Cell Biol 81:525–528

Caffaro CE, Hirschberg CB (2006) Nucleotide sugar transporters of the Golgi apparatus: from basic science to diseases. Acc Chem Res 39:805–812

Chan KW, Slotboom DJ, Cox S, Embley TM, Fabre O, van der Giezen M, Harding M, Horner DS, Kunji ERS, Leon-Avila G, Tovar J (2005) A novel ADP/ATP transporter in the mitosome of the microaerophilic human parasite *Entamoeba histolytica*. Curr Biol 15:737–742

Chavrier P, Goud B (1999) The role of ARF and Rab GTPases in membrane transport. Curr Opin Cell Biol 11:466–475

Clark CG, Roger AJ (1995) Direct evidence for secondary loss of mitochondria in *Entamoeba histolytica*. PNAS 92:6518–6521

Clark CG, Alsmark UC, Tazreiter M, Saito-Nakano Y, Ali V, Marion S, Weber C, Mukherjee C, Bruchhaus I, Tannich E, Leippe M, Sicheritz-Ponten T, Foster PG, Samuelson J, Noël CJ, Hirt RP, Embley TM, Gilchrist CA, Mann BJ, Singh U, Ackers JP, Bhattacharya S, Bhattacharya A, Lohia A, Guillén N, Duchêne M, Nozaki T, Hall N (2007) Structure and content of the *Entamoeba histolytica* genome. Adv Parasitol 65:51–190

Cruz-Castañeda A, Hernández-Sánchez J, Olivares-Trejo JJ (2009) Cloning and identification of a gene coding for a 26-kDa hemoglobin-binding protein from *Entamoeba histolytica*. Biochimie 91:383–389

Dacks JB, Doolittle WF (2004) Molecular and phylogenetic characterization of syntaxin genes from parasitic protozoa. Mol Biochem Parasitol 136:123–136

Dacks JB, Davis LA, Sjögren AM, Andersson JO, Roger AJ, Doolittle WF (2003) Evidence for Golgi bodies in proposed 'Golgi-lacking' lineages. Proc Biol Sci 270:S168–S171

Das SR, Ghoshal S (1976) Restoration of virulence to rat of axenically grown *Entamoeba histolytica* by cholesterol and hamster liver passage. Ann Trop Med Parasitol 70:439–443

Das S, Lohia A (2002) Delinking of S phase and cytokinesis in the protozoan parasite *Entamoeba histolytica*. Cell Microbiol 4:55–60

Dautry-Varsat A (1986) Receptor-mediated endocytosis: the intracellular journey of transferrin and its receptor. Biochimie 68:375–381

de la Vega H, Specht CA, Semino CE, Robbins PW, Eichinger D, Caplivski D, Ghosh S, Samuelson J (1997) Cloning and expression of chitinases of *Entamoebae*. Mol Biochem Parasitol 85:139–147

Descoteaux S, Yu Y, Samuelson J (1994) Cloning of *Entamoeba* genes encoding proteolipids of putative vacuolar proton-translocating ATPases. Infect Immun 62:3572–3575

Diamond LS, Phillips BP, Bartgis IL (1974) A comparison of the virulence of nine strains of axenically cultivated *Entamoeba histolytica* in hamster liver. Archivos de Investigacion Medica 5:423–426

Field H, Farjah M, Pal A, Gull K, Field MC (1998) Complexity of trypanosomatid endocytosis pathways revealed by Rab4 and Rab5 isoforms in *Trypanosoma brucei*. J Biol Chem 273:32102–32110

Fok AK, Lee Y, Allen RD (1982) The correlation of digestive vacuole pH and size with the digestive cycle in *Paramecium caudatum*. J Protozool 29:409–414

Fratti RA, Backer JM, Gruenberg J, Corvera S, Deretic V (2001) Role of phosphatidylinositol 3-kinase and Rab5 effectors in phagosomal biogenesis and mycobacterial phagosome maturation arrest. J Cell Biol 154:631–644

Geisow MJ, D'Arcy Hart P, Young MR (1981) Temporal changes of lysosome and phagosome pH during phagolysosome formation in macrohages: studies by fluorescence spectroscopy. J Cell Biol 89:645–652

Ghosh SK, Samuelson J (1997) Involvement of p21racA, phosphoinositide 3-kinase, and vacuolar ATPase in phagocytosis of bacteria and erythrocytes by *Entamoeba histolytica*: suggestive evidence for coincidental evolution of amebic invasiveness. Infect Immun 65:4243–4249

Ghosh SK, Field J, Frisardi M, Rosenthal B, Mai Z, Rogers R, Samuelson J (1999) Chitinase secretion by encysting *Entamoeba invadens* and Transfected *Entamoeba histolytica* Trophozoites: localization of secretory vesicles, endoplasmic reticulum, and Golgi apparatus. Infect Immun 67:3073–3081

Ghosh S, Field J, Rogers R, Hickman M, Samuelson J (2000) The *Entamoeba histolytica* mitochondrion-derived organelle (crypton) contains double-stranded DNA and appears to be bound by a double membrane. Infect Immun 68:4319–4322

Gillooly DJ, Simonsen A, Stenmark H (2001) Cellular functions of phosphatidylinositol 3-phosphate and FYVE domain proteins. Biochem J 355:249–258

Girard-Misguich F, Sachse M, Santi-Rocca J, Guillén N (2008) The endoplasmic reticulum chaperone calreticulin is recruited to the uropod during capping of surface receptors in *Entamoeba histolytica*. Mol Biochem Parasitol 157:236–240

Goldberg AV, Molik S, Tsaousis AD, Neumann K, Kuhnke G, Delbac F, Vivares CP, Hirt RP, Lill R, Embley TM (2008) Localization and functionality of microsporidian iron-sulphur cluster assembly proteins. Nature 452:624–628

Groves E, Dart AE, Covarelli V, Caron E (2008) Molecular mechanisms of phagocytic uptake in mammalian cells. Cell Mol Life Sci 65:1957–1976

Hackstein JHP, Tjaden J, Huynen M (2006) Mitochondria. hydrogenosomes and mitosomes: products of evolutionary tinkering. Curr Genet 50:225–245

Hartman DJ, Hoogenraad NJ, Condron R, Hoj PB (1992) Identification of a mammalian 10-kDa heat shock protein, a mitochondrial chaperonin 10 homologue essential for assisted folding of trimeric ornithine transcarbamoylase *in vitro*. PNAS 89:3394–3398

Hauri H, Kappeler F, Andersson H, Appenzeller C (2000) ERGIC-53 and traffic in the secretory pathway. J Cell Sci 113:587–596

Helenius A, Aebi M (2004) Roles of N-linked glycans in the Endoplasmic Reticulum. Annu Rev Biochem 73:1019–1049

Hirschberg CB, Robbins PW, Abeijon C (1998) Transporters of nucleotide sugars, ATP, and nucleotide sulfate in the endoplasmic reticulum and Golgi apparatus. Annu Rev Biochem 67:49–69

Hume AN, Collinson LM, Rapak A, Gomes AQ, Hopkins CR, Seabra MC (2001) Rab27a regulates the peripheral distribution of melanosomes in melanocytes. J Cell Biol 152:795–808

Huston CD, Houpt ER, Mann BJ, Hahn CS, Petri WA Jr (2000) Caspase 3-dependent killing of host cells by the parasite *Entamoeba histolytica*. Cell Microbiol 2:617–625

Huston CD, Boettner DR, Miller-Sims V, Petre WA Jr (2003) Apoptotic killing and phagocytosis of host cells by the parasite *Entamoeba histolytica*. Infect Immun 71:964–972

Jensen MS, Bainton DF (1973) Temporal changes in pH within the phagocytic vacuole of the polymorphonuclear neutrophilic leukocyte. J Cell Biol 56:379–388

Kane P (2006) The where, when, and how of organelle acidification by the yeast vacuolar H^+-ATPase. Microbiol Mol Biol Rev 70:177–191

Katiyar SK, Edlind T (1996) *Entamoeba histolytica* encodes a highly divergent β-tubulin. J Eukaryot Microbiol 43:31–34

Labruyère E, Zimmer C, Galy V, Olivo-Marin JC, Guillén N (2003) EhPAK, a member of the p21-activated kinase family, is involved in the control of *Entamoeba histolytica* migration and phagocytosis. J Cell Sci 116:61–71

Lal K, Field MC, Carlton JM, Warwicker J, Hirt RP (2005) Identification of a very large Rab GTPase family in the parasitic protozoan *Trichomonas vagnalis*. Mol Biochem Parasitol 143:226–235

Lang F, Gulbins E, Szabo I, Lepple-Wienhues A, Huber SM, Duranton C, Lang KS, Lang PA, Wieder T (2004) Cell volume and the regulation of apoptotic cell death. J Mol Recognit 17:473–480

Leippe M (1997) Amoebapores. Parasitol Today 13:178–183

Leippe M (1999) Antimicrobial and cytolytic polypeptides of amoeboid protozoa–effector molecules of primitive phagocytes. Dev Comp Immunol 23:267–279

Leippe M, Andra J, Nickel R, Rannich E, Muller-Eberhard HJ (1994) Amoebapores, a family of membranolytic peptides from cytoplasmic granules of *Entamoeba histolytica*: isolation, primary structure, and pore formation in bacterial cytoplasmic membranes. Mol Microbiol 15:895–904

Lejeune A, Gicquaud C (1987) Evidence for two mechanisms of human erythrocyte endocytosis by *Entamoeba histolytica*-like amoebae (Laredo strain). Biol Cell 59:239–245

Lejeune A, Gicquaud C (1992) Target cell deformability determines the type of phagocytic mechanism used by *Entamoeba histolytica*-like, Laredo strain. Biol Cell 74:211–216

León-Avila G, Tovar J (2004) Mitosomes of *Entamoeba histolytica* abundant mitochondrial-related remnant organelles that lack a detectable organellar genome. Microbiology 150:1245–1250

León-Sicairos N, Reyes-López M, Canizalez-Román A, Bermúdez-Cruz RM, Serrano-Luna J, Arroyo R, de la Garza M (2005) Human hololactoferrin: endocytosis and use as an iron source by the parasite *Entamoeba histolytica*. Microbiology 151:3859–3871

Libros-Ziv P, Villalobo E, Mirelman D (2005) *E. histolytica*: identification and characterization of an N-ethylmaleimide sensitive fusion protein homologue. Exp Parasitol 110:276–279

Lill R, Mühlenhoff U (2006) Iron-sulphur protein biogenesis in eukaryotes: components and mechanisms. Annu Rev Cell Dev Biol 22:457–486

Lill R, Mühlenhoff U (2008) Maturation of iron-sulfur proteins in eukaryotes: mechanisms, connected processes, and diseases. Annu Rev Biochem 77:669–700

Loftus B, Anderson I, Davies R, Alsmark UC, Samuelson J, Amedeo P, Roncaglia P, Berriman M, Hirt RP, Mann BJ, Nozaki T, Suh B, Pop M, Duchene M, Ackers J, Tannich E, Leippe M, Hofer M, Bruchhaus I, Willhoeft U, Bhattacharya A, Chillingworth T, Churcher C, Hance Z, Harris B, Harris D, Jagels K, Moule S, Mungall K, Ormond D, Squares R, Whitehead S, Quail MA, Rabbinowitsch E, Norbertczak H, Price C, Wang Z, Guillén N, Gilchrist C, Stroup SE, Bhattacharya S, Lohia A, Foster PG, Sicheritz-Ponten T, Weber C, Singh U, Mukherjee C, El-Sayed NM, Petri WA Jr, Clark CG, Embley TM, Barrell B, Fraser CM, Hall N (2005) The genome of the protist parasite *Entamoeba histolytica*. Nature 433:865–868

Lohia A (2003) The cell cycle of *Entamoeba histolytica*. Mol Cell Biochem 253:217–222

Lohia A, Mukherjee C, Majumder S, Dastidar PG (2007) Genome re-duplication and irregular segregation occur during the cell cycle of *Entamoeba histolytica*. Biosci Rep 27:373–384

López-Soto F, González-Robles A, Salazar-Villatoro L, León-Sicairos N, Piña-Vázquez C, Salazar EP, de la Garza M (2009) *Entamoeba histolytica* uses ferritin as an iron source and internalises this protein by means of clathrin-coated vesicles. Int J Parasitol 39:417–426

Lucht E, Evengård B, Skott J, Pehrson P, Nord CE (1998) *Entamoeba gingivalis* in human immunodeficiency virus type 1-infected patients with periodontal disease. Clin Infect Dis 27:471–473

Lushbaugh WB, Kairalla AB, Loadholt CB, Pittman FE (1978) Effect of hamster liver passage on the virulence of axenically cultivated *Entamoeba histolytica*. Am J Trop Med Hyg 27:248–254

Luzio JP, Pryor PR, Bright NA (2007) Lysosomes: fusion and function. Nat Rev Mol Cell Biol 8:622–632

Manning-Cela R, Marquez C, Franco E, Talamas-Rahana P, Meza I (2003) BFA-sensitive and insensitive exocytic pathways in *Entamoeba histolytica* trophozoites: their relationship to pathogenesis. Cell Microbiol 5:921–932

Marion S, Wilhelm C, Voigt H, Bacri C, Guillén N (2004) Overexpression of myosin IB in living *Entamoeba histolytica* enhances cytoplasm viscosity and reduces phagocytosis. J Cell Sci 117:3271–3279

Marion S, Laurent C, Guillén N (2005) Signalization and cytoskeleton activity through myosin IB during the early steps of phagocytosis in *Entamoeba histolytica*: a proteomic approach. Cell Microbiol 7:1504–1518

Martin J (1997) Molecular chaperones and mitochondrial protein folding. J Bioenerg Biomembr 29:35–43

Matouschek A, Pfanner N, Voos W (2000) Protein unfolding by mitochondria. The Hsp70 import motor. EMBO Rep 1:404–410

Mattern CF, Keister DB, Natovitz PC (1982) Virulence of *Entamoeba histolytica* upon continuous axenic cultivation. Archivos de Investigacion Nedica 13:185–190

Mayer A (2001) What drives membrane fusion in eukaryotes? Trends Biochem Sci 26:717–723

Mazzuco A, Benchimol M, De Souza W (1997) Endoplasmic reticulum and Golgi-like elements in *Entamoeba*. Micron 28:241–247

McGugan GC Jr, Temesvari LA (2003) Characterization of a Rab11-like GTPase, EhRab11, of *Entamoeba histolytica*. Mol Biochem Parasitol 129:137–146

McNew JA, Parlati F, Fukuda R, Johnston RJ, Paz K, Paument F, Söllner TH, Rothman JE (2000) Compartmental specificity of cellular membrane fusion encoded in SNARE proteins. Nature 407:153–159

Meléndez-Hernández MG, Barrios MLL, Orozco E, Luna-Arias JP (2008) The Vacuolar ATPase from *Entamoeba histolytica*: Molecular cloning of the gene encoding for the B subunit and subcellular localization of the protein. BMC Microbiol 8:235

Mellman I, Warren G (2000) The road taken: past and future foundations of membrane traffic. Cell 100:99–112

Meza I, Clarke M (2004) Dynamics of endocytic traffic of *Entamoeba histolytica* revealed by confocal microscopy and flow cytometry. Cell Motil Cytoskeleton 59:215–226

Meza I, Talamás-Rohana P, Vargas MA (2006) The cytoskeleton of *Entamoeba histolytica*: structure, function, and regulation by signaling pathways. Arch Med Res 37:234–243

Mitra BN, Yasuda T, Kobayashi S, Saito-Nakano Y, Nozaki T (2005) Differences in morphology of phagosomes and kinetics of acidification and degradation in phagosomes between the pathogenic *Entamoeba histolytica* and the non-pathogenic *Entamoeba dispar*. Cell Motil Cytoskeleton 62:84–99

Mitra BN, Kobayashi S, Saito-Nakano Y, Nozaki T (2006) *Entamoeba histolytica*: differences in phagosome acidification and degradation between attenuated and virulent strains. Exp Parasitol 114:57–61

Mitra BN, Saito-Nakano Y, Nakada-Tsukui K, Sato D, Nozaki T (2007) Rab11B small GTPase regulates secretion of cysteine proteases in the enteric protozoan parasite *Entamoeba histolytica*. Cell Microbiol 9:2112–2125

Moremen KW, Molinari M (2006) N-linked glycan recognition and processing: the molecular basis of endoplasmic reticulum quality control. Curr Opin Struct Biol 16:592–599

Mukherjee C, Clark CG, Lohia A (2008) *Entamoeba* shows reversible variation in ploidy under different growth conditions and between life cycle phases. PLoS Negl Trop Dis 2:e281

Mukherjee C, Majumder S, Lohia A (2009) Inter-cellular variation in DNA content of *Entamoeba histolytica* originates from temporal and spatial uncoupling of cytokinesis from the nuclear cycle. PLoS Negl Trop Dis 3:e409

Nakada-Tsukui K, Saito-Nakano Y, Ali V, Nozaki T (2005) A retromerlike complex is a novel Rab7 effector that is involved in the transport of the virulence factor cysteine protease in the enteric protozoan parasite *Entamoeba histolytica*. Mol Biol Cell 16:5294–5303

Nakada-Tsukui K, Okada H, Mitra BN, Nozaki T (2009) Phosphatidylinositol-phosphates mediate cytoskeletal reorganization during phagocytosis via a unique modular protein consisting of RhoGEF/DH and FYVE domains in the parasitic protozoon Entamoeba histolytica. Cell Microbiol 11:1471–1491

Nickel W, Weber T, McNew JA, Parlati F, Söllner TH, Rothman JE (1999) Content mixing and membrane integrity during membrane fusion driven by pairing of isolated v-SNAREs and t-SNAREs. PNAS 96:12571–12576

Nishi T, Forgac M (2002) The vacuolar (H^+)-ATPases-natures' most versatile proton pumps. Nat Rev Mol Cell Biol 3:94–103

Nothwehr SF, Bruinsma P, Strawn LA (1999) Distinct domains within Vps35p mediate the retrieval of two different cargo proteins from the yeast prevacuolar/endosomal compartment. Mol Biol Cell 10:875–890

Novick P, Zerial M (1997) The diversity of Rab proteins in vesicle transport. Curr Opin Cell Biol 9:496–504

Nozaki T, Nakada-Tsukui K (2006) Membrane trafficking as a virulence mechanisms of the protozoan parasite *Entamoeba histolytica*. Parasitol Res 98:179–183

Ogden CA, de Cathelineau A, Hoffmann PR, Bratton D, Ghebrehiwet B, Fadok VA, Henson PM (2001) C1q and mannose binding lectin engagement of cell surface calreticulin and CD91 initiates macropinocytosis and uptake of apoptotic cells. J Exp Med 194:781–795

Okada M, Huston CD, Mann BJ, Petri WA Jr, Kita K, Nozaki T (2005) Proteomic analysis of phagocytosis in the enteric protozoan parasite *Entamoeba histolytica*. Eukaryot Cell 4:827–831

Olivos-García A, Saavedra E, Martínez RD (2009) Molecular nature of virulence in *Entamoeba histolytica*. Infect Genet Evol. doi:10.1016/j.meegid.2009.04.005

Olkkonen VM, Stenmark H (1997) Role of Rab GTPases in membrane traffic. Int Rev Cytol 176:1–85

Orozco E, Guarneros G, Martinez-Palomo A, Sanchez T (1983) Entamoeba histolytica. Phagocytosis as a virulence factor. J Exp Med 158:1511–1521

Padilla-Vaca F, Ankri S, Bracha R, Koole LA, Mirelman D (1999) Down regulation of *Entamoeba histolytica* virulence by monoxenic cultivation with *Escherichia coli* O55 is related to a decrease in expression of the light (35-kilodalton) subunit of the Gal/GalNAc lectin. Infect Immun 67:2096–2102

Paul P, Kamisaka Y, Marks DL, Pagano RE (1996) Purification and characterization of UDP-glucose:ceramide glucosyltransferase from rat liver Golgi membranes. J Biol Chem 271:2287–2293

Petri WA, Haque R, Mann BJ (2002) The bittersweet interface of parasite and host: lectin-carbohydrate interactions during human invasion by the parasite *Entamoeba histolytica*. Annu Rev Microbiol 56:34–64

Phillips BP (1973) *Entamoeba histolytica*: concurrent irreversible loss of infectivity-pathogenicity and encystment potential after prolonged maintenance in axenic culture in vitro. Exp Parasitol 34:163–167

Picazarri K, Luna-Arias JP, Carrillo E, Orozco E, Rodriguez MA (2005) *Entamoeba histolytica*: Identification of EhGPCR-1, a novel putative G protein-coupled receptor that binds to EhRabB. Exp Parasitol 110:253–258

Pimenta PF, Diamond LS, Mirelman D (2002) *Entamoeba histolytica* Schaudinn, 1903 and *Entamoeba dispar* Brumpt, 1925: differences in their cell surfaces and in the bacteria-containing vacuoles. J Eukaryot Microbiol 49:209–219

Powell RR, Welter BH, Hwu R, Bowersox B, Temesvari LA (2006) FYVE-finger domains, phosphatidylinositol 3-phosphate biosensors associate with phagosomes but not fluid filled endosomes in *Entamoeba histolytica*. Exp Parasitol 112:221–231

Putignani L, Tait A, Smith HV, Horner D, Tovar J, Tetley L, Wastling JM (2004) Characterization of a mitochondrion-like organelle in *Cryptosporidium parvum*. Parasitology 129:1–18

Que X, Reed SL (2000) Cysteine proteinases and the pathogensis of amebiasis. Clin Microbiol Rev 13:196–206

Que X, Brinen SL, Perkins P, Herdman S, Hirata K, Torian BE, Rubin H, McKerrow JH, Reed SL (2002) Cysteine proteinases from distinct cellular compartments are recruited to phagocytic vesicles by *Entamoeba histolytica*. Mol Biochem Parasitol 119:23–32

Rabinowitz S, Horstmann H, Gordon S, Griffiths G (1992) Immunocytochemical characterization of the endocytic and phagolysomal compartments in peritoneal macrophages. J Cell Biol 116:95–112

Raiborg C, Bremnes B, Mehlum A, Gillooly DJ, D'Arrigo A, Stang E, Stenmark HJ (2001) FYVE and coiled-coil domains determine the specific localisation of Hrs to early endosomes. J Cell Sci 114:2255–2263

Ravdin JI (1988) Amebiasis: human infection by *Entamoeba histolytica*. Wiley, New York

Ray SS, Gangopadhyay SS, Pande G, Samuelson J, Lohia A (1997) Primary structure of *Entamoeba histolytica* γ-tubulin and localization of amoebic microtubule organizing center. Mol Biochem Parasitol 90:331–336

Reddy JV, Seaman MN (2001) Vps26p, a component of retromer, directs the interactions of Vps35p in endosome-to-Golgi retrieval. Mol Biol Cell 12:3242–3256

Reyes-López M, Serrano-Luna JJ, Negrete-Abascal E, León-Sicairos N, Guerrero-Barrera AL, de la Garza M (2001) *Entamoeba histolytica*: transferrin binding proteins. Exp Parasitol 99:132–140

Riordan CE, Ault JG, Langreth SG, Keithly JS (2003) *Cryptosporidium parvum* Cpn6 targets a relict organelle. Curr Genet 44:138–147

Rodriguez MA, Orozco E (1986) Isolation and characterization of phagocytosis- and virulence-deficient mutants of *Entamoeba histolytica*. J Infect Dis 154:27–32

Rodriguez MA, Garcia-Perez RM, Garcia-Rivera G, Lopez-Reyes I, Mendoz L, Ortiz-Navarrete V, Orozco E (2000) An *Entamoeba histolytica* Rab-like encoding gene and protein: function and cellular location. Mol Biochem Parasitol 108:199–206

Roy D, Lohia A (2004) Sequence divergence of Entamoeba histolytica tubulin is responsible for its altered tertiary structure. Biochem Biophys Res Commun 319:1010–1016

Rupper AC, Rodriguez-Paris JM, Grove BD, Cardellit JA (2001) p110-related PI 3-kinases regulate phagosome-phagosome fusion and phagosomal pH through a PKB/Akt dependent pathway in *Dictyostelium*. J Cell Sci 114:1283–1295

Saito-Nakano Y, Nakazawa M, Shigeta Y, Takeuchi T, Nozaki T (2001) Identification and characterization of genes encoding novel Rab proteins from *Entamoeba histolytica*. Mol Biochem Parasitol 116:219–222

Saito-Nakano Y, Yasuda T, Nakada-Tsukui K, Leippe M, Nozaki T (2004) Rab5-associated vacuoles play a unique role in phagocytosis of the enteric protozoan parasite *Entamoeba histolytica*. J Biol Chem 279:49497–49507

Saito-Nakano Y, Loftus BJ, Hall N, Nozaki T (2005) The diversity of Rab GTPases in *Entamoeba histolytica*. Exp Parasitol 110:244–252

Saito-Nakano Y, Mitra BN, Nakada-Tsukui K, Sato D, Nozaki T (2007) Two Rab7 isotypes, EhRab7A and EhRab7B, play distinct roles in biogenesis of lysosomes and phagosomes in the enteric protozoan parasite *Entamoeba histolytica*. Cell Microbiol 9:1796–1808

Sanchez MA, Peattie DA, Wirth D, Orozco E (1994) Cloning, genomic organization and transcription of the *Entamoeba histolytica* α-tubulin gene. Gene 146:239–244

Sanchez-Lopez R, Gama-Castro S, Ramos MA, Merino E, Lizardi PM, Alagón A (1998) Cloning and expression of the *Entamoeba histolytica* ERD2 gene. Mol Biochem Parasitol 92:355–359

Sanderfoot AA, Assaad FF, Raikhel NV (2000) The *Arabidopsis* genome. An abundance of soluble N-ethylmaleimide-sensitive factor adaptor protein receptors. Plant Physiol 124:1558–1569

Schulter OM, Khvotchev M, Jahn R, Sudhof TC (2002) Localization versus function of Rab3 proteins. Evidence for a common regulatory role in controlling fusion. J Biol Chem 277 (43):40919–40929

Scott CC, Dobson W, Botelho RJ, Coady-Osberg N, Chavrier P, Knecht DA, Heath C, Stahl P, Grinstein S (2005) Phosphatidylinositol-4,5-bisphosphate hydrolysis directs actin remodeling during phagocytosis. J Cell Biol 169:139–149

Seaman MN (2004) Cargo-selective endosomal sorting for retrieval to the Golgi requires retromer. J Cell Biol 165:111–122

Seaman NM (2005) Recycle your receptors with retromer. Trends Cell Biol 15:68–75

Seigneur M, Mounier J, Prevost MC, Guillén N (2005) A lysine- and glutamic acid-rich protein, KERP1, from Entamoeba histolytica binds to human enterocytes. Cell Microbiol 7:569–579

Serrano-Luna JJ, Negrete E, Reyes M, de la Garza M (1998) *Entamoeba histolytica* HM1:IMSS: hemoglobin-degrading neutral cysteine proteases. Exp Parasitol 89:71–77

Silverstein SC, Steinman RM, Cohn ZA (1977) Endocytosis. Annu Rev Biochem 46:669–722

Singer-Krüger B, Stenmark H, Düsterhöft A, Philippsen P, Yoo JS, Gallwitz D, Zerial M (1994) Role of three rab5-like GTPases, Ypt51p, Ypt52p, and Ypt53p, in the endocytic and vacuolar protein sorting pathways of yeast. J Cell Biol 125:283–298

Slapeta J, Keithly JS (2004) *Cryptosporidium parvum* mitochondrial-type HSP70 targets homologous and heterologous mitochondria. Eukaryot Cell 3:483–494
Stanley SL Jr (2003) Amoebiasis. Lancet 361:1025–1034
Stenmark H, Aasland R (1999) FYVE-finger proteins-effectors of an inositol lipid. J Cell Sci 112:4175–4183
Stenmark H, Olkkonen VM (2001) The Rab GTPase family. Genome Biol 2:REVIEWS3007
Stenmark H, Parton RG, Steele-Mortimer O, Lütcke A, Gruenberg J, Zerial M (1994) Inhibition of rab5 GTPase activity stimulates membrane fusion in endocytosis. EMBO J 13:1287–1296
Stenmark H, Aasland R, Driscoll PC (2002) The phosphatidylinositol 3-phosphate-binding FYVE finger. FEBS Lett 513:77–84
Stinchcombe JC, Barral DC, Mules EH, Booth S, Hume AN, Machesky LM, Seabra MC, Griffiths GM (2001) Rab27a is required for regulated secretion in cytotoxic T lymphocytes. J Cell Biol 152:825–834
Sutak R, Dolezal P, Fiumera HL, Hrdy I, Dancis A, Delgadillo-Correa M, Johnson PJ, Müller M, Tachezy J (2004) Mitochondrial-type assembly of FeS centers in the hydrogenosomes of the amitochondriate eukaryote *Trichomonas vaginalis*. PNAS 101:10368–10373
Tamaki TH, Yamashina S (2002) Structural integrity of the Golgi stack is essential for normal secretory functions of rat parotid acinar cells: effects of brefeldin A and okadaic acid. J Histochem Cytochem 50:1611–1623
Teixeira JE, Huston CD (2008) Participation of the serine-rich *Entamoeba histolytica* protein in amebic phagocytosis of apoptotic host cells. Infect Immun 76:959–966
Teixeira JE, Heron BT, Huston CD (2008) C1q- and Collectin-dependent phagocytosis of apoptotic host cells by the intestinal protozoan *Entamoeba histolytica*. J Infect Dis 198:1062–1070
Temesvari LA, Harris EN, Stanley SL, Cardelli JA (1999) Early and late endosomal compartments of *Entamoeba histolytica* are enriched in cysteine proteases acid phosphatase and several Ras-related Rab GTPases. Mol Biochem Parasitol 103:225–241
Tovar J, Fischer A, Clark CG (1999) The mitosome, a novel organelle related to mitochondria in the amitochondrial parasite *Entamoeba histolytica*. Mol Microbiol 32:1013–1021
Tovar J, León-Avila G, Sánchez LB, Sutak R, Tachezy J, van der Giezen M, Hernández M, Muller M, Lucocq JM (2003) Mitochondrial remnant organelles of *Giardia* function in iron-sulfur protein maturation. Nature 426:172–176
Tovar J, Cox SSE, van der Giezen M (2007) A mitosome purification protocol based on percoll density gradients and its use in validating the mitosomal nature of *E. histolytica* mitochondrial Hsp70. Methods Mol Biol 390:167–177
Trombetta ES, Parodi AJ (2003) Quality control and protein folding in the secretory pathway. Annu Rev Cell Dev Biol 19:649–676
Tsaousis AD, Kunji ER, Goldberg AV, Lucocq JM, Hirt RP, Embley TM (2008) A novel route for ATP acquisition by the remnant mitochondria of *Encephalitozoon cuniculi*. Nature 453:553–556
Ueda T, Nakano A (2002) Vesicular traffic: an integral part of plant life. Curr Opin Plant Biol 5:513–517
Ungermann C, Langosch D (2005) Functions of SNAREs in intracellular membrane fusion and lipid bilayer mixing. J Cell Sci 118:3819–3828
Ungermann C, Sato K, Wickner W (1998) Defining the functions of *trans*-SNARE pairs. Nature 365:543–548
Ungermann C, Price A, Wickner W (2000) A new role for a SNARE protein as a regulator of the Ypt7/Rab-dependent stage of docking. PNAS 97:8889–8891
van der Giezen M, Cox S, Tovar J (2004) The iron-sulfur cluster assembly genes iscS and iscU of *Entamoeba histolytica* were acquired by horizontal gene transfer. BMC Evol Biol 4:7
van der Giezen M, León-Avila G, Tovar J (2005) Characterization of chaperonin 10 (Cpn10) from the intestinal human pathogen *Entamoeba histolytica*. Microbiology 151:3107–3115
Vats D, Vishwakarma RA, Bhattacharya S, Bhattacharya A (2005) Reduction of cell surface glycosylphosphatidylinositol conjugates in *Entamoeba histolytica* by antisense blocking of

E. histolytica GlcNAc-phosphatidylinositol deacetylase expression: effect on cell proliferation, endocytosis, and adhesion to target cells. Infect Immun 73:8381–8392

Vayssié L, Vargas M, Weber C, Guillén N (2004) Double-stranded RNA mediates homology-dependent gene silencing of γ-tubulin in the human parasite *Entamoeba histolytica*. Mol Biochem Parasitol 138:21–28

Vieira OV, Harrison RE, Scott CC, Stenmark H, Alexander D, Liu J, Gruenberg J, Schreiber AD, Grinstein S (2004) Acquisition of Hrs, an essential component of phagosomal maturation, is impaired by mycobacteria. Mol Cell Biol 24:4593–4604

Voellmy R, Boellmann F (2007) Chaperone regulation of the heat shock protein response. Adv Exp Med Biol 594:89–99

Voigt H, Guillén N (1999) New insights into the role of the cytoskeleton in phagocytosis of *Entamoeba histolytica*. Cell Microbiol 1:195–203

Voigt H, Olivo JC, Sansonetti P, Guillén N (1999) Myosin IB from Entamoeba histolytica is involved in phagocytosis of human erythrocytes. J Cell Sci 112:1191–1201

Weber C, Blazquez S, Marion S, Ausseur C, Vats D, Krzeminski M, Rigothier MC, Maroun RC, Bhattacharya A, Guillén N (2008) Bioinformatics and functional analysis of an *Entamoeba histolytica* mannosyltransferase necessary for parasite complement resistance and hepatical infection. PLoS Negl Trop Dis 2:e165

Welter BH, Laughlin RC, Temesvari LA (2002) Characterization of a Rab7-like GTPase, EhRab7: a marker for the early stages of endocytosis in *Entamoeba histolytica*. Mol Biochem Parasitol 121:254–264

Welter BH, Powell RR, Leo M, Smith CM, Temesvari LA (2005) A unique Rab GTPase, EhRabA, is involved in motility and polarization of *Entamoeba histolytica* cells. Mol Biochem Parasitol 140:161–173

Welter BH, Powell RR, Laughlin RC, McGugan GC, Bonner M, King A, Temesvari LA (2006) *Entamoeba histolytica*: comparison of the role of receptors and filamentous actin among various endocytic processes. Exp Parasitol 113:91–99

Williams BA, Hirt RP, Lucocq JM, Embley TM (2002) A mitochondrial remnant in the microsporidian *Trachipleistophora hominis*. Nature 418:865–869

Yeung T, Ozdamar B, Paroutis P, Grinstein S (2006) Lipid metabolism and dynamics during phagocytosis. Curr Opin Cell Biol 18:429–437

Yi Y, Samuelson J (1994) Primary structure of the *Entamoeba histolytica* gene (EhvmaI) encoding the catalytic peptide of a putative vacuolar membrane proton-transporting ATPase (V-ATPase). Mol Biochem Parasitol 66:165–169

Zamarripa-Morales S, Villagómez-Castro JC, Calvo-Méndez C, Flores-Carreón A, López-Romero E (1999) *Entamoeba histolytica*: identification and properties of membrane-bound and soluble alpha-glucosidases. Exp Parasitol 93:109–115

Zerial M, McBride H (2001) Rab protein as membrane organizers. Nat Rev Mol Cell Biol 2:107–117

Secretory Organelles in Apicomplexa

Jean François Dubremetz

Contents

Abstract Apicomplexa interactions with their host are exquisitely dependent on unique secretory organelles that exocytose their contents during gliding motility, attachment, and host cell invasion. Recent developments boosted by molecular genetics and high throughput methods have unraveled a number of biological processes and allowed a better understanding of the role of these organelles in Apicomplexa biology. The major contribution of microneme proteins to the gliding motility, the cooperation of microneme and rhoptry neck proteins to the moving junction during invasion, and the major role played by rhoptry bulb proteins in

J.F. Dubremetz
UMR CNRS 5235, Bt 24, CC 107, Université de Montpellier 2, Place Eugène Bataillon, Montpellier cedex 05, 34095, France
e-mail: dubremet@univ-montp2.fr

W. de Souza (ed.), *Structures and Organelles in Pathogenic Protists*, Microbiology Monographs 17, DOI 10.1007/978-3-642-12863-9_8,

© Springer-Verlag Berlin Heidelberg 2010

high-jacking the host cell are the basis of new paradigms that underline the unique characteristics of intracellular parasitism developed by Apicomplexa.

1 Introduction

The phylum Apicomplexa comprises a vast array of protozoan parasites, likely deriving from free living marine alveolates that have gradually evolved toward intracellular parasitism by acquiring a sophisticated secretory system associated with a unique gliding motility apparatus, both involved in host cell invasion. These secretory organelles have been designated as rhoptries, micronemes and dense granules (Fig. 1). They are found in all Apicomplexa invasive stages with very few exceptions, and are among the most specific criteria of recognition of the members of the phylum. This chapter will deal essentially with these structures. Other secretory organelles are found in Apicomplexa, such as in female gametes, where they are involved in producing the highly resistant shell that protects the zygote during the extra-host part of the life cycle. They are not found in all members of the phylum however and will not be dealt with here.

Rhoptries, dense granules and micronemes were first described by electron microscopy and their morphological characteristics already allowed unambiguous identification across the phylum. The molecular characterization of their contents has been extensively studied more recently and has demonstrated both conserved and divergent molecular species, allowing for a better understanding of the basic, conserved functions shared by all members of the phylum, and specific differentiations developed by individual groups to serve their needs in their respective niches. These organelles cannot be introduced efficiently without mentioning two other Apicomplexa zoites specific structures, namely the inner membrane complex and

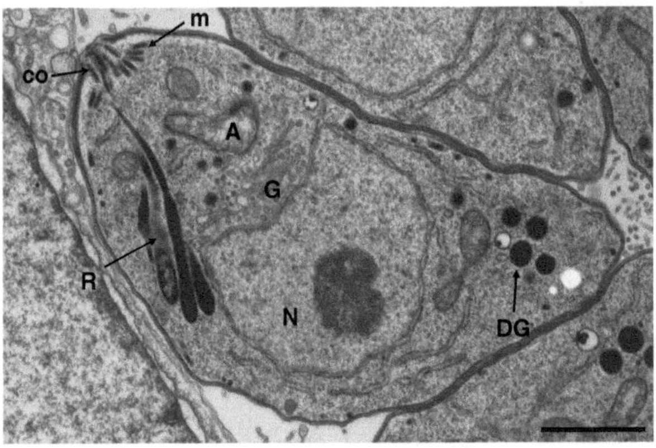

Fig. 1 *Toxoplasma gondii* tachyzoite showing the three types of secretory organelles: rhoptries (R), micronemes (m) and dense granules (DG). *co* conoid, *A* apicoplast, *N* nucleus. *Bar* = 1 μm

the apical cytoskeletal complex, both of which being functionally tightly associated with the exocytic organelles. The inner membrane complex is a system of contiguous flattened vesicles underlying the plasmalemma of the zoite except at the very apical end, constituting a three layered pellicle. The apical cytoskeletal complex comprises an array of longitudinal subpellicular microtubules anchored on an apical ring and that run longitudinally under the parasite pellicle; in some cases, the apical ring encircle a short hollow cylinder made of twisted tubulin fibers named the conoid.

2 Description

The organelles are too small to be seen by photonic microscopy and were therefore only discovered with the introduction of electron microscopy, starting in the late fifties; it took another 30 years before the molecular contents started being unraveled.

2.1 Identification

In the case of the rhoptries, these were given various names with the term "paired organelle" being extensively used because two club-shaped structures had been initially identified in *Plasmodium* sporozoites (Garnham et al. 1961). The name "rhoptries" was coined later by Senaud (1967) and eventually became the accepted term. The first identification of another apical organelle in addition to the rhoptries was again made by Garnham. Initially the short rod shaped structures were thought to be thin sections through long convoluted tubules, hence the name initially proposed (Garnham et al. 1961). They were later identified as being individual rod-shaped structures by Senaud (1967). Jacobs (1967) proposed the term micronemes to represent these rod-like organelles. In a review of the coccidian literature, Scholtyseck and Mehlhorn (1970) proposed retaining the term rhoptry for the club-like organelles and microneme for the rod-like organelles. The idea that rhoptries could be involved in invasion was first proposed by Gustafson et al. (1954) and Ludvik was the first researcher to propose a secretory function for these membrane-bound organelles (Ludvik 1956). The dense granules were first mentioned by Ogino and Yoneda (1966). Vivier described them as dense globules and demonstrated that they were not rhoptry sections, but spherical inclusions, by serial sectioning (Vivier and Petitprez 1972). Their isolation by subcellular fractionation from *Sarcocystis tenella* cystozoites provided the definitive evidence of these being a third set of specific secretory organelles in Apicomplexa zoites (Dubremetz and Dissous 1980). Last, a new set of secretory organelles, which might be a specific subset of dense granules, has been described as exonemes in *Plasmodium falciparum* merozoites (Yeoh et al. 2007).

2.2 Molecular Characterization

The molecular characterization of these organelles has been a long task which is far from being over, and that was dependent upon technical improvements in the characterization of proteins through either immunochemical, molecular genetic and/or biochemistry procedures.

2.2.1 Methods

The first step was to isolate these structures by subcellular fractionation which was obtained for the first time for micronemes and dense granules of *S. tenella* (Dubremetz and Dissous 1980); rhoptries were first isolated in *Eimeria necatrix* (Dubremetz et al. 1989) and soon later in *T. gondii* (Leriche and Dubremetz 1991). Then the corresponding organelles were isolated from *E. tenella* and *Plasmodium* spp. merozoites and ookinetes [reviewed by Blackman and Bannister (2001)]. This allowed the generation of monoclonal antibodies in addition to those obtained serendipidously through immunizing mice with crude parasite fractions. Another efficient way was also through expression libraries and affinity purification of specific antibodies from crude polyclonal sera.

A major step was achieved when mass spectroscopic analysis of proteins combined with whole genome sequencing allowed massive characterization of organelle fractions. The first was obtained from *T. gondii* rhoptries (Bradley et al. 2005), followed by *E. tenella* sporozoites (Bromley et al. 2003) and *Plasmodium* ookinete micronemes (Lal et al. 2009). These methods are highly efficient in generating large amounts of data, however, because subcellular fractions are inherently contaminated, each hit has to be individually confirmed by immunolocalization.

The lessons learned from these studies are that these organelles can be isolated intact, and are therefore bags filled up with proteins. In addition, in some cases, lipids have also been detected and analyzed (Foussard et al. 1991; Besteiro et al. 2008). Immunolocalization studies also showed that at least in rhoptries, two domains of distinct proteinic composition are found, one in the neck of the organelle, the other in the bulb shaped posterior part (Roger et al. 1988; Bradley et al. 2005): this later finding has proven highly fruitful on a functional point of view, as shown below.

2.2.2 Organelles Structure and Contents

The very characteristic shape of rhoptries and micronemes suggests strong structural constraints on their organization, of which we are entirely ignorant. If dense granules are clearly spherical bags needing no specific supporting structure, the elongated micronemes must carry a structural information and a molecular frame to maintain their shape in situ. This has been nicely shown in *Cryptosporidium* spp.

where a helical structure of unknown composition is found in the organelle by negative staining (Harris et al. 2003). The shape of rhoptries is even more complex, and yet nicely conserved across the phylum, meaning that it is undergoing a strong selective pressure and must be important for its function. Interestingly, upon exocytosis, the shape of the organelle is immediately lost (Nichols et al. 1983), suggesting that the shape is maintained by internal components, and not by an external skeleton.

Despite conserved morphological features, the organelle contents vary depending on the genera. Interestingly, micronemes and rhoptry neck proteins tend to be far more conserved across the phylum compared to the rhoptry bulb ones. This has been related to the respective functions of the proteins derived from these compartments. Indeed, schematically, microneme and rhoptry neck proteins are involved in the gliding motility, adhesion, invasion processes, highly conserved in Apicomplexa, whereas rhoptry bulb proteins would be more related to specific interactions with the host cell and therefore vary more depending on the niche occupied by the organism within the host.

Microneme proteins are dominated by polypeptides including motifs similar to known sugar or protein binding domains of higher eukaryotes. The most conserved protein is the TRAP (for thrombospondin related protein) (Robson et al. 1988) and its orthologs, shown to be involved in gliding motility, but others including EGF-like domains, integrin like domains, lectin domains, etc. are found in most Apicomplexa (Tomley and Soldati 2001). As shown below, this has been related to the role of micronemes in binding substrates. The characterization of the binding site of some of these proteins has been obtained at the atomic resolution level (Garnett et al. 2009). The reason why some organisms such as *T. gondii* have developed such an array of binding molecules in this organelle is not understood. In addition, micronemes contain proteases involved in pre and/or post exocytic processing of other microneme proteins (Miller et al. 2001; Dowse et al. 2005), and pore forming proteins (Ishino et al. 2004; Kadota et al. 2004; Kafsack et al. 2009) which are gaining more and more interest in being linked to egress and cell traversal.

Rhoptries have been shown to contain a vast array of proteins, with little conservation between Apicomplexa when considering the posterior bulgy part (those proteins are named ROPs in *T. gondii*, or RAPs or RHOPs in *Plasmodium*), whereas the ones found in the neck (named RONs) tend to be shared by most members of the phylum (Bradley et al. 2005). RONs have no orthologs out of the phylum, whereas some ROPs show structures and/or activities similar to known enzymes such as proteases, phosphatases or kinases, and probably also phospholipases. Under certain conditions, rhoptries have been shown to contain membrane like material (Stewart et al. 1985; Bannister et al. 1986), and subcellular fractionation has confirmed the presence of lipids in these organelles (Foussard et al. 1991).

When rhoptries and micronemes are almost universal in Apicomplexa (except with a few exceptions where rhoptries are missing such as *Plasmodium* ookinetes), the fate of dense granules is less clear. They are especially well represented in Sarcocystidae (*Sarcocystis* spp., *T. gondii*), but much less so in other groups and their exact importance remains to be demonstrated, needing more molecular details on their contents. They are best known in *Toxoplasma*, but most of them have no ortholog out of the Toxoplasmatidae and, when known, their functions are linked to the structure or the metabolic features of the parasitophorous vacuole (PV) (Mercier et al. 2005; Coppens et al. 2006). In *Plasmodium* spp., only two dense granule proteins have been described, one of which is RESA (for Ring infected erythrocyte surface antigen), which is actually translocated under the host cell surface after invasion (Culvenor et al. 1991).

3 Genesis

The schizogonic process characteristic of Apicomplexa leads to the production of infective zoites that are equipped with all the components of the invasion machinery; this is a complex morphogenetic process during which the secretory organelles are generated.

3.1 Morphological Studies

Electron microscope studies have shown very early that both micronemes and rhoptries were exclusively made during infective stages genesis, in close vicinity with the other components of the apical complex (i.e., conoid, inner membrane complex, subpellicular microtubules) during the last mitosis of the schizogonic process. Microneme genesis could not be tracked down, whereas rhoptries were shown to undergo a complex biogenetic process going through spherical vesicles (prerhoptries) before acquiring their definitive elongated shape. The pattern of zoite genesis suggests that rhoptries develop before micronemes but does not exclude overlap between the two. The definitive positioning of the organelles in the apical part of the zoites was also suggested from EM observation to be dependent on interaction with the subpellicular microtubules (Bannister et al. 2003). Concerning the dense granules, their biosynthetic process has not been unraveled.

3.2 Molecular Details

Once the protein contents of organelles started being identified, the intraparasite trafficking of organellar proteins could be tracked down and their fate could then be followed during the biogenesis of the organelles. Yet, the precise molecular details of rhoptry, microneme or dense granule genesis are still fairly elusive. One reason for this is that the classical endocytic–secretory machinery has not been fully identified in Apicomplexa, which makes the characterization of the vesicular intermediates difficult (Sheiner and Soldati-Favre 2008). Another reason is that zoite genesis features the quasi-simultaneous building of distinct compartments and the synchronization in vitro is not precise enough to follow the biosynthetic pattern.

Despite these drawbacks, the follow up of rhoptry and microneme proteins synthesis by IFA with specific probes during schizogony or endodyogeny also suggested successive building of the two organelles. In addition, like in many examples in higher eukaryotes, rhoptry and microneme proteins often undergo proteolytic processing en route from the golgi to the mature organelle, which remove a propeptide from the neosynthesized protein. Using antibodies generated against these prosequences provided the most convincing demonstration of the sequential trafficking of both organelles contents, meaning that there is no simultaneous transit of both sets of proteins in the secretory pathway and therefore no need to sort them specifically from one another to build the organelles (Besteiro et al. 2009). This does not answer the question of how the organelle specific proteins are sorted from others (and from dense granule proteins for example).

Works on *T. gondii* have suggested that rhoptry building could follow a rather classical endosomal route, supported in addition by the fact that prerhoptries and rhoptries are the only acidic compartment in these parasites (Shaw et al. 1998), and are also the site where the propeptide is cleaved. Concerning micronemes, no immature compartment is clearly identified, although again, the propeptide is only detected at the IFA level in a zone located between the post golgi and the apical end where mature micronemes are found. Trafficking from the golgi to mature organelles is believed to depend on interactions between cytosolic domains of transmembrane proteins and adaptor protein complexes (AP) involved in sorting into clathrin coated vesicles, the transmembrane proteins acting as escorter for soluble proteins (Baldi et al. 2000; Reiss et al. 2001). However, additional signals are found in propeptides or internal sequences, suggesting a multifactorial nature for the trafficking of rhoptry and microneme proteins [reviewed in Lebrun et al. (2007)].

Dense granule proteins are considered as being synthesized during all the replicative cycle (in *T. gondii* which is the only Apicomplexa for which clear data are available) and the path they follow before being packaged is unknown (again, no immature DG are known, but because there is no known processing for DG proteins, no probe exist that could be used to mark such a compartment).

4 Functions

Despite being almost exclusively devoted to invasion, the functions of the secretory organelles can be subdivided into several processes from gliding motility to highjacking the host cell.

4.1 Gliding Motility

In Apicomplexa zoites the amount of micronemes per cell can vary extensively from one genus to the other, between species, or even between stages of the life cycle: the *Plasmodium* spp. erythrocytic merozoites have only a few, whereas the sporozoite is filled up with these organelles up to about half its total volume. There is a rather good correlation between microneme richness and the capacity of parasites to exhibit gliding motility. This crude observation was explained when a microneme protein (named TRAP in *Plasmodium* spp., and conserved in all Apicomplexa analyzed so far) was found to be essential for sporozoite motility (Sultan et al. 1997). A large body of research followed, both in *T. gondii* and *Plasmodium* spp., leading to the concept of "glideosome" (Opitz and Soldati 2002), where the TRAP protein would be translocated on the zoite surface with its C-terminus facing the inner membrane complex and interacting with an actin–myosin motor anchored in the IMC. This unique mechanism of motility whereby transmembrane proteins ensure both binding to the substrate and transduction of the motor power followed by specific cleavage by rhomboid proteases at the posterior end of the organism is fascinating; whether this was designed by Apicomplexa or can be found elsewhere in unicellular organisms remains to be found and is an evolutionary enigma. This gliding motility cannot be clearly distinguished from the binding to substrates, which involves the extracellular part of microneme proteins, in addition to GPI anchored resident surface proteins. The Apicomplexa zoites glide on cells but also on inert surfaces such as glass slides, and despite some detailed studies on the specificity of binding of specific motifs in resident or transient surface proteins of some members of the phylum, little is known on the physical parameters of Apicomplexa binding to surfaces and on the relationship between binding and gliding.

Gliding motility has also been reported with gregarines at the trophozoite stage, which lacks an apical complex, but where small dense organelles of microneme appearance are found near the parasite surface and could serve the same role (Vivier 1968; King 1981).

4.2 Host Cell Invasion

The entry into the host cell comprises two simultaneous events, one being the PV formation, the other being the moving junction (MJ) that allows the parasite to enter the PV.

4.2.1 Moving Junction

Host cell invasion by intracellular apicomplexa has been extensively studied morphologically, but the molecular details of this process have only been discovered recently. Indeed, the moving junction described by Aikawa et al in *P. knowlesi* invasion (Aikawa et al. 1978) has been shown microscopically to be present in other genera (*T. gondii*, *Eimeria* spp., …) but no specific molecule had been found associated to this structure. A important step was done when proteins from the rhoptry neck where clearly shown to be translocated in the moving junction during *T. gondii* tachyzoite entry (Alexander et al. 2005; Lebrun et al. 2005); under certain conditions, a conserved transmembrane microneme protein named AMA1 could also be found in the same location. These observations were also reported for *Plasmodium* (Alexander et al. 2006), leading to the identification of a protein that had been described in the rhoptry neck 20 years before (Roger et al. 1988). These works led to discovering that the moving junction was created by a coordinated action between a microneme protein inserted in the parasite membrane and RON proteins translocated in the host cell membrane or cytosol, interacting together to create the close apposition between the two membranes, and probably linked to the submembranous actin–myosin gliding motor to propel the parasite into the vacuole (Besteiro et al. 2009) (Fig. 2).

This discovery provided a new role to microneme proteins (which had been suggested to participate in invasion before, but without any real support) and established a crucial role to proteins contained in the neck of the rhoptries, further enforcing the importance of the curious segregation of proteins within the organelle.

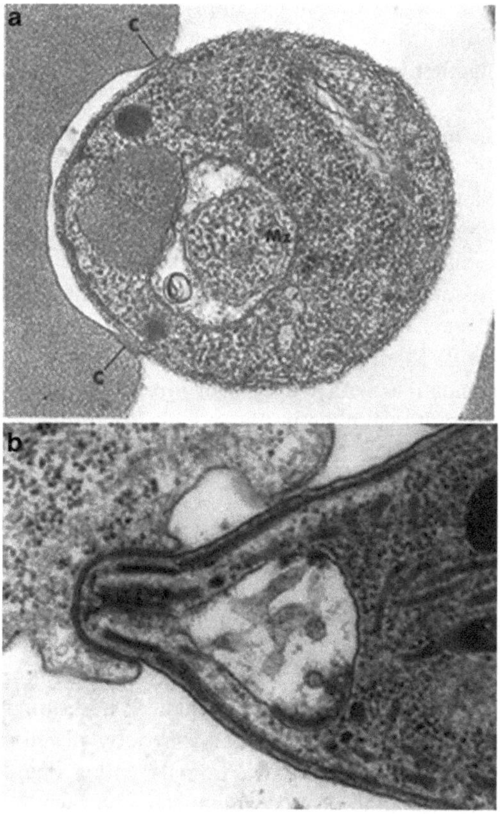

Fig. 2 (**a**) Moving junction (c) of a *Plasmodium knowlesi* merozoite (taken from Aikawa et al. (1978)). *Bar* = 0.2 μm. (**b**) Moving junction (c) of a *Toxoplasma gondii* tachyzoite; Host cell (HC), empty rhoptry (R), conoid (co). *Bar* = 0.5 μm

4.2.2 Parasitophorous Vacuole

The PV membrane (PVM) develops simultaneously with the entry of the parasite passing the MJ, is continuous with the plasma membrane, and is very distinct from the latter, both morphologically (Aikawa et al. 1981; Porchet-Hennere and Torpier 1983; Dubremetz et al. 1993) and biochemically (Mordue et al. 1999), having lost most of the host cell transmembrane proteins passing the MJ. Rhoptry proteins have been shown to be associated with the PVM in many studies, their topology with respect to this membrane being rather elusive.... The discovery of the association of RONs with the moving junction, together with new insights in the topology of rhoptry proteins after invasion has somehow clarified the way things happen during invasion. Nothing is known about how the PVM develops: whether the parasites pushes the lipid bilayer while the MJ glides backward or whether vesicles fuse with the developing PVM is unclear. When invasion is frustrated by using cytochalasin D, which blocks motility and stops the invasion process immediately after MJ initiation at the apical tip of the parasite, a cloud of vesicles that can be labeled with antibodies directed against rhoptry bulb proteins is found in the host cell cytosol near the parasite apex, suggesting the injection of rhoptry bulb material in the cell cytoplasm (Miller et al. 1979; Hakansson et al. 2001). This suggests that, once the MJ is built up (implying that rhoptry neck proteins have been injected in the HCM or cytosol and have associated with AMA1), rhoptry bulb proteins together with rhoptry lipids or lipids budding internally from the HCM are injected in the host cell cytosol. The Cyt B experiment suggests that in the normal situation, the material that is injected immediately fuses with the developing PVM, such that the e-vacuoles are usually not seen or not so conspicuous in this condition. As patch clamp experiments have clearly shown in *T. gondii* that the PVM is largely resulting from HPM invagination (Suss-Toby et al. 1996), the PVM building is most likely a composite between the two processes, but this remains to be clarified. Concerning the fate of rhoptry material, in *T. gondii*, where the phenomenon has been analyzed in much detail, rhoptry proteins are eventually found attached to the outer surface of the PVM, in the host cell cytosol, in the host cell nucleus, all these situations being compatible with HC cytosol injection, but also found in the PV, which is more difficult to explain in this model (Boothroyd and Dubremetz 2008).

4.3 Egress

At the end of intracellular development, usually after schizogony has led to the production of a number of daughter invasive stages (usually called merozoites), release of the parasites occur by lysis of the host cell. While in *T. gondii* the number of zoites formed within a vacuole is depending on the host cell size, in most other Apicomplexa, this number is strictly developmentally regulated, related to the number of mitosis (synchronous within one schizont) taking place before zoite differentiation occurs. This means that, while egress has long been considered in

T. gondii as due to overloading the host cell, this cannot be the case in most Apicomplexa, where egress must also be tightly regulated to enable release of mature parasites able to propagate the infection. How this regulation operates is unknown. Some insight came very recently concerning *P. falciparum* merozoites release from infected red cells, where it has been known for some time that proteolytic events are tightly linked to parasite egress: what was found is that a new set of apical organelles apparently distinct from the three canonical ones contained a protease required for egress and that these organelles, dubbed exonemes, were likely exocytosed in the PV before merozoite release (Yeoh et al. 2007). In addition to this, P falciparum egress also involves the red blood cell calpain, which would be activated by calcium upon PVM breakdown and would allow egress by digesting the red blood cell cytoskeletal proteins (Chandramohanadas et al. 2009).

In the case of *T. gondii* tachyzoites, egress has been extensively studied, especially concerning its triggering by altering the ionic composition of the host cell cytosol (Endo et al. 1982); the contribution of organelles in the process remains unclear. Micronemes are involved in exit and may contribute perforins that would help breaking down the PVM (Kafsack et al. 2009), but their major contribution is actually that their exocytosis allows triggering motility that allows exit of the cell. Concerning rhoptries, they are unlikely to be involved although a RON4 ring has been described in parasites triggered to egress by ionophore, suggesting that a MJ could operate also upon egress. This is topologically unsound, however, because the MJ would have to establish either with the PVM or with the inner side of the HCM, which in no case can lead to exit, and these preparations also contain invaders that may be confused with egressing parasites.

4.4 Apical Binding

Those Apicomplexa which remain extracellular (mostly gregarines), or extracytoplasmic (cryptosporidia) possess rhoptries and micronemes, these latter being likely involved in gliding motility. The function of rhoptries is less well defined, and no molecular detail is known concerning these organisms in this respect. However, morphological observations of the apical attachment zone in young gregarines (Schrevel 1968), or in early stages of Cyptosporidium-enterocyte interaction (Lumb et al. 1988), show a clear similarity with the early moving junction of *Plasmodium* spp. or coccidian, together with empty rhoptries attached to the apex, suggesting that in these cases, a stable junction could be built, sharing some properties with the one known in intracellular invaders, but which would not be moved backwards along the zoite, but would transform later in the complex attachment zone of the trophozoite. Early observations on empty rhoptries at the site of gregarine attachment led to the suggestion that these organelles were involved in feeding by endocytosing the host cell cytoplasm (Schrevel 1968). This would be a link with what are considered as free living representatives of apicomplexa ancestors, which possess a sort of apical complex and rhoptry-like organelles that engulf preys by their apical pole (Leander et al. 2003).

4.5 Cell Traversal

Another role played by the organelle proteins has received more attention in the recent years, when investigators working with *Plasmodium* ookinetes and sporozoites discovered that in both cases, parasites were able to glide across cells to reach an underlying target, and that they did so by disrupting cell membranes using pore forming proteins having homology to bacterial porins or complement related proteins. These proteins have been shown to be located in micronemes and exocytosed like other microneme proteins during gliding motility (Kadota et al. 2004; Ishino et al. 2004); in the same idea of traversal, chitinases involved in traversing peritrophic membrane have also been found in ookinete micronemes and they are required for successful development of the parasite in the mosquito gut (Langer and Vinetz 2001). A major issue concerning these membrane active substances is to understand how these microneme proteins can be inactivated when the zoite switches from traversal to invasion: this may be related to the exocytosis of rhoptry contents that may neutralize the perforating effect of these proteins.

4.6 High-jacking Host Cell Functions

One of the most exciting recent discoveries in the field has been that rhoptry proteins could be major virulence factors influencing the fate of the infection by *T. gondii* towards death or survival of the host and that this fate was fully determined at invasion by rhoptry protein exocytosis into the host cell. These data were obtained independently and simultaneously by genetic (Saeij et al. 2006; Taylor et al. 2006) and biochemical (El Hajj et al. 2007) studies determining the

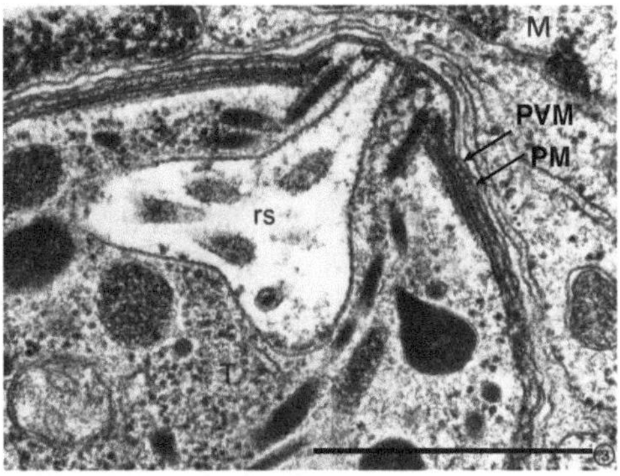

Fig. 3 *Toxoplasma gondii* tachyzoite captured during the process of invasion, showing a typical figure of rhoptry having undergone exocytosis (rs), with the duct open at the apical end of the parasite; *PVM* parasitophorous vacuole membrane, *PM* parasite plasmalemma. *Bar* = 0.5 μm [taken from Nichols et al. (1983)]

importance of a family of kinases and pseudo-kinases in *T. gondii*–host interaction. Although in the case of TgROP18 the protein does not seem to act on host cell gene expression but more on the role of the PV in the metabolic exchanges between host and parasite, another rhoptry protein is targeted to the host cell nucleus and clearly interfere with host cell transcription (Saeij et al. 2007). This latter is probably one among many others since early high throughput transcription studies of infected cells had shown a large number of changes in the very early times after invasion (Blader et al. 2001), which we now understand as derived from rhoptry protein dumping in the host cell during invasion.

5 Exocytosis of Apicomplexa Secretory Organelles

All the organelles described are exocytosed by invasive stages of apicomplexa. Micronemes and rhoptries are exclusively exocytosed when parasites are extracellular (or upon egress for micronemes). Dense granules are usually exocytosed when parasites are intracellular, sometimes at a very early stage (even before complete internalization) but can be exocytosed extracellularly by certain stages (*T. gondii* tachyzoites) under certain conditions.

Microneme exocytosis was suggested when the TRAP protein was discovered both in mics and on the sporozoite surface of *Plasmodium* (Rogers et al. 1992); but the exocytic process of this organelle was dissected in *T. gondii* where it has been shown that ion fluxes were responsible for the release of MIC proteins in the supernatant of motile parasites. The internal signaling pathway for MIC exocytosis seems to require release of Calcium ion from intraparasitic stores (Carruthers and Sibley 1999), through I3P/ryanodine sensitive channels (Lovett et al. 2002). But the environmental sensor responsible for transducing the external signal is not known: whether a contact with a surface or any chemical is needed is not known. In addition, there seems to exist a basic MIC exocytosis linked to motility, and a burst of exocytosis, this latter being mimicked by ionophores, upon host cell invasion.

Micronemes are exocytosed apically, but the precise location is unknown. Their release by fusion with the rhoptry duct has been suggested but never demonstrated and it is not compatible with the very different triggering mechanisms operating for the exocytosis of these organelles. The most likely way of microneme release is the surface located within the apical ring, but no specific site can be identified.

Rhoptries: although rhoptry exocytosis upon invasion was suggested as early as 1954, what occurs remains elusive. The main reason is that there is no way to trigger rhoptry exocytosis other than invading a host cell. In addition, what cytoplasmic signaling is required is entirely unknown.

Rhoptry exocytosis occurs at the center of the apical surface of the zoite (Nichols et al. 1983), where freeze fracture have shown a rosette of intramembranous particles (Porchet and Torpier 1977) identical to the one described for the exocytosis site for trichocysts of mucocysts in Paramecium or Tetrahymena (Satir et al. 1973), which are distantly related alveolates. Molecular details on this structure are unknown, but genetic studies in Paramecium have demonstrated that it is essential to exocytosis (Beisson et al. 1976). Artificial triggers of mucocysts or trichocyst exocytosis are known (Satir 1977), but they are without effect on rhoptry exocytosis (Dubremetz unpublished). As in Paramecium, images captured by freeze fracture during exocytosis show the opening of the organelle duct where the rosette was located (Dubremetz 2007), but whether the rosette is reconstituted after invasion is not known, as well as whether several rhoptries can be exocytosed in a row (Figs. 3, 4, 5).

Dense granules: dense granule exocytosis was first discovered in Sarcocystis (Entzeroth et al. 1986), and it happens laterally through a process that remains enigmatic since the location of the exocytosis site supposes traversing the inner membrane complex before fusing with the parasite plasma membrane. One must admit that the plates of the inner complex separate to allow DG exocytosis, but this has not been shown. Dense granule exocytosis was then shown to occur during invasion by *T. gondii* (Leriche and Dubremetz 1990), and a burst of exocytosis of this organelle indeed occurs at this stage. In the case of *T. gondii*, there is also a continuous "background noise" of release of dense granules which can occur extracellularly in serum supplemented medium, or during intracellular development (due to the fact that endodyogeny produces fully differentiated zoites upon every mitosis, which does not occur in schizogony). Whether this requires two different triggers is not known. Little is known about dense granule exocytosis in other apicomplexa: RESA is considered as exocytosed from merozoite dense granules upon red blood cell invasion by *P. falciparum* (Culvenor et al. 1991); but the compartment that trafficks the vast amount of proteins targeted to the red cell cytosol and plasma membrane through the Pexel motif does not seem to correspond to dense granules of Sarcocystidae. In Babesia, an equivalent of DG has been described, containing a protein of 225 kDa targeted to the rbc membrane after invasion (Hines et al. 1995) (Fig. 6).

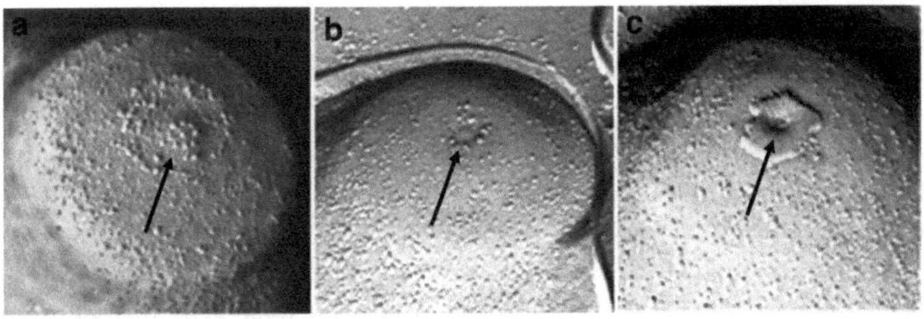

Fig. 4 Freeze fracture image of the apical end of: (**a**) an *Eimeria nieschulzi* sporozoite showing the apical rosette located at the exocytic site of rhoptries (*arrow*); (**b**, **c**) same area of *Toxoplasma gondii* tachyzoites captured during invasion and showing in (**b**) the rhoptry exocytosis pore (*arrow*), and in (**c**) the corresponding pore in the developing PVM (*arrow*). *Bar* = 0.1 μm

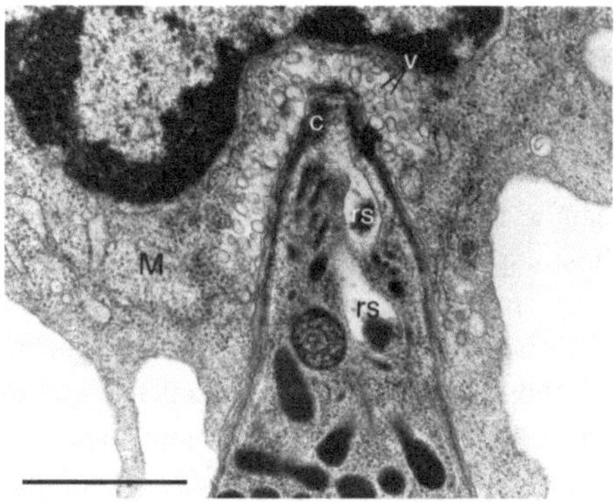

Fig. 5 *Toxoplasma gondii* tachyzoite invading a macrophage, showing a layer of vesicles (v) surrounding the apical area of the zoite and corresponding to e-vacuoles secreted by the rhoptries and which will fuse with the developing PVM. *M* macrophage; *rs* empty rhoptry, *c* conoid [taken from Nichols et al. (1983)]. *Bar* = 1 μm

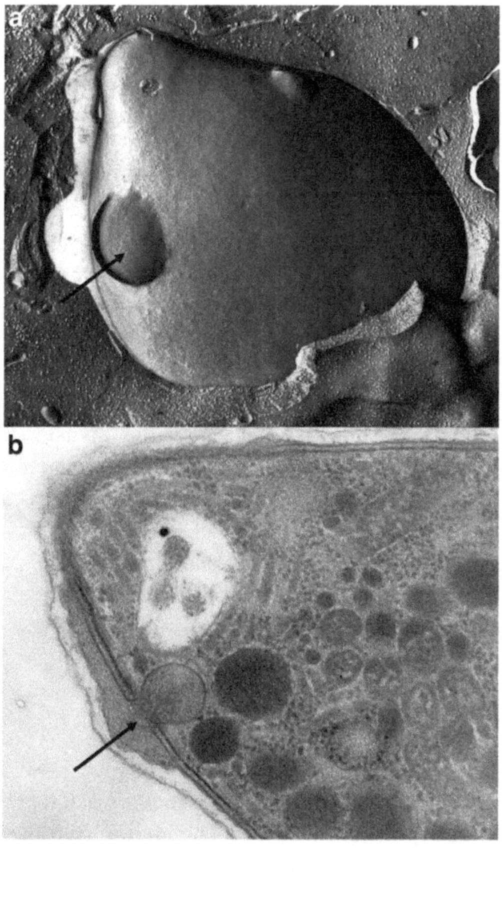

Fig. 6 Dense granule exocytosis (*arrow*) in the PV space newly formed by *Toxoplasma gondii* tachyzoites captured in section (**a**) or by freeze fracture as bumps in the PVM (**b**). *HCM* host cell membrane, *R* empty rhoptry; *DG* dense granule. *Bars* = 0.5 μm

6 Open Questions

All intracellular Apicomplexa use the invasion system at one stage or the other in their life cycle; however, in very rare cases, deviations from this universal scheme can be found: the only one clearly described is Theileria zoites in the intermediate host (sporozoites, merozoites) which have been shown to invade by using a zippering mechanism involving host cell surface adhesion, without organelle exocytosis. Microneme are said to be absent, coinciding with the absence of motility for these stages, rhoptries and dense granule exocytosis would occur after internalization in the vacuole, and rhoptries or dense granule contents are believed to be involved in dissolving the PVM; however, in the Theileria related parasite Babesia, rhoptries are exocytosed during invasion and the PVM disappears later, most likely through the activity of molecules released by dense granule exocytosis.

After invasion, Apicomplexa translocate proteins in the host cell, leading to structural transformation (*Plasmodium*) or reprogramming the host cell genome expression (*T. gondii*, *Theileria spp.*). When this occurs at invasion, rhoptries play this role, as shown in *T. gondii*; the dense granule contribution to this process is not known. When this translocation happens later during development, as for example in *P. falciparum* for export toward the red cell through the PEXEL machinery (Marti et al. 2004), the pathway cannot involve invasive stage organelles and is so far unknown.

The latest discovered invasive stage organelle is the exonemes, which is exocytosed in the Plasmodium PVM where it releases PfSUB1, a protease involved in merozoite release from the infected RBC. Is this newly described organelle a new category of DG, or of micronemes? The question is not solved and calls for a more general one which is about whether these organelles are or not heterogenous in contents, as already proposed (Entzeroth et al. 1991), and also suggested by recent demonstration of the presence of membranolytic components in micronemes used for egress or cell traversal, which is difficult to reconciliate with an organelle involved in an invasion process that preserves the host cell membrane. This remains to be investigated.

Little is known about how the invasion mechanism has evolved from the Apicomplexa free living ancestors (Leander et al. 2003), through extracellular parasites such as gregarines, to reach the present level of complexity and differentiation. A better knowledge of secretion processes in phylogenetically related extant protozoa would help understanding how this process has been selected and would also help exploring some of the related biochemical processes that are not easy to study with intracellular organisms.

7 Conclusion

The unique Apicomplexa–host cell invasion process is tightly linked to the invasive stages specific organelles. The high impact of Apicomplexa on human and animal health makes this crucial stage of these parasites life cycle a major target of therapeutic intervention. In addition to the biological interest of studying this phenomenon, this increases the need for more investigations in the field. Because of the high conservation of the process within the phylum, what is learned with one model Apicomplexa can shed light on all others, calling for exploring this world of parasites further, both in depth for the models under current study and in exploring more exotic ones, likely to bring new useful information on Apicomplexa biology.

References

Aikawa M, Miller LH, Johnson J, Rabbege J (1978) Erythrocyte entry by malarial parasites. A moving junction between erythrocyte and parasite. J Cell Biol 77:72–82

Aikawa M, Miller LH, Rabbege JR, Epstein N (1981) Freeze-fracture study on the erythrocyte membrane during malarial parasite invasion. J Cell Biol 91:55–62

Alexander DL, Mital J, Ward GE, Bradley P, Boothroyd JC (2005) Identification of the moving junction complex of *Toxoplasma gondii*: a collaboration between distinct secretory organelles. PLoS Pathog 1:e17

Alexander DL, Kapur SA, Dubremetz JF, Boothroyd JC (2006) *Plasmodium falciparum* AMA1 (PfAMA1) binds a rhoptry neck protein homologous to TgRON4, a component of the moving junction in *Toxoplasma*. Eukaryot Cell 5:1169–1173

Baldi DL, Andrews KT, Waller RF, Roos DS, Howard RF, Crabb BS, Cowman AF (2000) RAP1 controls rhoptry targeting of RAP2 in the malaria parasite *Plasmodium falciparum*. EMBO J 19:2435–2443

Bannister LH, Mitchell GH, Butcher GA, Dennis ED (1986) Lamellar membranes associated with rhoptries in erythrocytic merozoites of *Plasmodium knowlesi*: a clue to the mechanism of invasion. Parasitology 92:291–303

Bannister LH, Hopkins JM, Dluzewski AR, Margos G, Williams IT, Blackman MJ, Kocken CH, Thomas AW, Mitchell GH (2003) *Plasmodium falciparum* apical membrane antigen 1 (PfAMA-1) is translocated within micronemes along subpellicular microtubules during merozoite development. J Cell Sci 116:3825–3834

Beisson J, Lefort-Tran M, Pouphile M, Rossignol M, Satir B (1976) Genetic analysis of membrane differentiation in *Paramecium*. Freeze-fracture study of the trichocyst cycle in wild-type and mutant strains. J Cell Biol 69:126–143

Besteiro S, Bertrand-Michel J, Lebrun M, Vial H, Dubremetz JF (2008) Lipidomic analysis of *Toxoplasma gondii* tachyzoites rhoptries: further insights into the role of cholesterol. Biochem J 415:87–96

Besteiro S, Michelin A, Poncet J, Dubremetz JF, Lebrun M (2009) Export of a *Toxoplasma gondii* rhoptry neck protein complex at the host cell membrane to form the moving junction during invasion. PLoS Pathog 5:e1000309

Blackman MJ, Bannister LH (2001) Apical organelles of Apicomplexa: biology and isolation by subcellular fractionation. Mol Biochem Parasitol 117:11–25

Blader IJ, Manger ID, Boothroyd JC (2001) Microarray analysis reveals previously unknown changes in *Toxoplasma gondii*-infected human cells. J Biol Chem 276:24223–24231

Boothroyd JC, Dubremetz JF (2008) Kiss and spit: the dual roles of *Toxoplasma* rhoptries. Nat Rev Microbiol 6:79–88

Bradley PJ, Ward C, Cheng SJ, Alexander DL, Coller S, Coombs GH, Dunn JD, Ferguson DJ, Sanderson SJ, Wastling JM, Boothroyd JC (2005) Proteomic analysis of rhoptry organelles reveals many novel constituents for host-parasite interactions in *Toxoplasma gondii*. J Biol Chem 280:34245–34258

Bromley E, Leeds N, Clark J, McGregor E, Ward M, Dunn MJ, Tomley F (2003) Defining the protein repertoire of microneme secretory organelles in the apicomplexan parasite Eimeria tenella. Proteomics 3:1553–1561

Carruthers VB, Sibley LD (1999) Mobilization of intracellular calcium stimulates microneme discharge in *Toxoplasma gondii*. Mol Microbiol 31:421–428

Chandramohanadas R, Davis PH, Beiting DP, Harbut MB, Darling C, Velmourougane G, Lee MY, Greer PA, Roos DS, Greenbaum DC (2009) Apicomplexan parasites co-opt host calpains to facilitate their escape from infected cells. Science 324:794–797

Coppens I, Dunn JD, Romano JD, Pypaert M, Zhang H, Boothroyd JC, Joiner KA (2006) *Toxoplasma gondii* sequesters lysosomes from mammalian hosts in the vacuolar space. Cell 125:261–274

Culvenor JG, Day KP, Anders RF (1991) *Plasmodium falciparum* ring-infected erythrocyte surface antigen is released from merozoite dense granules after erythrocyte invasion. Infect Immun 59:1183–1187

Dowse TJ, Pascall JC, Brown KD, Soldati D (2005) Apicomplexan rhomboids have a potential role in microneme protein cleavage during host cell invasion. Int J Parasitol 35:747–756

Dubremetz JF (2007) Rhoptries are major players in *Toxoplasma gondii* invasion and host cell interaction. Cell Microbiol 9:841–848

Dubremetz JF, Dissous C (1980) Characteristic proteins of micronemes and dense granules from *Sarcocystis tenella* zoites (Protozoa, Coccidia). Mol Biochem Parasitol 1:279–289

Dubremetz JF, Ferreira E, Dissous C (1989) Isolation and partial characterization of rhoptries and micronemes from *Eimeria nieschulzi* zoites (Sporozoa, Coccidia). Parasitol Res 75:449–454

Dubremetz JF, Achbarou A, Bermudes D, Joiner KA (1993) Kinetics and pattern of organelle exocytosis during *Toxoplasma gondii*/host-cell interaction. Parasitol Res 79:402–408

El Hajj H, Lebrun M, Arold ST, Vial H, Labesse G, Dubremetz JF (2007) ROP18 Is a rhoptry kinase controlling the intracellular proliferation of *Toxoplasma gondii*. PLOS pathog 3:e14

Endo T, Sethi KK, Piekarski G (1982) *Toxoplasma gondii*: calcium ionophore A23187-mediated exit of trophozoites from infected murine macrophages. Exp Parasitol 53:179–188

Entzeroth R, Dubremetz JF, Hodick D, Ferreira E (1986) Immunoelectron microscopic demonstration of the exocytosis of dense granule contents into the secondary parasitophorous vacuole of *Sarcocystis muris* (Protozoa, Apicomplexa). Eur J Cell Biol 41:182–188

Entzeroth R, Konig A, Dubremetz JF (1991) Monoclonal antibodies identify micronemes and a new population of cytoplasmic granules cross-reacting with micronemes of cystozoites of *Sarcocystis muris*. Parasitol Res 77:59–64

Foussard F, Leriche MA, Dubremetz JF (1991) Characterization of the lipid content of *Toxoplasma gondii* rhoptries. Parasitology 102(Pt 3):367–370

Garnett JA, Liu Y, Leon E, Allman S, Friedrich N, Saouros S, Curry S, Soldati-Favre D, Davis B, Feizi T, Matthews S (2009) Detailed insights from microarray and crystallographic studies into carbohydrate recognition by microneme protein 1 (MIC1) of *Toxoplasma gondii*. Protein Sci 18:1935–1947

Garnham PC, Bird RG, Baker JR, Bray RS (1961) Electron microscope studies of motile stages of malaria parasites. II. The fine structure of the sporozoite of *Laverania (Plasmodium) falcipara*. Trans R Soc Trop Med Hyg 55:98–102

Gustafson PV, Agar HD, Cramer DI (1954) An electron microscope study of *Toxoplasma*. Am J Trop Med Hyg 3:1008–1021

Hakansson S, Charron AJ, Sibley LD (2001) *Toxoplasma* evacuoles: a two-step process of secretion and fusion forms the parasitophorous vacuole. EMBO J 20:3132–3144

Harris JR, Adrian M, Petry F (2003) Structure of the *Cryptosporidium parvum* microneme: a metabolically and osmotically labile apicomplexan organelle. Micron 34:65–78

Hines SA, Palmer GH, Brown WC, McElwain TF, Suarez CE, Vidotto O, Rice-Ficht AC (1995) Genetic and antigenic characterization of *Babesia bovis* merozoite spherical body protein Bb-1. Mol Biochem Parasitol 69:149–159

Ishino T, Yano K, Chinzei Y, Yuda M (2004) Cell-passage activity is required for the malarial parasite to cross the liver sinusoidal cell layer. PLoS Biol 2:E4

Jacobs L (1967) *Toxoplasma* and toxoplasmosis. Adv Parasitol 5:1–45

Kadota K, Ishino T, Matsuyama T, Chinzei Y, Yuda M (2004) Essential role of membrane-attack protein in malarial transmission to mosquito host. Proc Natl Acad Sci USA 101:16310–16315

Kafsack BF, Pena JD, Coppens I, Ravindran S, Boothroyd JC, Carruthers VB (2009) Rapid membrane disruption by a perforin-like protein facilitates parasite exit from host cells. Science 323:530–533

King CA (1981) Cell surface interaction of the protozoan *Gregarina* with concanavalin A beads – implications for models of gregarine gliding. Cell Biol Int Rep 5:297–305

Lal K, Prieto JH, Bromley E, Sanderson SJ, Yates JR 3rd, Wastling JM, Tomley FM, Sinden RE (2009) Characterisation of *Plasmodium* invasive organelles; an ookinete microneme proteome. Proteomics 9:1142–1151

Langer RC, Vinetz JM (2001) *Plasmodium* ookinete-secreted chitinase and parasite penetration of the mosquito peritrophic matrix. Trends Parasitol 17:269–272

Leander BS, Kuvardina ON, Aleshin VV, Mylnikov AP, Keeling PJ (2003) Molecular phylogeny and surface morphology of *Colpodella edax* (Alveolata): insights into the phagotrophic ancestry of apicomplexans. J Eukaryot Microbiol 50:334–340

Lebrun M, Michelin A, El Hajj H, Poncet J, Bradley PJ, Vial H, Dubremetz JF (2005) The rhoptry neck protein RON4 relocalizes at the moving junction during *Toxoplasma gondii* invasion. Cell Microbiol 7:1823–1833

Lebrun M, Carruthers VB, Cesbron MF (2007) *Toxoplasma* secretory proteins and their roles in cell invasion and intracellular survival. In: Louis W, Kami K (eds) *Toxoplasma gondii*: the model Apicomplexan perspectives and methods. Academic, London, pp 265–316

Leriche MA, Dubremetz JF (1990) Exocytosis of *Toxoplasma gondii* dense granules into the parasitophorous vacuole after host cell invasion. Parasitol Res 76:559–562

Leriche MA, Dubremetz JF (1991) Characterization of the protein contents of rhoptries and dense granules of *Toxoplasma gondii* tachyzoites by subcellular fractionation and monoclonal antibodies. Mol Biochem Parasitol 45:249–259

Lovett JL, Marchesini N, Moreno SN, Sibley LD (2002) Toxoplasma gondii microneme secretion involves intracellular Ca(2+) release from inositol 1,4,5-triphosphate (IP(3))/ryanodine-sensitive stores. J Biol Chem 277:25870–25876

Ludvik J (1956) Vergleichende elektronoptische untersuchungen an *Toxoplasma gondii* und *Sarcocystis tenella*. Zentralbl Bakteriol Abt 166:60–65

Lumb R, Smith K, O'Donoghue PJ, Lanser JA (1988) Ultrastructure of the attachment of *Cryptosporidium* sporozoites to tissue culture cells. Parasitol Res 74:531–536

Marti M, Good RT, Rug M, Knuepfer E, Cowman AF (2004) Targeting malaria virulence and remodeling proteins to the host erythrocyte. Science 306:1930–1933

Mercier C, Adjogble KD, Daubener W, Delauw MF (2005) Dense granules: are they key organelles to help understand the parasitophorous vacuole of all apicomplexa parasites? Int J Parasitol 35:829–849

Miller LH, Aikawa M, Johnson JG, Shiroishi T (1979) Interaction between cytochalasin B-treated malarial parasites and erythrocytes. Attachment and junction formation. J Exp Med 149:172–184

Miller SA, Binder EM, Blackman MJ, Carruthers VB, Kim K (2001) A conserved subtilisin-like protein TgSUB1 in microneme organelles of *Toxoplasma gondii*. J Biol Chem 276:45341–45348

Mordue DG, Desai N, Dustin M, Sibley LD (1999) Invasion by *Toxoplasma gondii* establishes a moving junction that selectively excludes host cell plasma membrane proteins on the basis of their membrane anchoring. J Exp Med 190:1783–1792

Nichols BA, Chiappino ML, O'Connor GR (1983) Secretion from the rhoptries of *Toxoplasma gondii* during host-cell invasion. J Ultrastruct Res 83:85–98

Ogino N, Yoneda C (1966) The fine structure and mode of division of *Toxoplasma gondii*. Arch Ophthalmol 75:218–227

Opitz C, Soldati D (2002) 'The glideosome': a dynamic complex powering gliding motion and host cell invasion by *Toxoplasma gondii*. Mol Microbiol 45:597–604

Porchet E, Torpier G (1977) Etude du germe infectieux de *Sarcocystis tenella* et *Toxoplasma gondii* par la technique du cryodecapage. Z Parasitenkd 54:101–124

Porchet-Hennere E, Torpier G (1983) Relations entre *Toxoplasma* et sa cellule-hote. Protistologica 19:357–370

Reiss M, Viebig N, Brecht S, Fourmaux MN, Soete M, Di Cristina M, Dubremetz JF, Soldati D (2001) Identification and characterization of an escorter for two secretory adhesins in *Toxoplasma gondii.* J Cell Biol 152:563–578

Robson KJ, Hall JR, Jennings MW, Harris TJ, Marsh K, Newbold CI, Tate VE, Weatherall DJ (1988) A highly conserved amino-acid sequence in thrombospondin, properdin and in proteins from sporozoites and blood stages of a human malaria parasite. Nature 335:79–82

Roger N, Dubremetz JF, Delplace P, Fortier B, Tronchin G, Vernes A (1988) Characterization of a 225 kilodalton rhoptry protein of *Plasmodium falciparum.* Mol Biochem Parasitol 27:135–141

Rogers WO, Malik A, Mellouk S, Nakamura K, Rogers MD, Szarfman A, Gordon DM, Nussler AK, Aikawa M, Hoffman SL (1992) Characterization of *Plasmodium falciparum* sporozoite surface protein 2. Proc Natl Acad Sci USA 89:9176–9180

Saeij JP, Boyle JP, Coller S, Taylor S, Sibley LD, Brooke-Powell ET, Ajioka JW, Boothroyd JC (2006) Polymorphic secreted kinases are key virulence factors in toxoplasmosis. Science 314:1780–1783

Saeij JP, Coller S, Boyle JP, Jerome ME, White MW, Boothroyd JC (2007) *Toxoplasma* co-opts host gene expression by injection of a polymorphic kinase homologue. Nature 445:324–327

Satir B (1977) Dibucaine-induced synchronous mucocyst secretion in *Tetrahymena.* Cell Biol Int Rep 1:69–73

Satir B, Schooley C, Satir P (1973) Membrane fusion in a model system. Mucocyst secretion in Tetrahymena. J Cell Biol 56:153–176

Scholtyseck E, Mehlhorn H (1970) Ultrastructural study of characteristic organelles (paired organelles, micronemes, micropores) of sporozoa and related organisms. Z Parasitenkd 34: 97–127

Schrevel J (1968) L'ultrastructure de la région antérieure de la grégarine *Sélénidium* et son intérêt pour l'étude de la nutrition chez les Sporozoaires. J Microsc 7:391–410

Senaud J (1967) Contribution à l'étude des Sarcosporidies et des Toxoplasmes (toxoplasmea). Protistologica 3:168–232

Shaw MK, Roos DS, Tilney LG (1998) Acidic compartments and rhoptry formation in *Toxoplasma gondii.* Parasitology 117:435–443

Sheiner L, Soldati-Favre D (2008) Protein trafficking inside *Toxoplasma gondii.* Traffic 9:636–646

Stewart MJ, Schulman S, Vanderberg JP (1985) Rhoptry secretion of membranous whorls by *Plasmodium berghei* sporozoites. J Protozool 32:280–283

Sultan AA, Thathy V, Frevert U, Robson KJ, Crisanti A, Nussenzweig V, Nussenzweig RS, Menard R (1997) TRAP is necessary for gliding motility and infectivity of *Plasmodium* sporozoites. Cell 90:511–522

Suss-Toby E, Zimmerberg J, Ward GE (1996) *Toxoplasma* invasion: the parasitophorous vacuole is formed from host cell plasma membrane and pinches off via a fission pore. Proc Natl Acad Sci U S A 93:8413–8418

Taylor S, Barragan A, Su C, Fux B, Fentress SJ, Tang K, Beatty WL, El Hajj H, Jerome M, Behnke MS, White M, Wootton JC, Sibley LD (2006) A secreted serine-threonine kinase determines virulence in the eukaryotic pathogen *Toxoplasma gondii.* Science 314:1776–1780

Tomley FM, Soldati DS (2001) Mix and match modules: structure and function of microneme proteins in apicomplexan parasites. Trends Parasitol 17:81–88

Vivier E (1968) L'organisation ultrastructurale corticale de la grégarine *Lecudina pellucida;* ses rapports avec l'alimentation et la locomotion. J Protozool 15:230–246

Vivier E, Petitprez A (1972) Données ultrastructurales complémentaires, morphologiques et cytochimiques, sur *Toxoplasma gondii.* Protistologica 8:199–221

Yeoh S, O'Donnell RA, Koussis K, Dluzewski AR, Ansell KH, Osborne SA, Hackett F, Withers-Martinez C, Mitchell GH, Bannister LH, Bryans JS, Kettleborough CA, Blackman MJ (2007) Subcellular discharge of a serine protease mediates release of invasive malaria parasites from host erythrocytes. Cell 131:1072–1083

Secretory Events During *Giardia* Encystation

Fernando D. Rivero, Dana Müller, and Hugo D. Lujan

Contents

Abstract *Giardia lamblia*, a flagellate protozoan that parasitizes the upper small intestine of humans, is one of the most common causes of diarrheal disease worldwide. *Giardia* has a simple life cycle, alternating between the disease-causing trophozoites and the infective cysts. *Giardia* is a true eukaryotic organism since it has two nuclei, an endomembranous system including the nuclear envelope/ endoplasmic reticulum, transport vesicles and lysosomes-like peripheral vacuoles, as well as a complex cytoskeleton. However, trophozoites possess several prokaryotic features, including bacterial metabolic pathways and the lack of organelles typical of higher eukaryotes, such as mitochondria, peroxisomes, and a recognizable

F.D. Rivero, D. Müller, and H.D. Lujan (✉)
Laboratory of Biochemistry and Molecular Biology, School of Medicine, Catholic University of Cordoba, Jacinto Rios 571, CP X5004ASK, Cordoba, Argentina
e-mail: hlujan@ucc.edu.ar

W. de Souza (ed.), *Structures and Organelles in Pathogenic Protists*,
Microbiology Monographs 17, DOI 10.1007/978-3-642-12863-9_9,
© Springer-Verlag Berlin Heidelberg 2010

Golgi apparatus. Despite these characteristics, *Giardia* carries out secretory events implying both constitutive and regulated trafficking pathways. Here we describe the secretory machinery employed by *Giardia* for intracellular transport of cyst wall materials, their exocytosis, and the extracellular assembly of the protective cyst wall. These processes are essential for both the survival of the parasite outside the host's intestine and transmission of the disease among susceptible individuals.

1 Introduction

1.1 General Characteristics of the Parasite

Giardia lamblia is an intestinal parasite that belongs to the earliest diverging branch of the eukaryotic linage of descent (Dacks et al. 2003; Morrison et al. 2007; Sogin et al. 1989). *Giardia* has only two stages in its life cycle that are remarkably well adapted for the survival of the parasite in very different and hostile environments: the ovoid, infective cysts, and the motile, pear-shaped trophozoites. Trophozoites colonize the upper parts of the small intestine of humans and several vertebrates and are responsible for the clinical manifestations of the disease (Fig. 1a, b) (Adam 2001; Gillin et al. 1996). This parasite is an important cause of diarrheal disease throughout the world. *Giardia* infections are initiated by ingestion of contaminated water or food or through direct fecal–oral contact. Besides chronic or acute diarrhea, additional manifestations include malabsorption, abdominal pain, flatulence, and desnutrition (Adam 2001). In addition to their medical importance, *Giardia* is considered an excellent model system to study the development of basic cellular processes because of its early position in eukaryotic evolution. Although *Giardia* was once considered a "missing link" between prokaryotes and eukaryotes (Sogin et al. 1989), more recent studies have demonstrated a different scenario (Embley and Hirt 1998; Lloyd and Harris 2002). Although several genes and proteins analyses used for phylogenic classification indicate that *Giardia* is in fact one of the earliest branching eukaryotes (Simpson et al. 2002; Sogin and Silberman 1998), some of its particular cellular characteristics are probably a result of the secondary loss of complex cell structures as a consequence of its parasitic life style rather than the primitive simplicity supposed for early diverging protists (Morrison et al. 2001; Samuelson et al. 2005). *Giardia* appears as a "living fossil" (Graczyk 2005) since it has neither typical mitochondria, peroxisomes, nor a classical Golgi apparatus (Gillin et al. 1996), but after the finding of mitochondrial genes and the discovering of a mitochondrial remnant, the so called "mitosome" (Tovar et al. 2003), together with the completion of the sequencing of the *Giardia* genome, new characteristics of this organism came to light, including the possibility of sexual reproduction (Caccio and Sprong 2010; Cooper et al. 2007; Poxleitner et al. 2008) and other cellular functions distinctive of more evolved organisms (Morrison et al. 2007).

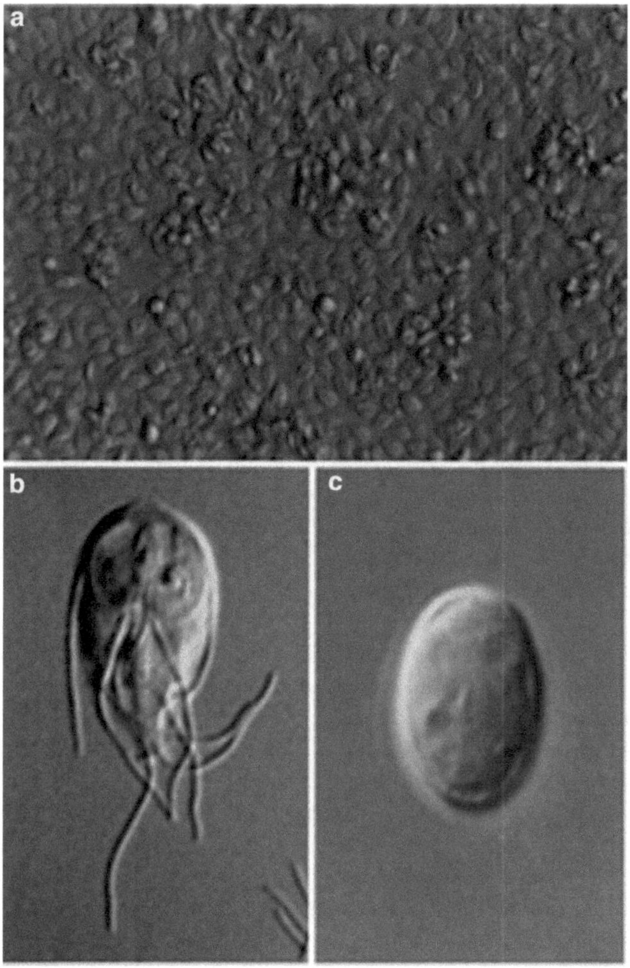

Fig. 1 *Giardia lamblia differentiation.* (**a**) Phase contrast image of a culture of encysting *Giardia*. Pear-shaped trophozoites form a monolayer attached to the glass wall of the culture tube. Newly formed cysts, detach and aggregate in the culture supernatant, from which are easily collected. (**b**) and (**c**) are DIC images of an isolated trophozoite and a cyst, respectively

G. lamblia trophozoites possess a fascinating secretory system in which a typical Golgi apparatus is not observable, although the packaging and sorting functions characteristic of this organelle are evident in *Giardia* (Lujan and Touz 2003; Marti et al. 2003a). For example, transport to the plasma membrane (PM) and release into the culture medium of variant-specific surface proteins (VSPs) (Lujan et al. 1995c; Reiner et al. 1990) and trafficking of both membrane and soluble enzymes to peripheral vacuoles (PVs) (Touz et al. 2003), which are thought to perform both lysosomal and endosomal activities (Lanfredi-Rangel et al. 1998), are evidence for constitutive protein transport (Lujan and Touz 2003; Marti et al. 2003a). Regulated secretion has been reported to take place only during trophozoite differentiation into cysts (Gillin et al. 1996), when large granules called “Encystation-specific secretory Vesicles or ESVs” are generated after the cells sensed the stimulus for differentiation (Faubert et al. 1991). Since encystation can be reproduced in vitro (Boucher and Gillin 1990; Lujan et al. 1996a) (Fig. 1a), it was possible to demonstrate that cyst formation comprises different stages, which include (a) the expression of encystation-specific genes, such as those necessary for the synthesis and processing of the cyst wall components (Lujan et al. 1997; Mowatt et al. 1995), (b) the biogenesis of electron-dense ESVs that transport cyst wall materials (Gillin et al. 1991), and (c) exocytosis of ESVs and extracellular assembly of the cyst wall (Gillin et al. 1991; Lujan et al. 1997).

Given that secretory granules form in the *trans*-Golgi Network (TGN) in higher eukaryotes (Orci et al. 1987; Rambourg et al. 1988) and that *Giardia* seems to lack this organelle in proliferating trophozoites (Marti et al. 2003b), important efforts by several laboratories have focused on this particular regulated secretory pathway that culminate with the formation of the cyst wall that protects the parasite outside the host’s intestine (Fig. 1a, c). Nevertheless, the current information is incomplete and, in many aspects, controversial.

In early studies, many molecules involved in *Giardia* secretory pathways have been identified and characterized, mainly by developing specific monoclonal antibodies to different subcellular organelles (Lujan et al. 1996b; Mowatt et al. 1995; Touz et al. 2002b) or by cloning genes encoding well-known proteins implicated in secretion in other organisms (Davids et al. 2006; Marti and Hehl 2003; Marti et al. 2003a; Sun et al. 2003), but the recent completion of the *Giardia* genome project (Morrison et al. 2007) has allowed the identification of an important number of molecules involved in protein trafficking in higher eukaryotes (http://www.GiardiaDB.org; see Table 1). Nevertheless, most of them have still not been characterized and several aspects of the regulated secretory pathway occurring during *Giardia* encystation remain enigmatic. Differentiation in *Giardia* is not only interesting from a cell biological point of view but the knowledge of the genetic and biochemical processes involved in the synthesis, posttranslational modifications, cross-linking, and fibrillar deposition of cell wall proteins might also lead to the design of new chemotherapeutic agents and of reagents for more sensitive and practical diagnostic tests. In addition, knowledge about cyst wall formation might be applicable to other cells as well and could afford information on the mechanisms of intracellular transport and biogenesis of the secretory organelles in eukaryotic cells.

In the next sections, we introduce the available knowledge of the intracellular endomembranous system of this parasite, describing in detail the morphological and biochemical evidence of the *Giardia*’s secretory compartments in proliferating trophozoites and in those undergoing encystation. In addition, we discuss contradictory results reported in the scientific literature, formulate questions about issues that need to be solved, and present new ideas and hypothesis based on the relevant information provided by the elucidation of the complete genome of this important human pathogen.

Table 1 Molecular components of the secretory pathway identified in the *Giardia lamblia* genome

Gene product	ORF	Function
ER components		
SRα	15156	Signal recognition particle receptor
SRβ	–	Signal recognition particle receptor
Sec61p-α	5744	Translocon
Sec61p-γ	3896	Translocon
Sec11p-1	9174	Signal peptide peptidase subunit
Sec11p-1	8429	Signal peptide peptidase subunit
BiP/GRP78	17121	ER chaperone
GRP94	15247	ER chaperone
PDI-1	29487	Protein disulfide isomerase
PDI-2	9413	Protein disulfide isomerase
PDI-3	14670	Protein disulfide isomerase
PDI-4	103713	Protein disulfide isomerase
PDI-5	8064	Protein disulfide isomerase
KDEL-receptor	4502	Retrieval of ER lumenal proteins
COPII		
Sar1p	7569	GTP binding protein
Sec13p	137698	Coatomer II component
Sec23p-1	17164	Coatomer II component
Sec23p-2	16520	Coatomer II component
Sec23p-3	9376	Coatomer II component
Sec24p	17065	Coatomer II component
Sec31p	2562	Coatomer II component
COPI		
ARF-1	7789	Coatomer I component
α-COP	11953	Coatomer I component
β-COP	88082	Coatomer I component
β′-COP	9593	Coatomer I component
δ-COP	6170	Coatomer I component
γ-COP	5603	Coatomer I component
ζ-COP	17345	Coatomer I component
ε-COP	–	Coatomer I component
ERP1	4509	Cargo receptor in COP I
Rer1p	15413	Possible COP I component
GTP regulator factors		
ARF-GAP-1	2834	Guanine-nucleotide activating protein
ARF-GAP-2	17561	Guanine-nucleotide activating protein
ARF-GAP-3	22454	Guanine-nucleotide activating protein
GEF-1	17192	Guanine-nucleotide exchange factor
GEF-2	112258	Guanine-nucleotide exchange factor
Clathrin coat		
AP large subunit β1	15339	Clathrin adaptor complex component
AP large subunit γ1	24122	Clathrin adaptor complex component
AP medium subunit μ1	8917	Clathrin adaptor complex component
AP small subunit σ1	5382	Clathrin adaptor complex component
AP large subunit β2	21423	Clathrin adaptor complex component
AP large subunit γ2	16364	Clathrin adaptor complex component
AP medium subunit μ2	89622	Clathrin adaptor complex component
AP small subunit σ2	91198	Clathrin adaptor complex component
Clathrin heavy chain	102108	Clathrin coat
Clathrin light chain	–	Clathrin coat

(*continued*)

Table 1 (continued)

Gene product	ORF	Function
Rabs		
Rab-1	9558	Vesicle docking
Rab-2	15567	Vesicle docking
Rab-3	9718	Vesicle docking
Rab-4	1695	Vesicle docking
Rab-5	16979	Vesicle docking
Rab-6	13109	Vesicle docking
Rab-7	12157	Vesicle docking
Rab-8	940	Vesicle docking
Rab-9	8497	Vesicle docking
Rab-10	8944	Vesicle docking
Rab-11	16636	Vesicle docking
Rab-12	9778	Vesicle docking
PPT-1	9382	Rab geranylgeranyltransferase
PPT-2	15820	Rab geranylgeranyltransferase
PPT-3	16675	Rab geranylgeranyltransferase
GGPT	17082	Rab geranylgeranyltransferase
Rab-GAP	11204	Guanine-nucleotide activating protein
Rab-GID	11495	Guanine-nucleotide inhibiting protein
Tethers		
Rer1p	15413	Retention of ER proteins
ARF-3	7562	ARF-like protein
ARF-4	13930	ARF-like protein
ARF-5	13478	ARF-like protein
ARF-6	4192	ARF-like protein
ARF-7	13523	ARF-like protein
Yif1p	7873	Interacting protein
Yip1p	8565	Interacting protein
Yip2p	12160	Interacting protein
Yip3p	14567	Hypothetical protein
SNAREs		
R-SNARE 1	9489	Vesicle-associated membrane protein
R-SNARE 2	7306	Vesicle-associated membrane protein
R-SNARE 3	14469	Vesicle-associated membrane protein
Qa-SNARE 1	7309	Syntaxin 1
Qa-SNARE 2	3869	Syntaxin 2
Qa-SNARE 3	96994	Syntaxin 3
Qa-SNARE 4	11220	Syntaxin 4
Qb-SNARE 5	10803	SNARE
Qb-SNARE 1	16054	Vti1p
Qb-SNARE 2	17464	SNARE
Qb-SNARE 3	5785	SNARE
Qc-SNARE 1	7590	SNARE
Qc-SNARE 2	19509	SNARE
Qc-SNARE 3	5927	SNARE
Qc-SNARE 4	10315	SNARE
Qc-SNARE 5	10013	SNARE
SNARE Master 1	16228	Regulation/Disassembly SNARE complex
SNARE Master 1	7718	Regulation/Disassembly SNARE complex

(*continued*)

Table 1 (continued)

Gene product	ORF	Function
α-SNAP	17224	Regulation/Disassembly SNARE complex
α-SNAP	16521	Regulation/Disassembly SNARE complex
NSF/Sec18p	10856	Regulation/Disassembly SNARE complex
NSF/Sec18p	114776	Regulation/Disassembly SNARE complex

Genes encoding different proteins involved in different aspects of the *Giardia* secretory machinery were identified by HMMsearch and HMMPfam with the HMMER program (Eddy 1998). SMART searches were performed with the SMART Web Server (Letunic et al. 2006). HHPred (Soding et al. 2005) was performed with varying options, against either PDB or COG/KOG databases. Cytoskeletal proteins are not included

2 Organelles and Machinery Involved in Protein Transport in *Giardia lamblia*

2.1 The Nuclear Envelope

It is well known that all eukaryotic cells present their genetic material surrounded by two nuclear membranes, forming the nuclear envelope (NE). The outer nuclear membrane is continuous with the endoplasmic reticulum (ER) and is normally studded with ribosomes, corresponding to the rough endoplasmic reticulum (RER). The inner nuclear membrane contains a meshwork of filamentous proteins forming the nuclear lamina that provides structural support for this membrane. The nuclear envelope of all eukaryotes is perforated by elaborate structures known as nuclear pore complexes (Alberts et al. 2007). Nuclear pore complexes are gated regions, where extensive traffic of materials between the nucleus and cytoplasm occurs (Alberts et al. 2007). Nevertheless, careful analyses of *Giardia* nuclei showed that the nuclear envelope in trophozoites presents unusual characteristics: For example, the frequent absence of ribosomes on parts of the outer nuclear membrane contrasts with the extensive RER of the parasite. In some studies, ribosomes are only occasionally observed on the nuclear envelope, which may indicate a more stringent functional separation between the NE and the ER in *Giardia* (Benchimol 2004a) (Fig. 2a).

Both layers of the *Giardia* nuclear envelope are fenestrated, and the pore complexes have the characteristic structure of those found in higher eukaryotic cells (Benchimol 2004a). Freeze-fracture assays of *G. lamblia* nuclei have shown large aggregations of closely packed pores and extensive pore-free areas. There is a marked heterogeneity in the number and distribution of nuclear pores between both nuclei of the same cell; consequently, Benchimol (2004a) suggested that the number and localization of nuclear pores might correlate with cell activity and protein synthesis. Thus, what is the functional significance of this observation? If both nuclei of the parasite seems to contain the same DNA content and both are transcriptionally active (Kabnick and Peattie 1990; Yu et al. 2002), why are nuclear pores differently distributed and how is this phenomenon related to the cell

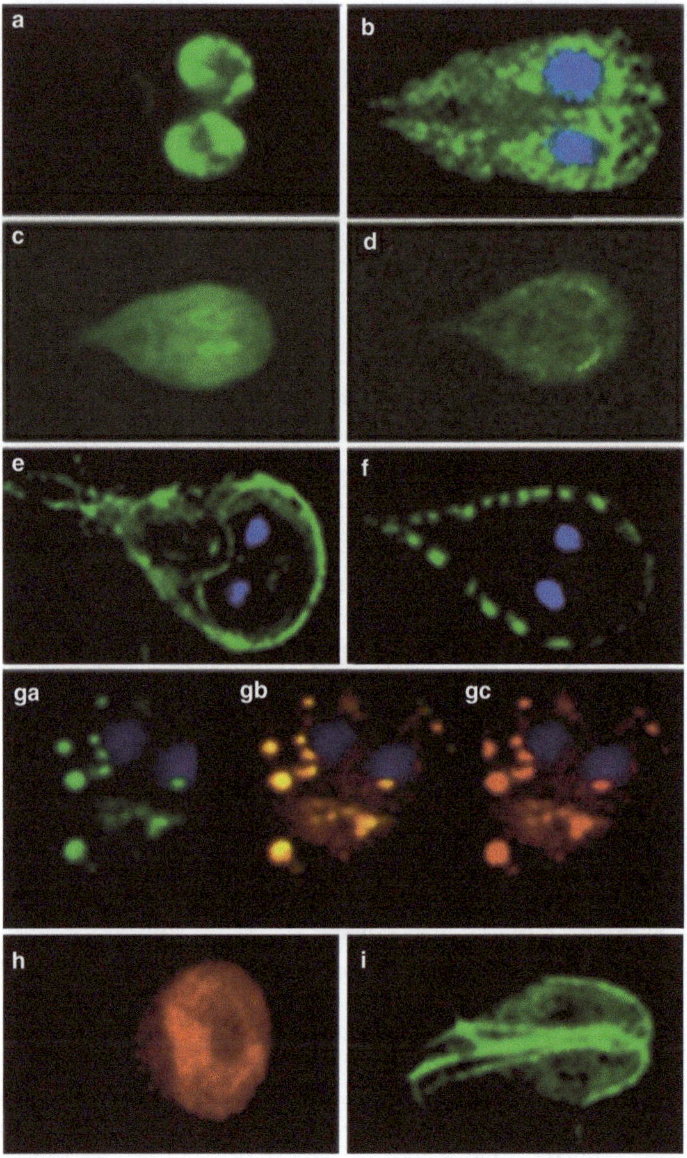

Fig. 2 *Markers of subcellular organelles in Giardia lamblia*. Immunofluorescence micrographs showing differential localization of several parasite molecules, as detected by specific monoclonal antibodies (mAbs). (**a**) Nuclear envelope (mAb 6E1 against an undetermined NE protein). (**b**) Endoplasmic reticulum (mAb 9C9 anti-BiP). (**c**) and (**d**) Potential organelles with Golgi-like function as labeled with HA-tagged ARF5 and ARF 6, respectively. (**e**) Plasma membrane (mAb 5C1 anti-VSP1267). (**f**) Peripheral vacuoles (mAb 5D2 anti-Cathepsin B). (**g**) Encystation-specific secretory vesicles (**a**, mAb 5-3C anti-CWP1; **c**, mAb 7D2 anti-CWP2; **b**, merged image showing colocalization of both CWPs into the ESVs). (**h**) Ventral disk (mAb 8C1 anti β-giardin). (**i**) Flagella (mAb 3H3 anti-α-tubulin)

activity? In cells where the secretory pathway is more polarized, such in *Toxoplasma* (Hager et al. 1999), the NE plays an important (and specialized) role in protein trafficking. It is therefore possible that the *Giardia* NE functions as a specialized organelle also required for protein transport (Benchimol 2004a; Elias et al. 2008). If so, the development of novel tools and techniques should allow further studies to provide insights about this neglected organelle.

During encystation, several changes in the nuclear envelope of both nuclei could be observed (Benchimol 2004a). These changes are the earlier ones observed during trophozoite differentiation into cysts. The nuclear envelope develops a sizable amount of nuclear pores, likely to allow a large amount of mRNA corresponding to components of the cyst wall to be transported to the cytoplasm. Other nuclear envelope process is the development of elaborate invaginations of its two membranes, surrounding some scarce cytoplasmic contents, as well as evaginations of the outer membrane that contain granular materials (Solari et al. 2003). Occasionally, using freeze-fracture, some structures can be observed inside the *Giardia* nuclei of encysting organisms as membranous profiles or even ESVs (Benchimol 2004a; see below). Some regions of the nuclear envelope present blebs which are morphologically distinct from the pore complexes. These blebs are larger than the regular pores, rounded or irregular structures in shape, and not usually found in both nuclei (Benchimol 2004a). Taken together, these results present compelling evidence that the NE is involved in a process (encystation) in which the cell needs to rapidly synthesize and transport an important number of molecules required for cyst wall formation, demonstrating that the NE in this early-branching protist is a highly dynamic organelle that respond to the changing conditions of the cell.

Regarding cell division, in higher eukaryotes the nucleus disassembles during mitosis, and the nuclear lamina, a meshwork of interconnected intermediate filaments, is involved in this process (Alberts et al. 2007). In *Giardia,* the inner nuclear membrane is covered by a dense material, probably the nuclear lamina (Benchimol 2004a). Throughout the cell cycle, *Giardia* displays the nuclear envelope without signs of any fragmentation. Even during mitosis, the nuclear envelope remains intact (Benchimol 2004a). Recently, Cande and coworkers suggested that the fusion between nuclei occurs in cyst but not in trophozoites. They suggest that somatic homologous recombination take place in cyst (Poxleitner et al. 2008).

2.2 *The Endoplasmic Reticulum*

In eukaryotes, secretory, organellar, and cell surface proteins are transported through a series of membranous compartments before their final destinations. The first (see above) of these compartments is generally the ER, where nascent polypeptides rely on molecular chaperones to facilitate their conformational maturation, which is essential for their biological activity (Haas 1994). As a rule, protein delivery to the next compartment, the Golgi apparatus, is tightly coupled to the acquisition of the correct protein structure (Bole et al. 1989). Misfolded polypeptides and unassembled protein subunits are usually subjected to ER-associated degradation, which concludes with the retro-translocation of substrates into the cytoplasm before elimination by the multicatalytic proteasome (Cabral et al. 2002).

Initially, because of morphological observations, it was believed that *Giardia* lacked an ER (Sheffield and Bjorvat 1977). The first hint that an ER was present in *Giardia* was obtained by Gupta et al. (1994) after isolating the gene encoding the *Giardia* immunoglobulin heavy chain-binding protein (BiP). BiP is an hsp70 homolog that resides in the lumen of the ER in eukaryotes (Munro and Pelham 1986). Later, Soltys et al. (1996) and Lujan et al. (1996b) using poly- and monoclonal antibodies in immunoelectron and immunofluorescence microscopic analyses were able to find BiP in a membranous meshwork distributed throughout the cytoplasm of the cell, including the NE (Fig. 2b). *Giardia* BiP contains a classic C-terminal KDEL ER-retention signal (Munro and Pelham 1987) implying that ER membranes are clearly present in this organism (Gupta et al. 1994; Lujan et al. 1996b; Soltys et al. 1996). In addition, BiP possesses an N-terminal signal peptide that is known to target the protein to the ER (Lujan and Touz 2003) and is apparently induced during encystation together with the expansion of the ER during differentiation into cysts (Lujan et al. 1996b). Subsequent studies demonstrated that the *Giardia* ER extends bilaterally through the cell body and is continuous with the nuclear envelope membrane (Hehl and Marti 2004). Similar to most other eukaryotes, it has discrete regions where small transport vesicles preferentially form (Marti et al. 2003a).

Electron microscopy (EM) studies demonstrated the localization of additional ER markers such as acid phosphatase (Feely and Dyer 1987; Lanfredi-Rangel et al. 1998), other ER chaperones (Knodler et al. 1999; McArthur et al. 2001), additionally to trophozoite and cyst secretory molecules (McCaffery et al. 1994; McCaffery and Gillin 1994).

Protein disulfide isomerases (PDIs) are ER-resident enzymes involved in disulphide bond formation of newly synthesized secretory proteins (Freedman et al. 1989). Generally, PDIs are retained in the ER via either KDEL or HDEL sequences at its C-terminus (Lujan and Touz 2003; Turano et al. 2002). Braakman et al. (1992) and Hwang et al. (1992) demonstrated that S–S bond formation in nascent secretory proteins occurs in the ER, the only sufficiently oxidizing cellular compartment in eukaryotic cells (Braakman et al. 1992; Hwang et al. 1992). Davids and coworkers found three PDIs in *Giardia* and showed that they also function as Ca^{2+}-dependent transglutaminases (TGases) (Davids et al. 2004). Unfortunately, at that time they missed two other PDIs that are now known to be present in the *Giardia* genome. When analyzing the localization of PDI 1–3, they found the presence of these molecules in the ER, the PVs, and the PM, and suggested that isopeptide bonds might help strengthen the cyst wall during differentiation. Nevertheless, in their studies they used polyclonal antibodies that might cross-react a PDI with each other due to their strong sequence similarity in the thioredoxin domain. Therefore, it is still unknown if all PDIs are located in all these subcellular compartments or individual ones function in different compartments of the secretory pathway.

Besides, only one of the five PDIs contains the KDEL motif at the C-terminus and other two have a transmembrane domain containing ER retention signals (see below).

It is well known that protein retention in (or targeting to) a certain organelle requires positive sorting information in proteins. Several amino acid motifs allowing the retrieval of proteins to the ER have been identified (e.g., KDEL, KXKXX, or KKXX) (Becker and Melkonian 1996). Analysis of the *Giardia* genome shows that very few genes encode proteins having some of these motifs (see Table 1), all of them known components of the lumen or membranes of the ER. These ER proteins are retrieved from the Golgi apparatus within COPI-coated vesicles (Letourneur et al. 1994). COPI is an essential component of this machinery, a heptameric protein complex that is recruited from cytoplasm to Golgi membrane before budding. Coatomer recruitment requires previous association of ARF1, a Ras-like GTPase that in its GTP-bound form initiates COPI coat assembly (Barlowe 2000; Donaldson and Lippincott-Schwartz 2000). The regulation of ARF function is controlled by the action of GEFs (guanine exchange factors) and GAPs (guanine activating proteins) (Randazzo et al. 2000). The *Giardia* genome encodes seven different ARFs or ARF-like proteins, three GEFs, and two GAPs (Table 1). Nonetheless, their function in ER-retrieval of any of these molecules has not been elucidated.

Retention of luminal ER-proteins is mediated by a salvage mechanism based on interaction with a specific receptor, whereby KDEL-bearing proteins are retrieved from post-ER compartment by recycling a membrane-bound receptor (Pelham 1995). pH differences between the ER and the Golgi have been proposed to account for the different affinities exhibited by the receptor toward ligands at both locations (Wilson et al. 1993). Within the lumen of the ER are a number of chaperones that bind to the polypeptide chain and assist the protein in forming the correct conformation. These chaperones include BiP (Gething 1999), lectins such Calnexin and Calreticulin (Helenius et al. 1997), which are absent in the *Giardia* genome (Morrison et al. 2007), and PDIs (Nigam et al. 1994). The issue of the retrieval of proteins back to the ER from the Golgi is particularly important in *Giardia* since in the absence of a morphologically identifiable Golgi, where does the retrieval of the KDEL-receptor take place? The *Giardia* genome possesses a single copy gene of the KDEL-receptor (ORF: GL50803_4502) and the expression of the tagged version of this molecule localizes the receptor at the ER (Stefanic et al. 2006). Then, is this receptor functional in *Giardia*? Since several KDEL-bearing chaperones can also be found in organelles later in the secretory pathway, it is clear that, at least, the KDEL-receptor is not able to retrieve all ER molecules back to the ER.

The early events triggered after the trophozoites have sensed the stimulus for encystation include the synthesis and export via ESVs of three cyst wall proteins (CWPs). CWP1–3 all form extensive intermolecular disulfide bonds that are essential for cyst wall (CW) formation and integrity, likely catalyzed by BiP and PIDs within the ER, ESVs, and, probably, at the PM (Gottig et al. 2006).

In mammalian cells, the protein traffic from ER to Golgi passes through membranous structures named ER–Golgi intermediate compartment (ERGIC) (Appenzeller-Herzog and Hauri 2006). These compartments are vesicular clusters defined by presence of the protein ERGIC-53 (Hauri et al. 2000; Klumperman et al. 1998). ERGICs have been proposed like the major site of anterograde and retrograde sorting, controlled by coat proteins (COP-I and COP-II), specific GTPases (RABs and ARFs), tethering components (SNAREs), and cytoskeletal networks (Appenzeller-Herzog and Hauri 2006). Nevertheless, in other eukaryotes, such as *Saccharomyses cerevisiae,* which lacks ERGIC clusters, a simple model for protein transport has been proposed. This raises a direct transport from the ER to Golgi, mediated by COP-II coated vesicles (Bonifacino and Glick 2004). In higher eukaryotes, protein transport from the ER is restricted to specialized domains named ERES (ER exit sites) which are evidenced by budding COP-II-coated vesicles (Bannykh et al. 1996; Hammond and Glick 2000). Marti and coworkers identified vesicular cluster similar to ERES in close proximity to ER and ESVs of *Giardia.* They observed that spherical vesicles of a diameter similar to typical COP-II vesicles (~70 nm), were closely clustered to ER networks and nascent ESVs in encysting cells. They suggest that those structures support the presence of a limited number of transitional ER sites during early encystation (Marti et al. 2003a). These observations are supported by the presence in the *Giardia* genome (http://www.GiardiaDB.org) of genes codifying to numerous proteins involved in traffic and sorting process, such as COPs, ARFs, RABs, and SNAREs (Morrison et al. 2007) (Table 1).

Soltys et al. (1996) labeled cryosections with antibodies to *Giardia* BiP. By this mean, they identified cytoplasmic clefts present throughout the cytoplasm (Soltys et al. 1996). These clefts are surrounded by membranes, suggesting that these structures derive from the ER. In addition, they observed numerous vesicles throughout the cytoplasm that could represent transport vesicles or cross-sections of ER tubules, and also stacked multilamellar membranes (Soltys et al. 1996). These cisterns and tubular forms of the ER were found throughout the cell body, often associated with other structural elements including the nucleus, PVs, and complex microtubule-based structures such as the axonemes and the adhesive disk. The tubular ER could be separated into two subregions, those with or without an associated cleft. Thus, the ER in *Giardia* could have subregions with different compositions, although they always contain BiP. The nuclear envelope is considered to be another subregion of ER. Cleft regions have been also found in the nuclear envelope, suggesting further compositional similarities (Lanfredi-Rangel et al. 1999; de Souza 2006). This is consistent with the current appreciation of the complex organization of ER in higher eukaryotes (Sitia and Meldolesi 1992). Interestingly, although BiP is essentially found in a reticular membranous network in vegetative and encysting trophozoites, it is also observed inside ESVs during encystation, probably still associated to CWPs during their transit to the plasma membrane (Lujan and Touz 2003).

2.3 The (elusive) Golgi Apparatus

The Golgi apparatus (GA) is an essential organelle for intracellular protein trafficking in higher eukaryotes. It carries out critical functions, generating posttranslational modifications of lipids and proteins, and producing packaging and sorting of secretory compounds (Jackson 2009). Despite its importance, it is known that some eukaryotes do not possess classical GA (Dacks et al. 2003). It was suggested that the lack of this organelle in primitive cells was a consequence of the inheritance from primitive ancestors (Cavalier-Smith 1987). However, new evidences have revealed that the loss of the classical GA described in some textbook of cell biology could be the result of late events during evolution, e.g., adaptation to parasitic life style (Dacks et al. 2003).

The typical flattened cisternal membranes of the GA have not been found in trophozoites by standard microscopic techniques (Becker and Melkonian 1996; de Souza 2006). Nevertheless, Golgi-like structures (GLS) have been observed in encysting cells (Becker and Melkonian 1996; Gillin et al. 1996; Lujan et al. 1995a). Using electron microscopy, a parallel array of smooth membranes (Golgi-like stacks) was observed in encysting trophozoites (Becker and Melkonian 1996). Nevertheless, the absence of GA markers precluded the correct assignment of these stacks as a Golgi in *Giardia*. GLS were also potentially evidenced in encysting (Lujan et al. 1995a) or both, encysting and vegetative trophozoites (Lanfredi-Rangel et al. 1999) using NBD-ceramide, a Golgi marker in higher eukaryotes (Lipsky and Pagano 1983). Those experiments showed that the fluorescent lipid localized in perinuclear areas after induction of encystation, usually over one of the two nuclei (Lujan et al. 1995a). However, the membranes labeled with this lipid analog were never linked to the stack of smooth membranes detected by EM, raising the question if NBD-ceramide only labels some NE/ER membranes containing a specific composition capable to incorporate the fluorescent marker. Our group has also reported that, during encystation, trophozoites undergo the induction of enzyme activities typical of the GA, which correlates with the appearance of NBD-ceramide-labeled structure (Lujan et al. 1995a). We thought at that time that these glycosyltransferases activities were proper GA enzymes, however, since we used artificial substrates, these enzymatic activities might have well corresponded to the still unidentified cyst wall synthase, a molecule that incorporated *N*-acetyl glucosamine and *N*-acetyl galactosamine to the complex carbohydrate portion of the cyst wall (Jarroll et al. 1989). Since it is still not clear where this carbohydrate polymer forms within the cell, our conclusions from these early experiments should be considered with care.

Interestingly, the treatment with BFA (Brefeldin A), a fungal toxin capable to block proteins sorting, induces dissociation of GLS to the cytoplasm and the nuclear envelope and redistribution of the activities of glycosyltransferases (Adam 2001; Lujan et al. 1995a). Nevertheless, other authors found GlcNAc transferase activity (normally associated to the Golgi) in the ESVs, suggesting that these organelles could play a role as the Golgi apparatus in this primitive cell (Marti and Hehl 2003). To demonstrate this theory, Marti and Hehl (2003) showed the presence of specific Golgi markers such as β-COP (a Coatomer subunit, Table 1) and ARF-1 (ADP-ribosylation factor 1) in ESVs. However, these results are controversial because the antibodies used in the experimental procedure were polyclonal sera against the heterologous mouse proteins (Marti et al. 2003a).

Conversely, Lujan´s group observed a different pattern using a specific polyclonal antibody generated against recombinant *Giardia* ARF. They showed that gARF and β-COP were associated with small structures scattered around the nuclei, which were also sensitive to BFA. Besides, the localization of these proteins was morphologically different from that shown by NBD-ceramide in encysting cells. The NBD-ceramide-labeled structures may represent a late Golgi compartment, as has been reported in mammalian cells or just an artifact as explained previously. If this compartment in *Giardia* is spatially segregated from early Golgi (where ARF and β-COP are present), this could explain the differential distribution of both β-COP and NBD-labeled structures (Lujan et al. 1995a).

It is evident that GLS play an important role in the encystation process, and probably are related with the sorting of some component of cyst wall (Hehl and Marti 2004; Lujan et al. 1995a; Marti and Hehl 2003). However, *Giardia* lacks the three large vesicles-tethering complexes (Marti et al. 2003b) called exocyst (a multimeric tethering complex involved in post-Golgi trafficking) (TerBush et al. 1996), Golgi-associated retrograde protein complex (GARP) (Liewen et al. 2005), and oligomeric Golgi complex (COG) (Smith and Lupashin 2008). In addition, the characterization of the complete set of SNAREs proteins in this organism (SNAREs comprise a family of membrane-associated proteins that confer the tight docking and subsequent fusion of membrane bilayers (Nickel et al. 1999)) did not evidence a typical Golgi apparatus (Elias et al. 2008). Moreover, SNAREs homologs to those found in the GA of higher eukaryotes localized in both the NE and ER in *Giardia* (Elias et al. 2008).

Based on the data shown above we hypothesize that *Giardia* do not have a typical Golgi apparatus, but probably their packaging and sorting functions are carried out by others organelles, such as the NE and the ER (Fig. 2c). The absence of the stack of flattened cisterns typical of the Golgi in higher eukaryotes must be due to the lack of complex protein glycosylation in *Giardia* (Elias et al. 2008; Gottig et al. 2006; Lujan et al. 1997). During the last years, it was reported that this parasite is only capable to add via dolichol-PP two GlcNAc to Asn residues of proteins. This incomplete glycosylation process appears to result from secondary loss of glycosyltransferases from a common ancestor that contained the complete set of N-linked glycan glycosyltransferases (Samuelson et al. 2005). Recently, analysis of the N-glycome of *Giardia* further supports this idea (Ratner et al. 2008).

2.4 The Peripheral Vacuoles

The PVs are small vesicles that are located underneath the plasma membrane (Fig. 2d), in the cell periphery, along the dorsal and ventral sides of trophozoites but not over the ventral disk (Fig. 2e). These are regular in size and ovoid to tubular in shape. The interior of individual PVs varies in electron opacity, which suggests the presence of different contents (Gillin et al. 1996; Lanfredi-Rangel et al. 1998). Despite that the interaction between the cellular membrane and the PVs has not been yet demonstrated, vesicles of 50–80 nm emerging from PVs toward the extracellular membrane were observed. It suggests a possible vesicle-mediated traffic between these organelles (Lanfredi-Rangel et al. 1998). The localization of ER markers such as glucose-6-phosphatase and BiP in portions of some PVs suggests continuity between both organelles (Adam 2001; Lanfredi-Rangel et al. 1998).

Activity of lysosomal soluble enzymes such as acid phosphatase, ribonucleases, and cathepsin-B and -C and cysteine proteases were also detected in PVs (Adam 2001; Lanfredi-Rangel et al. 1998; Touz et al. 2003). In addition, the sorting of an encystation-specific cysteine protease (ESCP) having a lysosomal tyrosine targeting motif was characterized in *Giardia* trophozoites (Touz et al. 2003, 2004). It suggests that PVs could play a role as a lysosomal compartment (Gillin et al. 1996; Lanfredi-Rangel et al. 1998; Touz et al. 2003). Although endocytosis has not been well characterized in *Giardia*, both the presence of genes encoding proteins related to this process (Clathrin, Adaptor proteins 1 and 2, etc.) and the uptake of transferrin from the extracellular environment toward the PVs indicates that PVs may perform functions similar to those of endosomes (Lanfredi-Rangel et al. 1998; Touz et al. 2003, 2004). Based on that evidence, the PVs are believed to correspond to an ancient endosomal–lysosomal system that later might evolve to early and late endosomes and lysosomes in higher eukaryotes (Lanfredi-Rangel et al. 1998). However, given that not all PVs look identical, it is also possible that some of them may act as endosomes and the others as lysosomes. Additionally, it was shown that PVs plays an important role during excystation (the process that allows the trophozoite to be released from the cyst to initiate a new infection (Ward et al. 1997)), being involved in the secretion of hydrolytic enzymes that facilitate the removal of the CW (Slavin et al. 2002). These results indicate that PVs may also constitute secretory organelles. Further studies regarding the endocytic and exocytic functions of the PVs are necessary to fully understand their involvement on the biology of the parasite.

2.5 The Encystation-Specific Secretory Vesicles

When *Giardia* is exposed to the encystation stimulus, it generates encystation-specific secretory vesicles (ESVs) containing cyst wall materials, which can constitute a regulated secretory pathway (Fig. 2f), functioning in parallel to the constitutive pathway transporting variant surface proteins (VSPs), to the plasma membrane and PV-resident enzymes (Marti et al. 2003a). The ESVs are the only large subcellular compartment containing cargo found and partially characterized in *Giardia* and, at the same time, is developmentally induced (Reiner et al. 1990) since these organelles are only present during encystation (Faubert et al. 1991). Reiner et al. (1990) were the first to demonstrate that the ESVs transport undetermined cyst antigens to the cell surface for the formation of the cyst wall (CW) (Reiner et al. 1990). However, years later, by using specific monoclonal antibodies

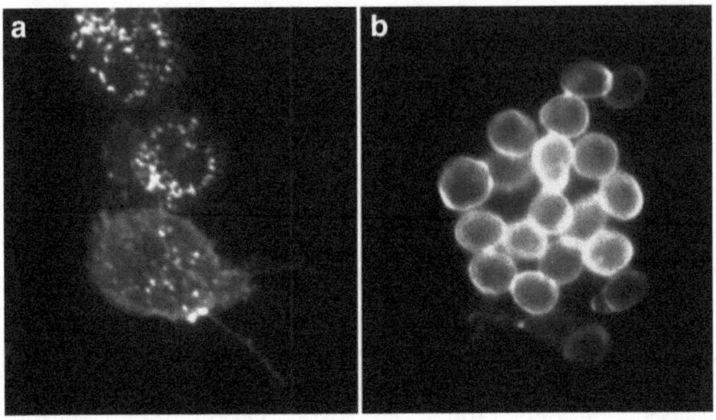

Fig. 3 *Dynamics of the process of encystation in Giardia lamblia.* Immunofluorescence assay using an anti-CWP2 monoclonal antibody during trophozoite encystation in vitro. (**a**) Trophozoites showing a variable number of ESVs of variable size and shape. One trophozoite in the latest stages of encystation shows cyst wall materials deposited on the plasma membrane, including the flagella, named "tailed cysts" (**b**) Mature cysts in the culture supernatant show rounded shape and the cyst walls heavily react with the specific monoclonal antibody

generated to the ESVs several components of these granules were identified (Lujan et al. 1995b).

In some trophozoites, the ESVs are extremely numerous and large in size, occupying much of the cytoplasm (Fig. 3a). Many are interconnected by a tubular network (Lanfredi-Rangel et al. 2003). Often the ESVs are perinuclear and, in occasional cases, they partly enclose the nucleus and appear to form near and in continuity with the nuclear envelope. Other ESVs can be seen at the cell periphery compressing or in close apposition to the PVs. Some ESVs are elongated and undulating, while others appear rather geometric (Faubert et al. 1991; McCaffery and Gillin 1994). Nevertheless, purified ESVs are spherical (Lujan et al. 1996b). Nothing is known about why ESVs are of different size and shapes, if this is a characteristic of these organelles, or if they are artifacts caused by fixation of the cells.

In higher eukaryotes, proteins destined for regulated secretion are folded, assembled and glycosylated as they are transported from the ER through the Golgi to secretory granules (Jackson 2009). Some cells are specialized to store secretory materials within these cytoplasmic organelles, which are generally round. The process involved in the formation of the secretory granule in higher eukaryotes can be summarized as three distinct events (a) the selective condensation of secretory proteins which aggregate to form a dense core, (b) the selection of the membranes which surround the aggregate, and (c) the budding and release of the nascent secretory granule (Orci et al. 1987; Rambourg et al. 1988). The biogenesis of the secretory granules in higher eukaryotes is difficult to study because cells specialized in regulated secretion contain secretory granules during their entire life (Orci et al. 1987; Rambourg et al. 1988). The ability to regulate the formation of these granules in *Giardia* just by changing the culture medium makes this parasite an excellent model to study how and where the secretory granules form, how proteins are sorted to and concentrated within these specialized organelles, and why other proteins are excluded from them. These issues have been analyzed in a recent work of Gottig et al. (2006) (see below).

Structurally, all ESVs have the characteristics of eukaryotic secretory granules (Lanfredi-Rangel et al. 2003). However, secretory granules in higher eukaryotes form in the TGN (Orci et al. 1987; Rambourg et al. 1988) which raises the question if *Giardia* really has a GA where these granules can be generated or the ESVs form directly from the ER (or specific portions of the ER) (Gottig et al. 2006). Several studies have shown that CWPs are synthesized in an area of the ER where the cistern is modified, forming a dilated region known as cleft (Gillin et al. 1996; Lanfredi-Rangel et al. 2003; Lujan et al. 1995b). The cleft is continuous with the ER and lacks an electron dense content. Gradually the cleft widens and becomes filled with a homogeneously dense material formed by the concentration of the CWPs. Although continuity of this structure with the ER is evident, glucose-6-phosphatase, a classical enzyme marker of the ER, is not found in these structures (Lanfredi-Rangel et al. 2003). Subsequently, the dense vesicle, which can be now designated as ESV, increases in density and migrates toward the periphery of the cell. In contrast, Marti and Hehl (2003) suggest that ER vesicles containing CWP fuse to each other to form the ESV (Marti and Hehl 2003).

It was recently demonstrated that the formation (biogenesis) of the ESVs is functionally linked to the regulated expression of exported cargo, i.e., CWPs (Gottig et al. 2006). In this work, it was shown that after *Giardia* senses the stimulus for encystation, CWPs are specifically expressed and concentrated within ESVs. Although CWP1, CWP2, and CWP3 are structurally similar, CWP2 distinguishes from CWP1 and CWP3 by the presence of a carboxy-terminal 121-amino acid basic extension. This basic C-terminal extension of CWP2 appears necessary but not sufficient to drive biogenesis of ESVs and that interaction/aggregation among secretory granule cargo proteins is also required. Additional results indicated that CWP2 is a key regulator of ESVs formation by functioning both as an aggregation factor for the other CWPs as well as a ligand for sorting. Sorting and cargo protein aggregation have been suggested to be important for secretory granule biogenesis in higher eukaryotes, but Gottig et al. (2006) provided evidence that both mechanisms (active and passive) are necessary. By expressing different CWP chimeras, either containing or not containing the basic tail of CWP2, they showed that those CWP containing the tail were able to form ESV in nonencysting trophozoites. Conversely, expression of a version of a CWP2 lacking the basic extension labeled the ER and was released constitutively to the culture medium. Over-expression of the tail alone did not form granules in nonencysting trophozoites, indicating that other regions of CWP2 are necessary to complex the additional CWPs. Due to the fact that the formation of ESVs just occurs during differentiation of *Giardia*, they were able to differentiate both mechanisms (sorting and aggregation) for the first time (Gottig et al. 2006).

Other authors, however, suggested that ESVs represent a novel Golgi equivalent in *Giardia* (Hehl et al. 2000; Marti et al. 2003a). These authors propose that ESVs appear to arise from the homotypic fusion of smaller transport intermediates, presumably ER-derived transport vesicles, as documented by transmission electronic microscopy (TEM) of encysting trophozoites (Marti and Hehl 2003). But again, this hypothesis is controversial since only stationary studies were performed and it is impossible to discriminate whether the small vesicles observed near the ESVs were going to fuse with the large ones, they were released from the large ESVs, or they were small ESVs becoming large after continuous addition of CW materials. In contrast, Gottig et al. (2006) presented results indicating that the sorting and packing functions of the Golgi apparatus can be provided by the ER in the absence of a morphologically identifiable Golgi complex, clearly supporting the hypothesis that the Golgi apparatus derives from the ER (Gottig et al. 2006).

Moreover, Gillin and coworkers propose that ESVs may have unusual pathways of formation and traffic. ESV have a very uniform density and are more electron dense than the cytoplasm, suggesting that if the proteins that are transported associate with the fibrous portion of the cyst wall, this association may occur after exocytosis. Gillin's group have observed ESV releasing their contents to the cyst wall by exocytosis (Reiner et al. 1990). However, since no specific markers for CWPs were utilized in those studies, it is not clear if the exocytosis corresponded to ESVs (see below). At the cell periphery, the ESV established contact both with the inner portion of the plasma membrane of the trophozoites as well as with the PVs (Lanfredi-Rangel et al. 1998). Because CWPs are processed by a cysteine proteinase localized in the PVs (Touz et al. 2002b), it suggests that fusion of the ESV with PVs takes place immediately before or simultaneously with the fusion of the ESV with the cell surface (de Souza 2006).

The fact that fluorescent proteins do not fluoresce in *Giardia* because it is a microaerophylic organism and these reporters require molecular oxygen to obtain the fluorescent conformation (Hehl et al. 2000), there is no possibility so far to perform dynamic studies in *Giardia* using those valuable approaches. This highly diminished our possibilities to get a better understanding of the secretory pathway in *Giardia* as compared to the knowledge obtained in other microorganisms (Regoes and Hehl 2005).

2.6 The Cytoskeleton

In *Giardia* there is an apparently strong link between the cytoskeleton and virulence, since trophozoites colonize the small intestine in the host by attaching to the intestinal cells, where they obtain the necessary nutrients by means of the ventral adhesive disk (Fig. 2g) and the flagella (Fig. 2h) (Elmendorf et al. 2003). *Giardia* must be able to swim and attach to the host intestinal epithelial layer to avoid being evacuated by peristalsis (Elmendorf et al. 2003). The four pairs of flagella allow the parasite to move around the lumen of the small intestine and the ventral disk mediates a mechanical attachment both to the intestinal wall. Thus, the cytoskeleton and the ventral disk play a key role in the survival of the organism within the host (Adam 2001). *Giardia* is a highly polarized protist, and the cytoskeleton is dominated by several structures, including eight flagella, the ventral disk, and the median body (Elmendorf et al. 2003). The flagella appear to be important for motility but not for attachment. Moreover, their early emergence through the cyst wall during the process of encystations suggests their importance in encystations (Buchel et al. 1987). The ventral disk contains the contractile proteins actinin, myosin, giardins, and tropomyosin (Feely et al. 1982) as the biochemical basis for the contraction of the disk involved in adherence (Adam 2001). The median body is a component of the cytoskeleton that is located in the midline and dorsal to the caudal flagella and consists of a group of microtubules in a tight bundle (Adam 2001). The function of the median body is not well established (Elmendorf et al. 2003) although possibly reserve cytoskeletal proteins for mobilization prior to mitosis or encystation (Gillin et al. 1996).

The *Giardia* genome shows a large repertoire or cytoskeletal proteins, including kinesins, dynamins, microtubule-associated proteins (MAPs), in addition to the components of microtubules and microfilaments (http://www.GiardiaDB.org). Nevertheless, the participation of the cytoskeleton in protein trafficking in *Giardia* has not been studied. The transport of the ESVs to the cell periphery is particularly relevant, since it is not clear how these vesicles (which are generated throughout all the cell body) contact and fuse with the cell membrane to release their content. Only few reports have claimed the observation of the release of ESV content to the cell exterior (Benchimol 2004b; Lanfredi-Rangel et al. 1998; Marti and Hehl 2003; Touz et al. 2002b). But, are all the ESVs releasing their content at once? Or are they fusing with the plasma membrane one after another? It was recently reported that a single dynamin homolog is recruited to the ESVs and involved in the formation and/or maturation of these organelles and indeed necessary for completion of cyst formation (Gaechter et al. 2008).

3 The Encystation Process

3.1 General Characteristics

G. lamblia alternates between the trophozoite stage in the vertebrate host and environmentally resistant infectious cyst, which is characterized by a thick extracellular matrix (the cyst wall) that protects the parasite outside the host's intestine (Adam 2001). Cyst formation begins when the trophozoites travel down the host's intestine and find an environment poor in cholesterol, which is known to be absorbed in the latest portions of the small intestine (Adam 2001).

Cyst diameter is about 5 by 7–10 μm and the trophozoite is enclosed by a wall that is 0.3–0.5-μm thick. The outer portion is covered by filamentous proteins associated to sugar compounds, predominantly *N*-acetyl galactosamine in the form of a complex polymer (Jarroll et al. 1989). This structure constitutes a resistant layer that allows the trophozoite to survive in hostile conditions such as osmotic shock, variable pHs, and temperatures changes, as well as the presence of chemical disinfectants (Lujan et al. 1998). The inner portion of the cyst wall is formed by two membranes enclosing the periplasmic space that permit interactions with the external media, exchanging different substances such oxygen and essential metabolites (Adam 2001). The presence of these two membranes is intriguing since one of them may represent the plasma membrane of the enclosed trophozoite because it contains VSPs and as far as the other one (on which the filamentous cyst wall is apposed) is concerned, its source is unknown since the antibodies against the VSPs cannot pass through the rigid cyst wall in mature cysts. One can speculate that during the release of the ESVs an excess of membranes occurs on the parasite surface. However, since VSPs are also observed in the PVs of encysting cells, a high level of membrane recycling must take place.

As stated before, the *Giardia* encystation process included three different stages (a) sensing the stimulus for the encystation and upregulation of encystation-specific genes; (b) biochemical and morphologic modifications involving the synthesis of cyst molecules and the biogenesis of secretory organelles implicated in trafficking of cyst wall components; and (c) release of secretory granules content and assembly of the extracellular cyst wall (Lujan et al. 1997; Lujan and Touz 2003; Touz et al. 2002b). Entry of cells into encystation is not synchronic. Once the stimulus is received, expression of molecules related to this process such as CWPs show a high increase at the level of mRNAs and proteins, clearly indicating that encystation-specific genes are transcriptionally regulated (Lujan et al. 1997; Touz et al. 2002a). Similarly, other molecules that are involved in transport and secretion of cyst wall materials such us BiP/GRP78, protein disulfide isomerase 2 (PDI-2), cathepsin C, SNARE proteins (Sintaxin 1 and 2), NSF, SNAP, and VAMP are also upregulated (Touz et al. 2002a; Wang et al. 2007).

3.2 Encystation and Cyst Specific Molecules

Biochemical analysis indicated that the cyst wall consist of both carbohydrate (43%) and protein (57%) components (Jarroll et al. 2001, 1989). Only three structural CWP are known (CWP1, CWP2, and CWP3). These CWP are related leucine-rich repeat containing proteins (Lujan et al. 1995b; Sun et al. 2003). Recently, a novel High Cysteine Non-VSP molecule has been reported to be present on the ESVs and in the mature cyst wall. This protein resembles trophozoite's VSPs, but possesses a different cytoplasmic tail (Lauwaet et al. 2007). CWP1 (26 kDa) and CWP2 (39 kDa) contain 18 and 19 cysteine residues, respectively, and form disulfide-bonded heterodimers and oligomers soon after synthesis (Lujan et al. 1995b). CWP3 (27 kDa) also form heterodisperse molecular weight disulfide-bonded complexes (Sun et al. 2003). CWP1–3 has a hydrophobic N-terminal signal peptide that targets them to the secretory pathway. The major difference between CWP1 and 3 with CWP2 is the presence of a 121 amino acid carboxy-terminal extension in CWP2, which predicts a high positive charge in this domain at physiological pH (Lujan et al. 1997). This C-terminal region, in CWP2, is present within ESVs, but is protolytically cleaved before cyst wall assembly (Gottig et al. 2006). It was demonstrated that this basic extension in CWP2 is necessary but not sufficient to trigger the formation of the ESVs (Gottig et al. 2006).

In contrast to CWP1–3, whose exclusive destination is the ESVs and the CW (Fig. 3), the *H*igh *C*ysteine *N*onvariant *C*yst *p*rotein, HCNCp, which is also upregulated during differentiation, differs in its behavior (Davids et al. 2006). This protein is detected in trophozoites and it colocalizes with CWP to the ESVs during encystation. Although HCNCp is in the wall of mature cyst, much of it remains in the cell body. In addition, this protein lacks LRR and has 14% cysteine with many "CXXC" or "CXC" motifs and a divergent, VSP-like C-terminal transmembrane domain (Lauwaet et al. 2007). It is possible that this integral protein is a structural component of the ESV membrane. Its presence in the cyst wall should be better studied to determine if it is in the outer membrane of the periplasmic space or integrated into the cyst wall. If HCNCp is a component of the ESV membrane and then is retained during exocytosis on the cell surface, this protein may constitute a fundamental tool to determine the biogenesis of the double membrane that is arranged underneath the cyst wall.

After induction of encystation, large amount of CWPs and glycans are produced. Morphologically recognizable ESVs are observed early during encystation and constitute the first morphological evidence that trophozoites are undergoing encystation (Faubert et al. 1991; Reiner et al. 1990). All the newly synthesized materials are exported from the ER to nascent, which are leading to the cell surface for release and assembly of the extracellular cyst wall (Marti and Hehl 2003). Although the exact mechanism for this process remains unclear, the transport of the CWPs constituted a regulated pathway. Independently, several studies have shown that ESVs are central to cyst wall formation as many genetic or chemical manipulation that interferes with ESV formation block all downstream events (Lauwaet et al. 2007). Based in all the data about the biogenesis of ESVs, two possible models have been postulated (a) aggregation and concentration of cyst wall material in specialized ER subcompartments, (b) export of cyst wall materials in COPII coated transport vesicles that give rise to ESVs by homotypic fusion (Stefanic et al. 2006).

The first model suggests that ESV biogenesis is a direct consequence of CWP synthesis (Gottig et al. 2006). This theory supports the idea that the ESV form de novo because of the accumulation of CWPs within membrane bound clefts corresponding to the RER. They appear without content and have glucose-6-phosphatase activity but lower than in the ER. Gradually, the clefts widen and become filled with homogeneous denser material, later shown to consist of CWPs (Lanfredi-Rangel et al. 2003). By direct addition of these CWPs, these aggregations of recently synthesized CWPs are increasing, giving rise to larger ESV where the glucose-6-phosphatase activity is lost. The electron-dense nature of these vesicles indicates a tightly packed or highly condensed arrangement of their contents. Some mechanisms for preventing premature formation of such filamentous pH, molecular chaperones, calcium ions, must exist within ESV because no filamentous structures are in the ESV (Lujan et al. 1995b). As described earlier, ESV formation entails active and passive mechanisms because it is rigorously dependent on the complex interactions between granule material (namely CWPs) and granule membrane receptors (Gottig et al. 2006).

The second model implies that the ESVs arise from homotypic fusion of smaller intermediates of transport, apparently ER-derived vesicles which are most likely COPII-coated vesicles (Hehl and Marti 2004). Marti et al. (2003a) suggest that this process is carried out by recruitment of only one or just a few factors that also stabilize the nascent cisternae until exocytosis, without the complex neogenesis of an entire secretory system (Marti et al. 2003b). Immunofluorescence analysis and subcellular fractionation studies performed by those authors have shown that ESVs are not generated de novo but form from an existing compartment in trophozoites, which clearly contrast the results of Gottig et al. 2006. The hypothesis of Hehl's group imply that *Giardia* accommodates the export of a great amount of cyst wall materials via ESV trough reorganization and extension of an unknown preexisting machinery, rather than by the biogenesis of compartments during encystation (Marti and Hehl 2003). According to these results, exported proteins are sorted according to their targeting signal to distinct COPII-vesicles, possibly already present at the ER exit sites. The soluble CWPs are dependent on cotransport with other *trans*-membrane proteins providing this function. Only transport intermediates containing CWPs are able to fuse homotypically and give rise to the larger ESV compartments (Marti et al. 2003a).

The CWPs are transported by ESV from the ER to the cell periphery for subsequent cyst wall assembly (Benchimol 2004b), but the exact mechanism of this process is until undefined. There are at least three possible pathways for the protein secretion during cyst wall formation:

(A) The secretory granules (ESV) are fragmented into smaller transport vesicles before their release on cell surface, suggesting that the ESV are the latest portions of the GA, the TGN of higher eukaryotic cells (Benchimol 2004b). Scanning EM studies have shown large numbers of small blobs of cyst wall material on the outer face of the plasma membrane of encysting cells at the beginning of the process, arguing for secretion of the cyst wall material in small portions (Erlandsen et al. 1996).
(B) Direct contact between ESV and the plasma membrane. Close to the plasma membrane, the ESV establishes contact with the inner portion of the plasma membrane of the trophozoites. The ESV membranes fuse directly with the plasma membrane and release their content (Hehl and Marti 2004). This fusion is incomplete, because fragments are formed during the vesicle release. These remnant membrane fragments are sealed and empty vesicles are formed (Benchimol 2004b).
(C) Interaction between PVs and ESV followed by the exocytosis of CWPs. This model has been supported for electron microphotograph which showed an

intimate contact between ESVs and PVs (Lujan et al. 1997). At the cell periphery, the ESV make contact with the PVs, acidic organelles that correspond to an endosome–lysosome system (de Souza 2006), whose function could be related to the storage or processing of cyst wall materials (Lujan et al. 1997). Touz et al. (2002b) observed that CWPs are processed by a cysteine proteinase localized in the PVs, suggesting that fusion of the ESV with the PVs takes place immediately before or simultaneously with the fusion of the ESV with the cell surface (de Souza 2006; Touz et al. 2002b). Because of the acidic nature of the PVs, they contain proteolytic enzymes, which may participate in CWP2 cleavage after interaction between the two organelles. This is a central step, as inhibition of CWP2 proteolytic processing also inhibits cyst wall formation (Touz et al. 2002b).

3.3 Cyst Wall Assembly and Maturation

The biosynthesis and assembly of eukaryotic extracellular superstructures such as the plant and fungal cell walls (Klis 1994; Reiss et al. 1992), and the cyst or spore wall of medically important intestinal pathogens (Mehlotra 1993) are poorly known processes. How *Giardia* generates a highly ordered supramolecular cyst wall is an interesting problem that can be taken as a model to study cell wall morphogenesis in general. Synthesis of the cyst wall poses topological problems since precursors must be synthesized intracellularly, but deposition of the components and all macromolecular organization occur outside the permeability barrier of the cell. Therefore, there must be special mechanisms to convey the precursors and the machinery for cyst wall assembly to the cell surface, avoiding their polymerization within the cell.

After induction in vitro (Boucher and Gillin 1990; Lujan et al. 1996a), the first morphological evidence of ESV release is that the cells became rounded and unattached. Subsequently, the flagella are gradually internalized while the cell is folded due to the increased size of the flange (Midlej and Benchimol 2009). The flagella and the ventral disk are irrelevant in the cyst form; cytoskeletal structures are broken down in several and large fragments and stored in the cyst's cytoplasm until they become necessary to rebuild the ventral disk and the flagella during excystation (Elmendorf et al. 2003). The process of encystation has been shown to consist of two morphologically different stages (intracellular and extracellular) requiring approximately 16 h for completion (Erlandsen et al. 1996).

It is not yet fully understood how the carbohydrate and protein portions of the cyst wall are released and how they interact extracellularly to organize the wall architecture. The formation and processing of the resistant cyst in *Giardia* comprises a complex series of coordinated events involving components of the endomembrane system, as well the activation of several synthetic processes, including profound alterations of the cell shape and intracellular organellar modifications (Midlej and Benchimol 2009). The CWPs are transported by the

ESV to the plasma membrane in stable and insoluble complex that avoids fibril formation inside the cell (Lujan and Touz 2003). This suggests that the formation of this compartment, and also probably their cargo, undergo a maturation process in which CWPs are posttranslationally modified (Stefanic et al. 2006). The late phase of encystation consist of the appearance on the trophozoite membrane of several sites for initiation of the assembly of cyst wall filaments followed by the assembly of the filamentous portion of the cyst wall (Adam 2001). The congregation of materials that constitute the CW is a direct consequence of exocytosis of ESV (de Souza 2006).

Once the CWP are released from the ESVs and secreted to the cell surface, they will undergo further modifications like disulphide bond formation and rearrangements in order to reach their final conformation as resistant fibrils. Evidence for cargo modification is inferred because proteins like BiP and PDIs have been observed that change their localization (Lujan and Touz 2003). These enzymes colocalize with the CWPs in the ER and ESVs, acting as chaperones that assist CWPs folding and probably CWPs complex formation. PDIs also have been found on the cell surface, and probably have *trans*-glutaminase activity forming isopeptide protein cross-links that are resistant to degradation (Lauwaet et al. 2007). Consequently, these data suggest that chaperones may also be participating at later times in the secretory pathway that concludes with CW formation (Lujan and Touz 2003; Reiner et al. 2001).

The secretory granules of encysting trophozoites contain large amounts of Ca^{2+} (Touz et al. 2002a). Calcium function has also been attributed to the packaging and processing of the contents of the secretory granules in higher eukaryotes (Pozzan et al. 1994). A *Giardia* granule-specific-protein (GSP) has been identified in the lumen of ESVs. This enzyme is proteolytically cleaved before cyst wall formation and is able to bind Ca^{2+}. In the regulated secretory pathway that develop when *Giardia* encyst, GSP might function as a calcium sensor within the ESVs and, therefore, might regulate granule discharge at the time of cyst wall formation. Thus, GSP may also play an essential role avoiding the premature assembly of the cyst protein inside the secretory vesicles and controlling the exocytic mechanism by Ca^{2+} regulation, since knock-down of this molecule blocks ESV release (Touz et al. 2002a). Besides, a single dynamin-related protein (GIDRP) has been identified in *Giardia*. It is primarily localized at the cell periphery where the PVs underlie the plasma membrane. A direct role of GIRDP in the late process of encystation has been proposed in relation with the recruitment of clathrin to ESVs during their last stage of maturation to secretor-competent vesicles. Therefore, GIRDP seems to be necessary for the secretion of the cyst wall materials playing an essential role in the last phase of encystation (Gaechter et al. 2008).

Regarding cyst wall composition, the filamentous cyst walls contain carbohydrate and proteins. The assembly begin with the deposition of fibrillate material as irregularly shaped bundles (fibril patches) of variable extension ($0.1 - > 2.0\ \mu m$) over the dorsal and ventral surface. These materials first appear like small protrusions and become larger patches. The fusion of fibril patches apparently advanced from the anterior to the caudal portion of the cells, leading to flagella entrapping.

These structures grow randomly forming multiple areas of coalescence that overlap progressively arising at the cyst wall with a constant thickness of about 0.25–0.30 μm. At that moment, the cell turns into rounded form completely covered by the filamentous meshwork and acquires a smooth texture to its surface. A flagellum-like structure remained visible seemingly a "tailed-cyst" and when the whole "tail" disappeared from the cell body, it became oval and acquired a nonmotile cystic form (Arguello-Garcia et al. 2002) (Fig. 3a).

The filamentous structure of the cyst wall has sufficient flexibility for the trophozoites to move within the cyst, but they are rigid enough for maintaining the viability of trophozoites. This particularity could be supported by the LRR (leucine-rich repeat) domains found in the CWP. The LRR motif could constitute a point of interaction between the cyst wall components providing stability to this superstructure (Lujan et al. 1997). Despite CWP 1 and CWP2 each one has a possible site of N-glycosylation; (Lujan et al. 1997), carbohydrates-bound to the CWPs have not been demonstrated to date. In addition, both the presence of numerous potential sites of O-glycosylation and the induction of the galactosamine and *N*-acetyl-galactosamine (GalNac) transferase activities during encystation suggest that the CWPs may be O-glycosylated (Lujan et al. 1995c). But again, no direct evidence neither for glycosylation nor attachment to the carbohydrate polymer that compose the CW has been reported for the CWPs.

The major monosaccharide constituent of the cyst wall is *N*-acetyl-galactosamine (GalNac). This is synthesized de novo from endogenous glucose through a pathway of inducible enzymes which are both transcriptionally and allosterically regulated, and the GalNac is fixed into an insoluble polysaccharide by the action of "cyst wall synthase." Because of the abundance of the GalNac and the insolubility of the cyst wall, it is possible that the filaments are composed of a polysaccharide-protein complex. A possible model for the inclusion of the saccharides on the cyst wall is their interaction with preexisting cyst wall polypeptides that are finally incorporated within the mature filaments (Erlandsen et al. 1990, 1996). Regarding this issue, it has been demonstrated that the filaments of the cyst wall contain a novel (β1–3)-linked GalNac homopolymer, but not chitin.

The generation of the carbohydrate homopolymer is unclear. The activity of the enzyme that generates this compound has been established, but neither their primary sequence nor their subcellular localization is unknown. The precursors of GalNac are synthesized by cytoplasmic enzymes that are upregulated during encystation. Therefore, where is the homopolymer assembled and how is it transported and released to the cell exterior? Benchimol (2002) described some novel vesicles that are able to be stained for carbohydrates and are different from the ESVs transporting CWPs (Benchimol 2002). If there is another set of encystation-specific vesicles whose function is the transport of the carbohydrate portion of the cyst wall is controversial. If they exist, how is the polymer synthesized in the cytoplasm translocated to the interior of these vesicles? What are the characteristics of these organelles and where do they come from? How do they secrete their content? Several important questions regarding this important issue remain unanswered.

4 Concluding Remarks

During the last two decades, important advances have been obtained in our knowledge of the secretory machinery of *Giardia*. Nevertheless, many relevant issues still remain unclear. Several elegant morphological EM studies have been performed and provided key information about the structural characteristics of the constitutive and regulated secretory pathway of *Giardia*. On the other hand, many groups have focused on the molecular aspects of these mechanisms, showing particular characteristics of the *Giardia* secretory pathways as well as insights on processes not well understood in other pathogenic and not pathogenic organisms. Unfortunately, studies linking molecules with morphology have precluded a better understanding of the secretory events occurring during parasite proliferation and differentiation. *Giardia* possesses simple organellar characteristics and a compact and small genome, probably caused by secondary loss of molecular machineries due to its parasitic life style. These features make this cell an excellent model to decipher molecular mechanisms difficult to explore in more complex cells. Further studies in which the molecular processes are complemented with in deep morphological analysis will provide the answers to many relevant questions still unsolved. In addition, it is essential to develop novel tools and techniques directed to study dynamic processes in organisms living in anaerobic or microaerophilic conditions such as *Giardia*, *Entamoeba* and *Trichomonas*. If it is possible, one should expect in the future a great development in the knowledge of the secretory pathways involved in many aspect of virulence of these important human pathogens.

Acknowledgments This work was supported by grants from the Agencia Nacional para la Promoción de la Ciencia y la Tecnología (ANPCYT), Consejo Nacional de Investigaciones Científicas y Técnicas (CONICET), Universidad Católica de Córdoba (UCC), and the Howard Hughes Medical Institute (HHMI). H.D.L. is an HHMI International Research Scholar and a Member of the Scientific Investigator's Career of the CONICET.

References

Adam RD (2001) Biology of Giardia lamblia. Clin Microbiol Rev 14:447–475

Alberts B, Johnson A, Lewis J, Raff M, Roberts K, Walter P (2007) Molecular biology of the cell, 5th edn. Garland Science, New York

Appenzeller-Herzog C, Hauri HP (2006) The ER-Golgi intermediate compartment (ERGIC): in search of its identity and function. J Cell Sci 119:2173–2183

Arguello-Garcia R, Arguello-Lopez C, Gonzalez-Robles A, Castillo-Figueroa AM, Ortega-Pierres MG (2002) Sequential exposure and assembly of cyst wall filaments on the surface of encysting Giardia duodenalis. Parasitology 125:209–219

Bannykh SI, Rowe T, Balch WE (1996) The organization of endoplasmic reticulum export complexes. J Cell Biol 135:19–35

Barlowe C (2000) Traffic COPs of the early secretory pathway. Traffic 1:371–377

Becker B, Melkonian M (1996) The secretory pathway of protists: spatial and functional organization and evolution. Microbiol Rev 60:697–721

Benchimol M (2002) A new set of vesicles in Giardia lamblia. Exp Parasitol 102:30–37
Benchimol M (2004a) Giardia lamblia: behavior of the nuclear envelope. Parasitol Res 94: 254–264
Benchimol M (2004b) The release of secretory vesicle in encysting Giardia lamblia. FEMS Microbiol Lett 235:81–87
Bole DG, Dowin R, Doriaux M, Jamieson JD (1989) Immunocytochemical localization of BiP to the rough endoplasmic reticulum: evidence for protein sorting by selective retention. J Histochem Cytochem 37:1817–1823
Bonifacino JS, Glick BS (2004) The mechanisms of vesicle budding and fusion. Cell 116:153–166
Boucher SE, Gillin FD (1990) Excystation of in vitro-derived Giardia lamblia cysts. Infect Immun 58:3516–3522
Braakman I, Helenius J, Helenius A (1992) Manipulating disulfide bond formation and protein folding in the endoplasmic reticulum. EMBO J 11:1717–1722
Buchel LA, Gorenflot A, Chochillon C, Savel J, Gobert JG (1987) In vitro excystation of Giardia from humans: a scanning electron microscopy study. J Parasitol 73:487–493
Cabral CM, Liu Y, Moremen KW, Sifers RN (2002) Organizational diversity among distinct glycoprotein endoplasmic reticulum-associated degradation programs. Mol Biol Cell 13:2639–2650
Caccio SM, Sprong H (2010) Giardia duodenalis: genetic recombination and its implications for taxonomy and molecular epidemiology. Exp Parasitol 124:107–112
Cavalier-Smith T (1987) The origin of eukaryotic and archaebacterial cells. Ann N Y Acad Sci 503:17–54
Cooper MA, Adam RD, Worobey M, Sterling CR (2007) Population genetics provides evidence for recombination in Giardia. Curr Biol 17:1984–1988
Dacks JB, Davis LA, Sjogren AM, Andersson JO, Roger AJ, Doolittle WF (2003) Evidence for Golgi bodies in proposed "Golgi-lacking" lineages. Proc Biol Sci 270(Suppl 2):S168–S171
Davids BJ, Mehta K, Fesus L, McCaffery JM, Gillin FD (2004) Dependence of Giardia lamblia encystation on novel transglutaminase activity. Mol Biochem Parasitol 136:173–180
Davids BJ, Reiner DS, Birkeland SR, Preheim SP, Cipriano MJ, McArthur AG, Gillin FD (2006) A new family of giardial cysteine-rich non-VSP protein genes and a novel cyst protein. PLoS ONE 1:e44
de Souza W (2006) Secretory organelles of pathogenic protozoa. An Acad Bras Cienc 78:271–291
Donaldson JG, Lippincott-Schwartz J (2000) Sorting and signaling at the Golgi complex. Cell 101:693–696
Eddy SR (1998) Profile hidden Markov models. Bioinformatics 14:755–763
Elias EV, Quiroga R, Gottig N, Nakanishi H, Nash TE, Neiman A, Lujan HD (2008) Characterization of SNAREs determines the absence of a typical Golgi apparatus in the ancient eukaryote Giardia lamblia. J Biol Chem 283:35996–36010
Elmendorf HG, Dawson SC, McCaffery JM (2003) The cytoskeleton of Giardia lamblia. Int J Parasitol 33:3–28
Embley TM, Hirt RP (1998) Early branching eukaryotes? Curr Opin Genet Dev 8:624–629
Erlandsen SL, Bemrick WJ, Schupp DE, Shields JM, Jarroll EL, Sauch JF, Pawley JB (1990) High-resolution immunogold localization of Giardia cyst wall antigens using field emission SEM with secondary and backscatter electron imaging. J Histochem Cytochem 38:625–632
Erlandsen SL, Macechko PT, van Keulen H, Jarroll EL (1996) Formation of the Giardia cyst wall: studies on extracellular assembly using immunogold labeling and high resolution field emission SEM. J Eukaryot Microbiol 43:416–429
Faubert G, Reiner DS, Gillin FD (1991) Giardia lamblia: regulation of secretory vesicle formation and loss of ability to reattach during encystation in vitro. Exp Parasitol 72:345–354
Feely DE, Dyer JK (1987) Localization of acid phosphatase activity in Giardia lamblia and Giardia muris trophozoites. J Protozool 34:80–83
Feely DE, Schollmeyer JV, Erlandsen SL (1982) Giardia spp.: distribution of contractile proteins in the attachment organelle. Exp Parasitol 53:145–154

Freedman RB, Bulleid NJ, Hawkins HC, Paver JL (1989) Role of protein disulphide-isomerase in the expression of native proteins. Biochem Soc Symp 55:167–192

Gaechter V, Schraner E, Wild P, Hehl AB (2008) The single dynamin family protein in the primitive protozoan Giardia lamblia is essential for stage conversion and endocytic transport. Traffic 9:57–71

Gething MJ (1999) Role and regulation of the ER chaperone BiP. Semin Cell Dev Biol 10:465–472

Gillin FD, Reiner DS, McCaffery M (1991) Organelles of protein transport in Giardia lamblia. Parasitol Today 7:113–116

Gillin FD, Reiner DS, McCaffery JM (1996) Cell biology of the primitive eukaryote Giardia lamblia. Annu Rev Microbiol 50:679–705

Gottig N, Elias EV, Quiroga R, Nores MJ, Solari AJ, Touz MC, Lujan HD (2006) Active and passive mechanisms drive secretory granule biogenesis during differentiation of the intestinal parasite Giardia lamblia. J Biol Chem 281:18156–18166

Graczyk TK (2005) Is Giardia a living fossil? Trends Parasitol 21:104–107

Gupta RS, Aitken K, Falah M, Singh B (1994) Cloning of Giardia lamblia heat shock protein HSP70 homologs: implications regarding origin of eukaryotic cells and of endoplasmic reticulum. Proc Natl Acad Sci USA 91:2895–2899

Haas IG (1994) BiP (GRP78), an essential hsp70 resident protein in the endoplasmic reticulum. Experientia 50:1012–1020

Hager KM, Striepen B, Tilney LG, Roos DS (1999) The nuclear envelope serves as an intermediary between the ER and Golgi complex in the intracellular parasite Toxoplasma gondii. J Cell Sci 112(Pt 16):2631–2638

Hammond AT, Glick BS (2000) Dynamics of transitional endoplasmic reticulum sites in vertebrate cells. Mol Biol Cell 11:3013–3030

Hauri HP, Kappeler F, Andersson H, Appenzeller C (2000) ERGIC-53 and traffic in the secretory pathway. J Cell Sci 113(Pt 4):587–596

Hehl AB, Marti M (2004) Secretory protein trafficking in Giardia intestinalis. Mol Microbiol 53:19–28

Hehl AB, Marti M, Kohler P (2000) Stage-specific expression and targeting of cyst wall protein-green fluorescent protein chimeras in Giardia. Mol Biol Cell 11:1789–1800

Helenius A, Trombetta ES, Hebert DN (1997) Simons JF. Calnexin, calreticulin and the folding of glycoproteins Trends in Cell Biology 7:193–200

Hwang C, Sinskey AJ, Lodish HF (1992) Oxidized redox state of glutathione in the endoplasmic reticulum. Science 257:1496–1502

Jackson CL (2009) Mechanisms of transport through the Golgi complex. J Cell Sci 122:443–452

Jarroll EL, Manning P, Lindmark DG, Coggins JR, Erlandsen SL (1989) Giardia cyst wall-specific carbohydrate: evidence for the presence of galactosamine. Mol Biochem Parasitol 32:121–131

Jarroll EL, Macechko PT, Steimle PA, Bulik D, Karr CD, van Keulen H, Paget TA, Gerwig G, Kamerling J, Vliegenthart J, Erlandsen S (2001) Regulation of carbohydrate metabolism during Giardia encystment. J Eukaryot Microbiol 48:22–26

Kabnick KS, Peattie DA (1990) In situ analyses reveal that the two nuclei of Giardia lamblia are equivalent. J Cell Sci 95(Pt 3):353–360

Klis FM (1994) Review: cell wall assembly in yeast. Yeast 10:851–869

Klumperman J, Schweizer A, Clausen H, Tang BL, Hong W, Oorschot V, Hauri HP (1998) The recycling pathway of protein ERGIC-53 and dynamics of the ER-Golgi intermediate compartment. J Cell Sci 111(Pt 22):3411–3425

Knodler LA, Noiva R, Mehta K, McCaffery JM, Aley SB, Svard SG, Nystul TG, Reiner DS, Silberman JD, Gillin FD (1999) Novel protein-disulfide isomerases from the early-diverging protist Giardia lamblia. J Biol Chem 274:29805–29811

Lanfredi-Rangel A, Attias M, de Carvalho TM, Kattenbach WM, de Souza W (1998) The peripheral vesicles of trophozoites of the primitive protozoan Giardia lamblia may correspond to early and late endosomes and to lysosomes. J Struct Biol 123:225–235

Lanfredi-Rangel A, Kattenbach WM, Diniz JA Jr, de Souza W (1999) Trophozoites of Giardia lamblia may have a Golgi-like structure. FEMS Microbiol Lett 181:245–251
Lanfredi-Rangel A, Attias M, Reiner DS, Gillin FD, de Souza W (2003) Fine structure of the biogenesis of Giardia lamblia encystation secretory vesicles. J Struct Biol 143:153–163
Lauwaet T, Davids BJ, Reiner DS, Gillin FD (2007) Encystation of Giardia lamblia: a model for other parasites. Curr Opin Microbiol 10:554–559
Letourneur F, Gaynor EC, Hennecke S, Demolliere C, Duden R, Emr SD, Riezman H, Cosson P (1994) Coatomer is essential for retrieval of dilysine-tagged proteins to the endoplasmic reticulum. Cell 79:1199–1207
Letunic I, Copley RR, Pils B, Pinkert S, Schultz J, Bork P (2006) SMART 5: domains in the context of genomes and networks. Nucleic Acids Res 34:D257–D260
Liewen H, Meinhold-Heerlein I, Oliveira V, Schwarzenbacher R, Luo G, Wadle A, Jung M, Pfreundschuh M, Stenner-Liewen F (2005) Characterization of the human GARP (Golgi associated retrograde protein) complex. Exp Cell Res 306:24–34
Lipsky NG, Pagano RE (1983) Sphingolipid metabolism in cultured fibroblasts: microscopic and biochemical studies employing a fluorescent ceramide analogue. Proc Natl Acad Sci USA 80:2608–2612
Lloyd D, Harris JC (2002) Giardia: highly evolved parasite or early branching eukaryote? Trends Microbiol 10:122–127
Lujan HD, Touz MC (2003) Protein trafficking in Giardia lamblia. Cell Microbiol 5:427–434
Lujan HD, Marotta A, Mowatt MR, Sciaky N, Lippincott-Schwartz J, Nash TE (1995a) Developmental induction of Golgi structure and function in the primitive eukaryote Giardia lamblia. J Biol Chem 270:4612–4618
Lujan HD, Mowatt MR, Conrad JT, Bowers B, Nash TE (1995b) Identification of a novel Giardia lamblia cyst wall protein with leucine-rich repeats. Implications for secretory granule formation and protein assembly into the cyst wall. J Biol Chem 270:29307–29313
Lujan HD, Mowatt MR, Wu JJ, Lu Y, Lees A, Chance MR, Nash TE (1995c) Purification of a variant-specific surface protein of Giardia lamblia and characterization of its metal-binding properties. J Biol Chem 270:13807–13813
Lujan HD, Mowatt MR, Byrd LG, Nash TE (1996a) Cholesterol starvation induces differentiation of the intestinal parasite Giardia lamblia. Proc Natl Acad Sci U S A 93:7628–7633
Lujan HD, Mowatt MR, Conrad JT, Nash TE (1996b) Increased expression of the molecular chaperone BiP/GRP78 during the differentiation of a primitive eukaryote. Biol Cell 86:11–18
Lujan HD, Mowatt MR, Nash TE (1997) Mechanisms of Giardia lamblia differentiation into cysts. Microbiol Mol Biol Rev 61:294–304
Lujan HD, Mowatt MR, Nash TE (1998) The molecular mechanisms of *Giardia* encystation. Parasitol Today 14:446–450
Marti M, Hehl AB (2003) Encystation-specific vesicles in Giardia: a primordial Golgi or just another secretory compartment? Trends Parasitol 19:440–446
Marti M, Li Y, Schraner EM, Wild P, Kohler P, Hehl AB (2003a) The secretory apparatus of an ancient eukaryote: protein sorting to separate export pathways occurs before formation of transient Golgi-like compartments. Mol Biol Cell 14:1433–1447
Marti M, Regos A, Li Y, Schraner EM, Wild P, Muller N, Knopf LG, Hehl AB (2003b) An ancestral secretory apparatus in the protozoan parasite Giardia intestinalis. J Biol Chem 278:24837–24848
McArthur AG, Knodler LA, Silberman JD, Davids BJ, Gillin FD, Sogin ML (2001) The evolutionary origins of eukaryotic protein disulfide isomerase domains: new evidence from the Amitochondriate protist Giardia lamblia. Mol Biol Evol 18:1455–1463
McCaffery JM, Gillin FD (1994) Giardia lamblia: ultrastructural basis of protein transport during growth and encystation. Exp Parasitol 79:220–235
McCaffery JM, Faubert GM, Gillin FD (1994) Giardia lamblia: traffic of a trophozoite variant surface protein and a major cyst wall epitope during growth, encystation, and antigenic switching. Exp Parasitol 79:236–249

Mehlotra RK (1993) Cultivation, encystation and excystation of Entamoeba histolytica: present status and future prospects. Biol Membr 19:59–84

Midlej V, Benchimol M (2009) Giardia lamblia behavior during encystment: how morphological changes in shape occur. Parasitol Int 58:72–80

Morrison HG, Roger AJ, Nystul TG, Gillin FD, Sogin ML (2001) Giardia lamblia expresses a proteobacterial-like DnaK homolog. Mol Biol Evol 18:530–541

Morrison HG, McArthur AG, Gillin FD, Aley SB, Adam RD, Olsen GJ, Best AA, Cande WZ, Chen F, Cipriano MJ, Davids BJ, Dawson SC, Elmendorf HG, Hehl AB, Holder ME, Huse SM, Kim UU, Lasek-Nesselquist E, Manning G, Nigam A, Nixon JE, Palm D, Passamaneck NE, Prabhu A, Reich CI, Reiner DS, Samuelson J, Svard SG, Sogin ML (2007) Genomic minimalism in the early diverging intestinal parasite Giardia lamblia. Science 317:1921–1926

Mowatt MR, Lujan HD, Cotten DB, Bowers B, Yee J, Nash TE, Stibbs HH (1995) Developmentally regulated expression of a Giardia lamblia cyst wall protein gene. Mol Microbiol 15:955–963

Munro S, Pelham HR (1986) An Hsp70-like protein in the ER: identity with the 78 kd glucose-regulated protein and immunoglobulin heavy chain binding protein. Cell 46:291–300

Munro S, Pelham HR (1987) A C-terminal signal prevents secretion of luminal ER proteins. Cell 48:899–907

Nickel W, Weber T, McNew JA, Parlati F, Sollner TH, Rothman JE (1999) Content mixing and membrane integrity during membrane fusion driven by pairing of isolated v-SNAREs and t-SNAREs. Proc Natl Acad Sci U S A 96:12571–12576

Nigam SK, Goldberg AL, Ho S, Rohde MF, Bush KT, Sherman M (1994) A set of endoplasmic reticulum proteins possessing properties of molecular chaperones includes Ca(2+)-binding proteins and members of the thioredoxin superfamily. J Biol Chem 269:1744–1749

Orci L, Ravazzola M, Amherdt M, Perrelet A, Powell SK, Quinn DL, Moore HP (1987) The trans-most cisternae of the Golgi complex: a compartment for sorting of secretory and plasma membrane proteins. Cell 51:1039–1051

Pelham HR (1995) Sorting and retrieval between the endoplasmic reticulum and Golgi apparatus. Curr Opin Cell Biol 7:530–535

Poxleitner MK, Carpenter ML, Mancuso JJ, Wang CJ, Dawson SC, Cande WZ (2008) Evidence for karyogamy and exchange of genetic material in the binucleate intestinal parasite Giardia intestinalis. Science 319:1530–1533

Pozzan T, Rizzuto R, Volpe P, Meldolesi J (1994) Molecular and cellular physiology of intracellular calcium stores. Physiol Rev 74:595–636

Rambourg A, Clermont Y, Hermo L (1988) Formation of secretion granules in the Golgi apparatus of pancreatic acinar cells of the rat. Am J Anat 183:187–199

Randazzo PA, Nie Z, Miura K, Hsu VW (2000) Molecular aspects of the cellular activities of ADP-ribosylation factors. Sci STKE 2000:RE1

Ratner DM, Cui J, Steffen M, Moore LL, Robbins PW, Samuelson J (2008) Changes in the N-glycome, glycoproteins with Asn-linked glycans, of Giardia lamblia with differentiation from trophozoites to cysts. Eukaryot Cell 7:1930–1940

Regoes A, Hehl AB (2005) SNAP-tag mediated live cell labeling as an alternative to GFP in anaerobic organisms. Biotechniques 39:809–810, 812

Reiner DS, McCaffery M, Gillin FD (1990) Sorting of cyst wall proteins to a regulated secretory pathway during differentiation of the primitive eukaryote, Giardia lamblia. Eur J Cell Biol 53:142–153

Reiner DS, McCaffery JM, Gillin FD (2001) Reversible interruption of Giardia lamblia cyst wall protein transport in a novel regulated secretory pathway. Cell Microbiol 3:459–472

Reiss E, Hearn VM, Poulain D, Shepherd MG (1992) Structure and function of the fungal cell wall. J Med Vet Mycol 30(Suppl 1):143–156

Samuelson J, Banerjee S, Magnelli P, Cui J, Kelleher DJ, Gilmore R, Robbins PW (2005) The diversity of dolichol-linked precursors to Asn-linked glycans likely results from secondary loss of sets of glycosyltransferases. Proc Natl Acad Sci U S A 102:1548–1553

Sheffield HG, Bjorvat B (1977) Ultrastructure of the cyst of Giardia lamblia. Am J Trop Med Hyg 26:23–30

Simpson AG, Roger AJ, Silberman JD, Leipe DD, Edgcomb VP, Jermiin LS, Patterson DJ, Sogin ML (2002) Evolutionary history of "early-diverging" eukaryotes: the excavate taxon Carpediemonas is a close relative of Giardia. Mol Biol Evol 19:1782–1791

Sitia R, Meldolesi J (1992) Endoplasmic reticulum: a dynamic patchwork of specialized subregions. Mol Biol Cell 3:1067–1072

Slavin I, Saura A, Carranza PG, Touz MC, Nores MJ, Lujan HD (2002) Dephosphorylation of cyst wall proteins by a secreted lysosomal acid phosphatase is essential for excystation of Giardia lamblia. Mol Biochem Parasitol 122:95–98

Smith RD, Lupashin VV (2008) Role of the conserved oligomeric Golgi (COG) complex in protein glycosylation. Carbohydr Res 343:2024–2031

Soding J, Biegert A, Lupas AN (2005) The HHpred interactive server for protein homology detection and structure prediction. Nucleic Acids Res 33:W244–W248

Sogin ML, Silberman JD (1998) Evolution of the protists and protistan parasites from the perspective of molecular systematics. Int J Parasitol 28:11–20

Sogin ML, Gunderson JH, Elwood HJ, Alonso RA, Peattie DA (1989) Phylogenetic meaning of the kingdom concept: an unusual ribosomal RNA from Giardia lamblia. Science 243:75–77

Solari AJ, Rahn MI, Saura A, Lujan HD (2003) A unique mechanism of nuclear division in Giardia lamblia involves components of the ventral disk and the nuclear envelope. Biocell 27:329–346

Soltys BJ, Falah M, Gupta RS (1996) Identification of endoplasmic reticulum in the primitive eukaryote Giardia lamblia using cryoelectron microscopy and antibody to Bip. J Cell Sci 109 (Pt 7):1909–1917

Stefanic S, Palm D, Svard SG, Hehl AB (2006) Organelle proteomics reveals cargo maturation mechanisms associated with Golgi-like encystation vesicles in the early-diverged protozoan Giardia lamblia. J Biol Chem 281:7595–7604

Sun CH, McCaffery JM, Reiner DS, Gillin FD (2003) Mining the Giardia lamblia genome for new cyst wall proteins. J Biol Chem 278:21701–21708

TerBush DR, Maurice T, Roth D, Novick P (1996) The Exocyst is a multiprotein complex required for exocytosis in Saccharomyces cerevisiae. EMBO J 15:6483–6494

Touz MC, Gottig N, Nash TE, Lujan HD (2002a) Identification and characterization of a novel secretory granule calcium-binding protein from the early branching eukaryote Giardia lamblia. J Biol Chem 277:50557–50563

Touz MC, Nores MJ, Slavin I, Carmona C, Conrad JT, Mowatt MR, Nash TE, Coronel CE, Lujan HD (2002b) The activity of a developmentally regulated cysteine proteinase is required for cyst wall formation in the primitive eukaryote Giardia lamblia. J Biol Chem 277:8474–8481

Touz MC, Lujan HD, Hayes SF, Nash TE (2003) Sorting of encystation-specific cysteine protease to lysosome-like peripheral vacuoles in Giardia lamblia requires a conserved tyrosine-based motif. J Biol Chem 278:6420–6426

Touz MC, Kulakova L, Nash TE (2004) Adaptor protein complex 1 mediates the transport of lysosomal proteins from a Golgi-like organelle to peripheral vacuoles in the primitive eukaryote Giardia lamblia. Mol Biol Cell 15:3053–3060

Tovar J, Leon-Avila G, Sanchez LB, Sutak R, Tachezy J, van der Giezen M, Hernandez M, Muller M, Lucocq JM (2003) Mitochondrial remnant organelles of Giardia function in iron-sulphur protein maturation. Nature 426:172–176

Turano C, Coppari S, Altieri F, Ferraro A (2002) Proteins of the PDI family: unpredicted non-ER locations and functions. J Cell Physiol 193:154–163

Wang CH, Su LH, Sun CH (2007) A novel ARID/Bright-like protein involved in transcriptional activation of cyst wall protein 1 gene in Giardia lamblia. J Biol Chem 282:8905–8914

Ward W, Alvarado L, Rawlings ND, Engel JC, Franklin C, McKerrow JH (1997) A primitive enzyme for a primitive cell: the protease required for excystation of Giardia. Cell 89:437–444

Wilson DW, Lewis MJ, Pelham HR (1993) pH-dependent binding of KDEL to its receptor in vitro. J Biol Chem 268:7465–7468

Yu LZ, Birky CW Jr, Adam RD (2002) The two nuclei of Giardia each have complete copies of the genome and are partitioned equationally at cytokinesis. Eukaryot Cell 1:191–199

The Remarkable Mitochondrion of Trypanosomes and Related Flagellates

Julius Lukeš, Hassan Hashimi, Zdeněk Verner, and Zdeňka Čičová

Contents

Abstract While the single mitochondrion of trypanosomatid flagellates contains many of the hallmarks that are known from mitochondria in other conventional model organisms, it also possesses several unique features, making it a subject of intense research. Here, we summarize current knowledge of the (1) structure, maintenance and replication of the extensive kinetoplastid DNA network, (2) byzantine organellar RNA metabolism, including insertion/deletion RNA editing,

---

J. Lukeš (✉), H. Hashimi, Z. Verner, and Z. Čičová
Biology Centre, Institute of Parasitology, Czech Academy of Science and Faculty of Science, University of South Bohemia, České Budějovice (Budweis), Czech Republic
e-mail: jula@paru.cas.cz

W. de Souza (ed.), *Structures and Organelles in Pathogenic Protists*,
Microbiology Monographs 17, DOI 10.1007/978-3-642-12863-9_10,
© Springer-Verlag Berlin Heidelberg 2010

(3) translation of mitochondrial-encoded proteins, and, finally, (4) distinctive aspects of energy metabolism of the organelle. While we focus on the mitochondrion of *Trypanosoma brucei*, particularly in the context of its radical physiological and biochemical changes during the life cycle of the parasite, in order to get a more complete picture of the workings of this fascinating organelle, we also discuss significant findings obtained from other trypanosomatids.

The single mitochondrion of trypanosomes and their relatives is a remarkable organelle, containing almost all of its hallmarks as well as some unique features. Among the latter are exceedingly complex mitochondrial DNA and RNA editing. Furthermore, the organelle is very different in the two principal stages of the life cycle, as it is metabolically active in the procyclic stage transmitted by the insect vector (Fig. 1d), while its morphology and metabolism are highly reduced in the bloodstream stage (Fig. 1e). Thanks to initiatives such as description of the mitochondrial proteome of the procyclic stage, our knowledge of the organelle of *Trypanosoma brucei* and *Leishmania* species increased substantially within the last decade. In this chapter, we focus on mitochondrial DNA and its transcription and translation and conclude with a brief description of the function of mitochondrial-encoded proteins and energy metabolism. Within the last decade, these topics were subject to several authoritative reviews (Besteiro et al., Trends Parasitol 21:185–191, 2005; Bringaud et al., Mol Biochem Parasitol 149:1–9, 2006; Liu et al., Trends Parasitol 21:363–369, 2005; Lukeš et al., Eukaryot Cell 1:495–502, 2002, Curr Genet 48:277–299, 2005; Rubio and Alfonzo, Top Curr Genet 12:71–86, 2005; Shlomai, Curr. Mol. Med. 4:623–647, 2004; Schnaufer et al., Int J Parasitol 32:1071–1084, 2002; Schneider, Int J Parasitol 31:1403–1415, 2001; Simpson et al., RNA 10:159–170, 2004, Trends Parasitol 22, 168–174, 2006; Stuart et al., Trends Biochem Sci 30, 97–105, 2005). In this chapter, we briefly summarize our present knowledge with somewhat more detailed treatise of the findings obtained mostly within the last 5 years.

1 The Kinetoplast DNA

Protists of the class Kinetoplastea derive their name from the mitochondrial genome, termed kinetoplast (k) DNA, for a good reason. Thanks to its enormous size (and complexity, as we shall see later), it is likely the first organellar DNA observed (Ziemann 1898), which until present arguably belongs to the best studied organellar genomes. It also represents a unifying feature that these protists carry in their mitochondrion. Indeed the presence of an extranuclear DNA, easily stainable with the Giemsa solution, is a strong hint that the cell in question belongs to the kinetoplastid flagellates.

The kDNA of trypanosomes and related flagellates belongs arguably to the most complex DNA known. Since its description by light and electron microscopy, it is the advent of molecular biology methods that enables dissection of this fascinating structure. Principal progress in our understanding of the kDNA structure and replication was recently achieved mainly via functional analyses using RNA interference in *Trypanosoma brucei*.

1.1 kDNA: Its In vivo Structure and Diversity

Initial studies by transmission electron microscopy in the 1960s revealed the presence of a compact electron-dense structure invariably located in a region of the single reticulated mitochondrion that is adjacent to the flagellar basal body (Fig. 1a, b). This structure represents the kDNA, which in all trypanosomatid flagellates exists in the form of a disk-like structure, with DNA strands aligned in parallel to the axis of the disk (Fig. 1f) (Shapiro and Englund 1995). Numerous transmembrane filaments attach the kDNA to the flagellar basal body (Zhao et al. 2008). During

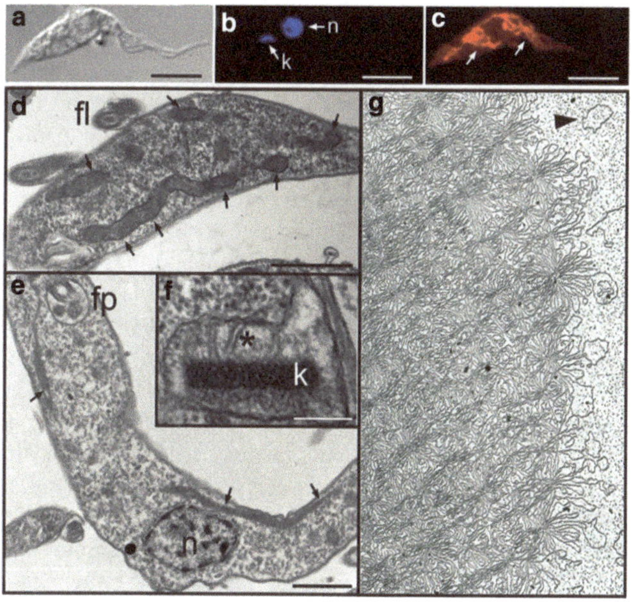

Fig. 1 *Light and electron microscopy of Trypanosoma brucei* (**a**) Procyclic cell of strain 29-13; (**b**) The same cell stained with 4′,6-diamidino-2-phenylindole (DAPI) revealing the large oval nucleus (n) and distinct kinetoplast (k); (**c**) Tubular reticulated mitochondrion stained in the same cell by mitotracker red (indicated by arrows); (**d**) Transmission electron microscopy of a procyclic cell, showing multiple cross-sections thru the peripherally located mitochondrial network (*arrows*). Cross-sectioned flagella at the periphery are indicated (fl); (**e**) Transmission electron microscopy of a bloodstream cell, revealing the reduced thin mitochondrion of this stage (*arrows*). Nucleus (n) and flagellar pocket (fp) are also indicated; (**f**) Longitudinal section thru the kinetoplast (k) in the extended periflagellar portion of the mitochondrion. Peripherally located tubular cristae are indicated with an *asterisk*. (**g**) Electron microscopy of the kinetoplast DNA network and free minicircles (*arrowhead*). *Bars*, 3 μm (**a–c**), 1 μm (**d**), 0.5 μm (**e, f**)

the cell cycle, kDNA division follows after the formation of a new flagellum, which is among the first and most conspicuous morphological signs of cell division (Woodward and Gull 1990). Duplication of the nuclear DNA in the form of closed mitosis, with the formation of an intranuclear spindle in the nucleus with an intact envelope terminates the cell cycle (Ogbadoyi et al. 2003).

While kDNA of all the studied members of the family Trypanosomatidae exists in the form of a disk (Fig. 1f), the homologous structure assumes a variety of forms in the sister family Bodonidae. In *Bodo* species, which are considered to be free-living predecessors of obligatory parasitic trypanosomatids (Simpson et al. 2006), kDNA is also confined to the periflagellar position, yet it is formed by a bundle of DNA strands, comprising an arrangement termed pro-kDNA (Lukeš et al. 2002). In other bodonids, the kDNA is either evenly or unevenly distributed throughout the mitochondrial lumen. In the former case, first described from the fish parasite *Trypanoplasma borreli*, the so-called pan-kDNA seems to contain a comparable amount of nucleic acids as the nucleus, at least as judged by staining with the DNA-binding dyes. The free-living and commensalic bodonids of the genera *Dimastigella* and *Cruzella* also contain a huge amount of DNA in their mitochondrion. However, it is present in the form of multiple foci evenly distributed throughout organellar lumen, in an arrangement termed poly-kDNA (Lukeš et al. 2002). The morphologically most unusual structure is the globular kDNA found in the highly diverged kinetoplastid *Perkinsella*, a parasite of amoebae found on the gills of fish (Dyková et al. 2003). The interesting finding that bodonids are eukaryotes with the largest amount of DNA in their mitochondrion deserves thorough research, which is unfortunately not forthcoming, so we know close to nothing about its organization, gene content and function.

Initial characterization of the morphologically prominent kDNA commenced with the studies of the nonpathogenic insect parasite *Crithidia fasciculata*, which can be cultivated in a cheap medium and to high cell densities. Treatment of the purified kDNA with topoisomerase II showed that it is composed of circular DNA molecules that are mutually catenated into a single large network (Englund 1979). Following restrictions with different endonucleases demonstrated that there are two classes of DNA circles, termed maxicircles and minicircles (Fig. 2a) (Shapiro and Englund 1995). As it turned out, these circular molecules have strikingly different functions (see below). Subsequent studies have shown a very similar arrangement of kDNA in other trypanosomatid species, such as *Leishmania tarentolae*, the plant pathogen *Phytomonas serpens*, and *T. brucei*, *T. cruzi* and *Leishmania* spp., the causative agents of African sleeping sickness, Chagas disease and Leishmaniasis.

1.2 Maxicircles

Maxicircles are homologs of classical mitochondrial DNA of other eukaryotes, as they contain many typical protein-coding genes, a single subunit of mitochondrial

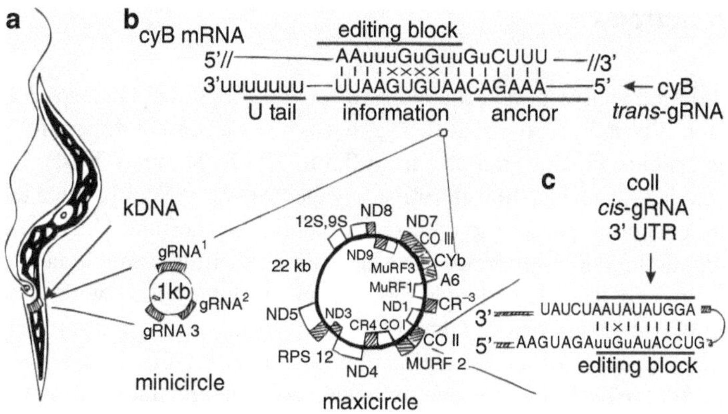

Fig. 2 *Integration of kinetoplast (k) DNA with RNA editing* (**a**) Genomic organization of kDNA maxi- and minicircles of *T. brucei*. The ~1 kb minicircle contains on average three guide (g) RNA genes. The ~22 kb maxicircle contains the complement of mitochondrial genes encoding subunits of the respiratory chain and mitoribosomal components. Genes undergoing RNA editing are shaded. Never-edited genes are left open. The genes are abbreviated as follows: mitoribosomal rRNAs 9 S (9S) and 12 S (12 S); ATPase subunit 6 (A6); cytochrome oxidase subunits 1 (COI) 2 (COII) and 3 (COIII); cytochrome reductase subunit b (cyB); maxicircle unknown reading frames 1 (MURF1), 2 (MURF2), and 3 (MURF3); NADH dehydrogenase subunits 1(ND1), 3 (ND3), 4 (ND4), 5 (ND5), 6 (ND6), 7 (ND7), 8 (ND8), and 9 (ND9); ribosomal protein S12 (RPS12); unknown open reading frames also labeled by CR3 and CR4. (**b**) CyB gRNA:mRNA duplex, with features of the gRNA molecule *highlighted*. Noncanonical U:G pairings are depicted as *crosses*, and inserted Us are shown in *lowercase*. (**c**) The *cis*-acting gRNA of COII, located in the 3′-UTR. Noncanonical U:G pairings are depicted as *crosses*, and inserted Us are shown in *lowercase*

ribosome, and the 9S and 12S mitoribosomal RNAs (Fig. 2a) (Estevez and Simpson 1999). They are composed of a gene-coding region or conservative region and a variable or divergent region, in which putative replication origin is located (Liu et al. 2005; Lukeš et al. 2005). Transcripts of some genes are readily translatable, however, most has to undergo (extensive) uridine insertion and deletion type of RNA editing (see section on RNA editing) to be rendered translatable (Stuart et al. 2005). It was proposed that approximately a dozen maxicircles, which along with minicircles constitute the kDNA disk, are mutually interlocked, forming a network within a network (Shapiro and Englund 1995). Maxicircles were considered to be homogeneous in sequence, but recent evidence from *Leishmania major* indicates that a cell may contain several maxicircle classes that differ in the variable region (Flegontov et al. 2009). The size of maxicircles has been established only in a handful of species, with a median size of 20 kb. In *T. brucei evansi* and *T. brucei equiperdum*, maxicircles are subject to deletions eventually leading to their complete loss. As a consequence, the mitochondrion loses the potential to produce key subunits of the respiratory complexes, turning it effectively into a petite mutant of *T. brucei* that cannot be transmitted via the tse-tse fly (Lai et al. 2008).

1.3 Minicircles

Initial restriction analyses and sequencing of minicircles did not bring any clue as to their "raison d'etre," yet revealed their extensive sequence heterogeneity, with the size being species-specific, ranging from 0.5 to 10 kb (Shlomai 2004). In fact the size of minicircles can also be determined by electron microscopy, since their DNA strands are packed in parallel to the axis of the disk (Lukeš and Votýpka 2000).

An estimated 5,000 minicircles per kinetoplast constitute a single large network (Fig. 1g), in which each and every minicircle is interlocked with three of its neighbors (Chen et al. 1995). Characteristic features of the minicircles include lack of supercoiling, conserved replication origins and unique sequence features such as the bent helix formed by evenly spaced polyadenine tracks (Liu et al. 2005). The seminal discovery of small genes encoding guide (g) RNAs on minicircles (Blum et al. 1990) uncovered their functional integration with maxicircles via the process of RNA editing (see Sect. 2). The coding capacity varies from a single to three gRNAs per minicircle (the assignment of some regions for gRNA is only tentative) (Fig. 2a), depending on the trypanosomatid species. Sequence heterogeneity of minicircles differs between species, but it is reasonable to assume that each trypanosomatid encodes hundreds of gRNAs in its minicircle kDNA.

In *T. brucei evansi* and *T. brucei equiperdum*, the (partial) deletion of maxicircles triggers loss of minicircle heterogeneity, although initially their abundance remains unaltered. However, further diminution of the kDNA is characterized by homogenization down to a single minicircle sequence class (dyskinetoplastic strains), which is eventually lost altogether, turning the cells into the akinetoplastic form (Lai et al. 2008; Jensen et al. 2008).

1.4 Replication and Maintenance of kDNA

It is apparent that the extremely complicated network composed of mini- and maxicircles (Fig. 1g) has to faithfully divide, in order to equip both daughter cells with a full kDNA complement. For that purpose, a sophisticated and likely also a highly exact mechanism evolved, which is expected to entail the participation of more than a hundred different proteins for this task (Liu et al. 2005).

In a nonreplicating network, all DNA circles are covalently closed and interlocked with their neighbors. By the action of topoisomerase II, individual minicircles are released from the network into a region between the mitochondrial membrane and the kDNA disk, or so-called kinetoflagellar zone, marking the initiation of replication (Drew and Englund 2001). It is in this region where DNA primase, two DNA polymerases (Klingbeil et al. 2002), and universal minicircle sequence-binding protein (UMSBP) (Abu-Elneel et al. 2001) perform replication of the free minicircles, in addition to the several other anticipated protein yet to be identified. Interaction of UMSBP with the minicircle replication origin is uniquely affected by its redox state. While oxidation drives oligomerization of UMSBP and eliminates its binding to DNA, binding is activated under reduced conditions, which favors monomers (Sela and Shlomai 2009). Next, the newly replicated minicircles migrate into antipodal sites, which are two protein-rich regions flanking the kDNA disk 180° apart. Specific antibodies have been used to characterize the content of these well-defined protein centers, which contain topoisomerase II, structure-specific endonuclease 1 (SSE1), DNA polymerase β (Torri and Englund 1995), DNA ligase (Downey et al. 2005) and p38 (Liu et al. 2006). The antipodal sites were recently visualized also by electron microscopy (Gluenz et al. 2007). The reattachment of newly replicated minicircles occurs within the antipodal centers via the action of topoisomerase II, and all replicated minicircles are marked with the retention of at least one gap (Wang and Englund 2001). Proteins identified thus far leave no doubt about the extreme complexity of the system, as the mitochondrion of *T. brucei* contains at least six DNA polymerases (Klingbeil et al. 2002; Saxowsky et al. 2003), six DNA helicases (Lindsay et al. 2008) and two DNA ligases (Sinhaet al. 2006). Moreover, there are at least four histone-like proteins, called kinetoplastid-associated proteins (KAPs) 1–4, that have been implicated in stabilizing the kDNA disk in its compacted structure (Avliyakulov et al. 2004).

Downregulation, mostly by RNAi, of proteins associated with the kDNA leads to colorful phenotypes, many of which illuminate the intricate mechanisms behind its maintenance and replication. After topoisomerase II was ablated, the kinetoplast shrinks (Wang and Englund 2001) and accumulates the holes that remain after the release of minicircles from the dividing kDNA (Lindsay et al. 2008). In the absence of SSE1, the attachment of minicircles is altered (Liu and Englund 2007), while the elimination of HsIVU protease causes over-replication of minicircles, triggering growth of the kDNA disk to an enormous size (Li et al. 2008). Giant kinetoplasts were observed also in cells that interfered against some other proteins. Their presence in cells lacking p166 is explained as a consequence of the disrupted tripartite attachment complex (Zhao et al. 2008), whereas missegregation of minicircles in the absence of UMSBP triggers a similar outcome (Milman et al. 2007). The advent of methods of forward and reverse genetics, to which trypanosomes and related protists are well amenable, promises to eventually disentangle the functions of dozens, perhaps hundreds of proteins engaged in the faithful replication and maintenance of kDNA.

Comparative analysis of the distribution of newly replicated minicircles in the kDNA networks of *T. brucei* and *C. fasciculata* produced an unexpected conundrum, which might be solved by the recent data. While in *T. brucei* and some other flagellates, the newly replicated minicircles accumulate close to the two antipodal sites, labeling of the gapped minicircle progeny with [^{3}H]thymidine in *C. fasciculata* and *L. tarentolae* revealed their ring-like distribution around the network periphery. In order to explain this distribution, a mechanism has been suggested by which the kDNA of the latter species rotates between the antipodal sites (Liu et al. 2005). Therefore, it was postulated that two dramatically different mechanisms of kDNA replication evolved independently in trypanosomatids. Using fluorescent microscopy, Liu and Englund (2007) have shown that the kinetoplast either rotates (*C. fasciculata*) or oscillates (*T. brucei*), resulting in the strikingly different patterns of distribution of replicated minicircles.

2 RNA Editing

2.1 Mechanism of RNA Editing

In 1986, Benne and colleagues made the seminal discovery that four uridine (U) residues not encoded in the gene are posttranscriptionally inserted into specific sites in the cytochrome oxidase subunit (co) 2 mRNA. This process was named RNA editing. A surge of reports followed indicating that mRNAs from 12 of the 20 genes residing on *T. brucei* maxicircles required this process for their maturation (Fig. 2a). Translatable open reading frames are created by insertions and/or deletions of hundreds of uridine (U) residues into/from maxicircle transcripts. Molecules undergoing this kind of maturation are conceptually grouped as pre-, partially- and fully edited RNAs, depending on their current stage in the process, while RNA molecules that bypass this route are referred to as never-edited (for review in RNA editing see Simpson et al. 2004; Stuart et al. 2005).

Another breakthrough in the field was the discovery of small RNA molecules (50–70 nts long) almost entirely encoded on the minicircles of the kDNA network. These primary transcripts were called guide (g) RNAs because they provide the genetic information defining the editing sites on a given pre- and/or partially-edited mRNA (Blum et al. 1990; Sturm and Simpson 1990). Examination of the three regions that make up the primary structure of a gRNA suggests how they act as blueprints for RNA editing events (Fig. 2b). The 5′-positioned anchor domain is a small stretch of ~10 nts that hybridizes to a complementary sequence on the mRNA, just downstream of the editing sites. The information domain starts at the first base mismatch, providing a template for the appropriate U insertion/deletion. The transfer of information the gRNA to mRNA relies on both Watson–Crick and noncanonical G:U base pairing between the two molecules (Sturm and Simpson 1990). After completion of editing, the edited part of the mRNA, called the editing block, is complementary to the information domain. The third part of the gRNA molecule, the 3′-oligo(U) tail, which is added to the molecule posttranscriptionally and demonstrated to interact with the purine-rich sequences upstream of the editing block (McManus et al. 2000). Almost all gRNAs act in *trans* in the described fashion. A notable exception is the editing of cox2, which utilizes a *cis*-acting gRNA in its 3′-UTR (Golden and Hajduk 2005) (Fig. 2c).

While gRNAs represent the informational component of RNA editing, a cascade of enzymatic activities is also required (reviewed in Simpson et al. 2004; Lukeš et al. 2005; Stuart et al. 2005) (Fig. 3). An endonucleolytic cleavage occurs at the editing site, dividing the mRNA into 5′ and 3′ fragments that are bridged by the bound gRNA. What occurs next depends on whether a U insertion or deletion event

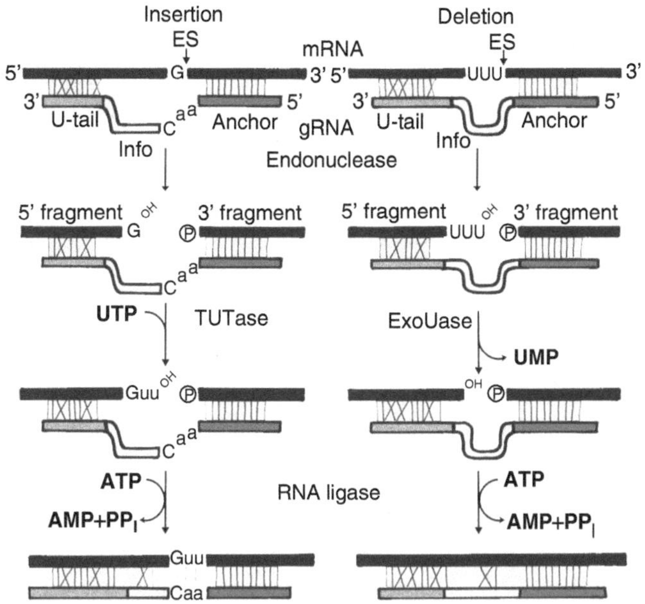

Fig. 3 *Mechanism of RNA editing*. The first base pair mismatch between mRNA (*top*) and anchor domain (*dark gray*) defines the editing site (ES), which is cleaved by an endonuclease. A 3′ terminal uridylyltransferase (TUTase) adds Us to the 5′ fragment while an exonuclease (exoUase) deletes extra Us. The two fragments are joined back together by an RNA ligase. The 3′ oligo(U) tail of the gRNA is indicated by light grey shading

is designated by the gRNA. In the case of the former, free UTP is added to the 3′ hydroxyl group of the 5′ fragment, the number of which is dictated by A and G residues in the information domain. A 3′→5′ exonuclease prunes away 3′-protruding U(s) from the 5′ fragment in the case of deletion. Once the processing step results in a fully complementary duplex, the two mRNA fragments are rejoined by an RNA ligase. After an editing block is completed, the current gRNA in unwound from the mRNA to allow the upstream hybridization of a subsequent gRNA for the next round. As a consequence, editing of pan-edited mRNAs proceeds with a 3′ to 5′ polarity (Maslov and Simpson 1992).

2.2 The RNA Editing Core Complex

The macromolecular RNA Editing Core Complex (RECC) confers the core editing activities required for mitochondrial biogenesis (Simpson et al. 2010). It has also been called in the literature the (20S) editosome, a name reflecting the settling rate in Svedberg units of the active complex in glycerol gradient ultracentrifugation experiments (Stuart et al. 2005), and the L-complex (Simpson et al. 2004).

The three-dimensional structure of the complex that has been resolved recently shares a similar structure with a larger version that sediments at 35–40S and contains extra elements such as substrate RNAs (Golas et al. 2009; Li et al. 2009). The ~20 protein subunits comprising the complex often occur in sets or pairs sharing motifs, domains and/or functions. Some of these related subunits are further divided into the two subcomplexes of RECC, each of which confers *in vitro* insertion or deletion editing activities (Schnaufer et al. 2003). Three distinct types of RECCs exist that differ in their incorporation of one of the three endonucleases that catalyze the initial mRNA cleavage step (Carnes et al. 2008). The residing endonuclease determines whether the complex has the capacity to cleave RNA editing substrates with U deletion or insertion editing sites (Carnes et al 2005; Trotter et al. 2005), or processes cox2 editing mediated by its *cis*-gRNA (Carnes et al. 2008).

The finding that RNA editing ligase is essential in the bloodstream stage was surprising (Schnaufer et al. 2001), as the existence of dyskinetoplastic trypanosomes lacking the gRNA repertoire for processing mRNAs, ostensibly implied that RNA editing is not required during this stage of its life cycle (Schnaufer et al. 2002). However, further studies clearly demonstrated that editing is essential for survival of flagellates both in the tse-tse fly and mammalian host (Fisk et al. 2008; Hashimi et al. 2009). The solved crystal structure of RNA editing ligase (Deng et al. 2004) has primed a study for drug-like inhibitors of its function (Amaro et al. 2008), which may potentially be developed for drug treatment of the various diseases caused by trypanosomes.

2.3 Other Proteins Involved in RNA Editing and/or Processing

Several other proteins and complexes that have a role in RNA editing aside from imparting the core enzymatic activities in this process have been described. A DExD/H-box RNA helicase found unstably associated with RECC was proposed to have gRNA-unwinding role (Missel et al. 1997). Another 3′ terminal uridylyltransferase that acts independently of RECC is responsible for the posttranscriptional addition of the 3′-oligo(U) tail to gRNA molecules, and is essential for RNA editing (Aphasizhev et al. 2003a).

The participation of RNA binding proteins has always been an expected feature of RNA editing. RBP16 is an example of such a protein, which has a demonstrated gRNA/mRNA annealing activity (Ammerman et al. 2008). Its silencing by RNAi had a pleimorphic phenotype, affecting some never-edited transcripts as well as those undergoing this process (Pelletier and Read 2003). The mitochondrial RNA binding proteins MRP1 and 2 associate in a heterotetrameric complex that was shown to have in vitro RNA matchmaking activity (Schumacher et al. 2006; Zíková et al. 2008b). Although RNAi-knockdowns of these proteins also affected a subset of both never-edited and edited mRNAs (Vondrušková et al. 2005), the crystal structure of this complex, in which positively-charged amino acids on its surface bind gRNAs, is consistent with a role in gRNA:mRNA duplex formation (Schumacher et al. 2006; Zíková et al. 2008a). TbRGG1 has affinity for poly(U) (Vanhamme et al. 1998), and its RNAi-silencing affects steady state levels of edited mRNAs but not gRNAs (Hashimi et al. 2008).

TbRGG1 associates with a macromolecular complex in a RNA that mediates fashion (Hashimi et al. 2008). This putative complex was provisionally named the mitochondrial RNA binding complex 1 (MRB1) since many of its constituents contain motifs and domains involved in RNA binding and metabolism. The composition of MRB1 shares a significant degree of overlap with other isolated complexes, including the recently discovered mitochondrial poly(A) polymerase complex (Etheridge et al. 2008; Panigrahi et al. 2008; Weng et al. 2008). Several of the interactions within MRB1 appear to be also mediated by RNA interactions (Fisk et al. 2008; Weng et al. 2008) and there is a diverse array of RNA phenotypes, lending credence to the idea that MRB1 represents a collection of smaller complexes and/or monomers that assemble around RNA (Weng et al. 2008; Hashimi et al. 2009). The so-called gRNA binding complex (GBRC) contains two orthologs, known as the gRNA associated proteins (GAPs) 1 and 2 that appear to have a role in gRNA stability and/or processing (Hashimi et al. 2008; Weng et al. 2008).

2.4 The raison d'etre *of RNA Editing*

Shortly after its discovery, it was proposed that kinetoplastid RNA editing may be a relic of an ancient "RNA world," when only these molecules existed. The lack of catalytic activity of the substrate RNAs and the participation of a sophisticated protein complex has negated this idea. In addition, a possible link of seemingly cumbersome process to parasitism has been invalidated by its existence in free-living bodonids. Several hypotheses have suggested the evolutionary advantages bestowed by RNA editing, including (1) extra level of regulation of mitochondrial gene expression, (2) fixing mutations that have accumulated in a nonfunctional mitochondrion, (3) accelerated evolution by creating more genetic variation, (4) multiple proteins coded by one gene, and several other hypotheses (for review see Speijer 2008).

The persistent editing of some transcripts in the bloodstream stage, despite the fact that proteins encoded by these mRNAs are obviously not required at this stage (see below), has spurred the exploration of the idea that RNA editing contributes to protein diversity. An interesting study has recently provided evidence for a protein product of an alternatively edited cox3 mRNA that has a role in kDNA maintenance (Ochsenreiter et al. 2008a). Although more alternatively edited RNAs of other mt gene transcripts have been described (Ochsenreiter et al. 2008b), hard evidence of their translation is needed to confirm this exciting theory.

3 The Mitochondrial RNA Metabolism

RNA editing is integrated into what is emerging as a byzantine RNA metabolism (Fig. 4). While only a single mitochondrial RNA polymerase appears to be required for mini- and maxicircle RNA synthesis (Fig. 4) (Grams et al. 2002; Hashimi et al. 2009), their transcripts undergo different maturation pathways before the gRNAs are duplexed with their cognate preedited mRNAs. Minicircles are thought to be transcribed polycistronically and cleaved by a 19S protein complex into one or more gRNAs (Grams et al. 2000), before being polyuridylylated by the terminal uridylyl transferase 1 (RET1) (Fig. 4) (Aphasizhev et al. 2003a). These molecules are believed to assume a secondary structure with two hairpin loops, perhaps as a way of being recognized by the protein machinery of editing (Schumacher et al. 2006). The two mitoribosomal rRNAs (9S and 12S) also undergo posttranscriptional modification, forming their short 3′ oligo(U) tails (Fig. 4) (Adler et al. 1991).

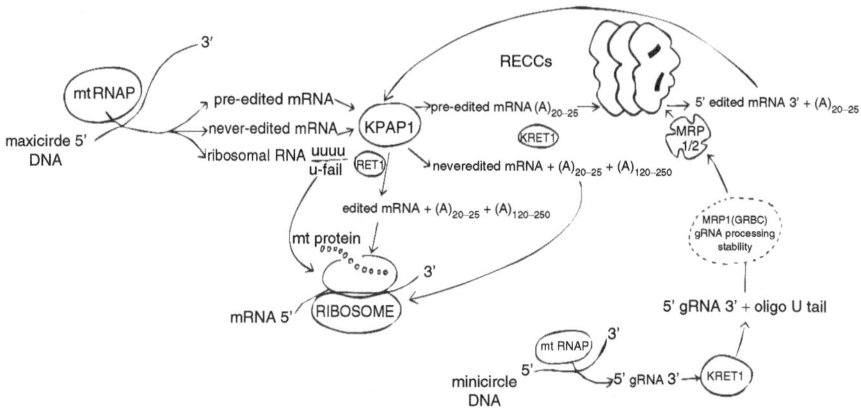

Fig. 4 *Mitochondrial RNA metabolism in procyclic T. brucei* Maxicircle kDNA is transcribed by the mitochondrial RNA polymerase (mtRNAP) into three types of transcripts: preedited mRNAs, never edited mRNAs and mitoribosomal RNAs. Preedited mRNA is equipped with a short 20–25 (A) tail in the kinetoplast poly (A) polymerase 1 complex (KPAP1), and subsequently undergoes editing in one of the RECCs involved in insertion, deletion or *cis*-gRNA mediated editing. 120–250 (A/U)-tail is appended to the fully edited transcript, which is then transcriptionally competent, perhaps with the involvement of the kinetoplastid RNA editing 3′- terminal uridylyl transferase 1 (RET1). Never-edited mRNA is equipped with both short and long tails in the KPAP1 complex and is transported to ribosome to be translated. Minicircle kDNA is transcribed by mtRNAP into gRNA to which oligo-U tail is added by KRET1. gRNAs are then probably stabilized and/or processed by mitochondrial RNA binding complex 1 (MRB1) (also termed GRBC). The heterotetrameric complex comprised of mitochondrial RNA binding proteins 1 and 2 (MRP1/2) stabilize the gRNA molecule in unfolded conformation suitable for pre-mRNA–gRNA hybridization in the initial stage of RNA editing in the 20S editosome. This model depicts a possible sequence of steps a mitochondrial RNA species requires for their maturation, although the precise order of these events remains to be elucidated. The depictions of the 20S editosome and the MRP1/2 complex are based on their resolved three-dimensional structures

The dense gene structure of the T. brucei maxicircle (Fig. 2) indicates that it is transcribed polycistronically. Editing mediated by *trans*-gRNAs occurs independently of cleavage of these precursors into monocistronic transcripts, sometimes even preceding this event (Koslowsky and Yahampath 1997). The resulting mRNAs are polyadenylated by kinetoplast poly(A) polymerase (Fig. 4) (Etheridge et al. 2008). Interestingly, this enzyme appears to be in association with some of the subunits of the MRB1 complex (Fig. 4) and mitochondrially-targeted pentatricopeptide proteins (Pusnik et al. 2007), although the nature of these interactions remains uncertain.

In mitochondria, polyadenylation either stabilizes mRNAs, as in humans, or marks them for degradation, as is the case in plants. However, it turns out that the role of polyadenylation is more complex in *T. brucei*, in which the length of the poly(A) tail appears to be a key determinant of the fate of the molecule. Preedited mRNAs have short poly$(A)_{20-25}$ tails, while never- and fully-edited transcripts have either short poly$(A)_{20-25}$ or long poly$(A/U)_{120-200}$ extensions, in which oligo(U) tracts are interspersed among the poly(A) (Fig. 4) (Militello and Read 2000; Ryan and Read 2005; Etheridge et al. 2008). While the short tail destabilizes preedited molecules, it has the opposite effect in edited mRNAs (Kao and Read 2005). Moreover, another mitochondrial mRNA degradation pathway seems to exist that is independent of the poly(A) tail or UTP (Militello and Read 2000). The exact mechanism of all processing events has to be further explored since some reports claim A/U-tails are destabilizing elements, in which the RET1 enzyme marks the RNA for decay by polyuridylylation (Militello and Read 2000; Ryan and Read 2005).

The 12S rRNA and NADH dehydrogenase subunit 3 genes represent the 5′ ends of the major and minor strands of the maxicircle, respectively, as they are adjacent to the variable sequence domain (Fig. 2a). TbDSS-1, which is a homolog to the eponymous yeast mitochondrial degradosome exonuclease (Penschow et al. 2004), targets aberrant byproducts of these loci, which still contain their unprocessed 5′-ends (Mattiacio and Read 2008). Thus, this enzyme has a role in surveillance of the mitochondrial transcriptome for improperly processed RNAs.

4 Mitochondrial Transfer RNAs

Transfer (t) RNA genes are lacking in the kDNA and thus a complete set of mitochondrial tRNAs has to be imported from the cytosol (Rubio and Alfonzo 2005). The tRNA import system remains, despite intense study, poorly understood. While the requirement for elongation factor 1α and aminoacylation of tRNA for import were rigorously proven (Bouzaidi-Tiali et al. 2007), it was also proposed that thiolation of tRNAs acts as a negative determinant for organellar import in *L. tarentolae* (Kaneko et al. 2003). However, recent data indicate that this is not the case in *T. brucei* (Paris et al. 2009). In a similar vein, the postulated essentiality of the putative tRNA import complex in *L. tropica* (Mukherjee et al. 2007) was not

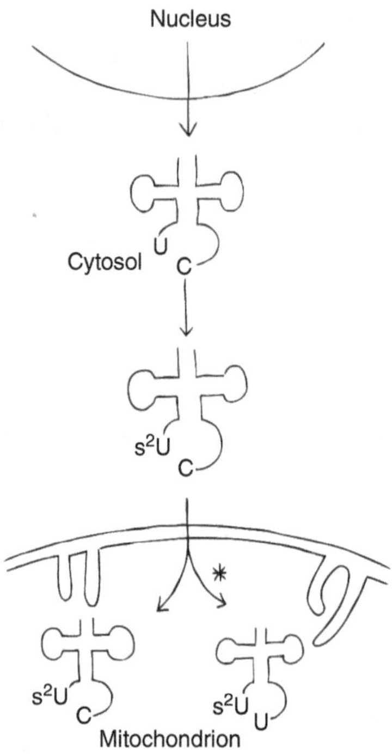

Fig. 5 *Import of tRNAs from the nucleus in T. brucei*. All mitochondrial tRNAs are coded in the nuclear genome and imported into the single mitochondrion from the cytosol. During mitochondrial import, tRNAs are thiolated (s^2U) (at position 33) and subsequently a fraction of the molecules (indicated with an *asterisk*) undergo C to U editing (at position 34)

confirmed in related *T. brucei*, where tRNA import is also independent on membrane potential (Paris et al. 2009). A unique feature of the trypanosomatid tRNA system is C to U editing in the anticodon of a single tryptophanyl tRNA, which allows it to decode the predominantly mitochondrial tryptophan codons (Alfonzo et al. 1999). We have shown recently that thiolation and editing of the neighboring bases in the tRNA molecule are intertwinned processes (Fig. 5) (Wohlgamuth-Benedum et al. 2009).

5 Mitochondrial-Encoded Proteins

Having started with kDNA we browsed through transcription and editing to end the story with translation. Since kDNA codes for components of mitoribosome and subunits of respiratory chain complexes, we are going to limit our protein description mainly to these two areas. For genes encoded on a maxicircle see Fig. 2a.

5.1 Mitochondrial Translation in T. brucei

Mitochondrial ribosomes are of the prokaryotic-type and their rRNA component has been minimized during evolution. In *T. brucei*, mitochondrial ribosomes contain extremely reduced rRNAs (Zíková et al. 2008a). The protein composition of these mitoribosomes was determined by tandem affinity purification followed by mass spectrometry. The large ribosomal subunit sediments at 50S, while the small subunit sediments at 30S. The analysis identified 133 proteins, of which 77 were associated with the large subunit and 56 with the small subunit (Zíková et al. 2008a, b). Quite a similar protein composition was described in the related trypanosomatid *L. tarentolae* (Maslov et al. 2006). Comparisons of this set of proteins with the bacterial and mammalian mitoribosomal proteins identified a number of homologues of both large and small subunits, although the degree of conservation varied widely. Sequence characteristics of some of the component proteins indicated apparent functions in rRNA modification and processing, protein assembly and mitochondrial metabolism, implying possible additional roles for these proteins.

5.2 Composition of Mitochondrial Respiratory Complexes

The eukaryotic respiratory chain is canonically composed of five multisubunit complexes, commonly termed I thru V (Fig. 6a). As in most eukaryotes, the *T. brucei* complexes contain at least one mitochondrial-encoded subunit, with the exception of complex II (FAD-dependent succinate:ubiquinone oxidoreductase). However, in trypanosomes there are several important departures from this general arrangement, which will be discussed below.

Most of the genes encoded by the kDNA maxicircles belong to complex I (NADH:ubiquinone oxidoreductase) (Fig. 2a). Thus, the core complex, as defined by subunits also present in the bacterial homolog of complex I, is composed of nine mitochondrial-encoded subunits, all of which bind Fe–S clusters as cofactors, and six nuclear-encoded subunits. During the evolution of eukaryotes, complex I gradually acquired more and more proteins, reaching in humans a huge complex of over 40 subunits (Gabaldón et al. 2005). In the nuclear and mitochondrial genomes of *T. brucei*, different authors identified a total of 19 subunits (Opperdoes and Michels 2008; Pagliarini et al. 2008). In frame of an extensive mitochondrial proteome study, 17 obvious homologs of complex I subunits and 12 additional hypothetical conserved proteins associated with this complex were identified (Panigrahi et al. 2009).

No subunits of complex II are encoded in the mitochondrial genome, yet we still mention it here in order to sequentially describe the respiratory chain. This complex has a core composed of four subunits that are shared with prokaryotes. So far, a proteomic study revealed two subunits in *T. brucei* (Panigrahi et al. 2009). This number is certainly not final, since the same complex from *T. cruzi* is composed of at least 12 subunits (Morales et al. 2009).

Following the flow of electrons, the next complex to carry a mitochondrial-encoded subunit is complex III (cytochrome c reductase or complex bc1) (Fig. 6a).

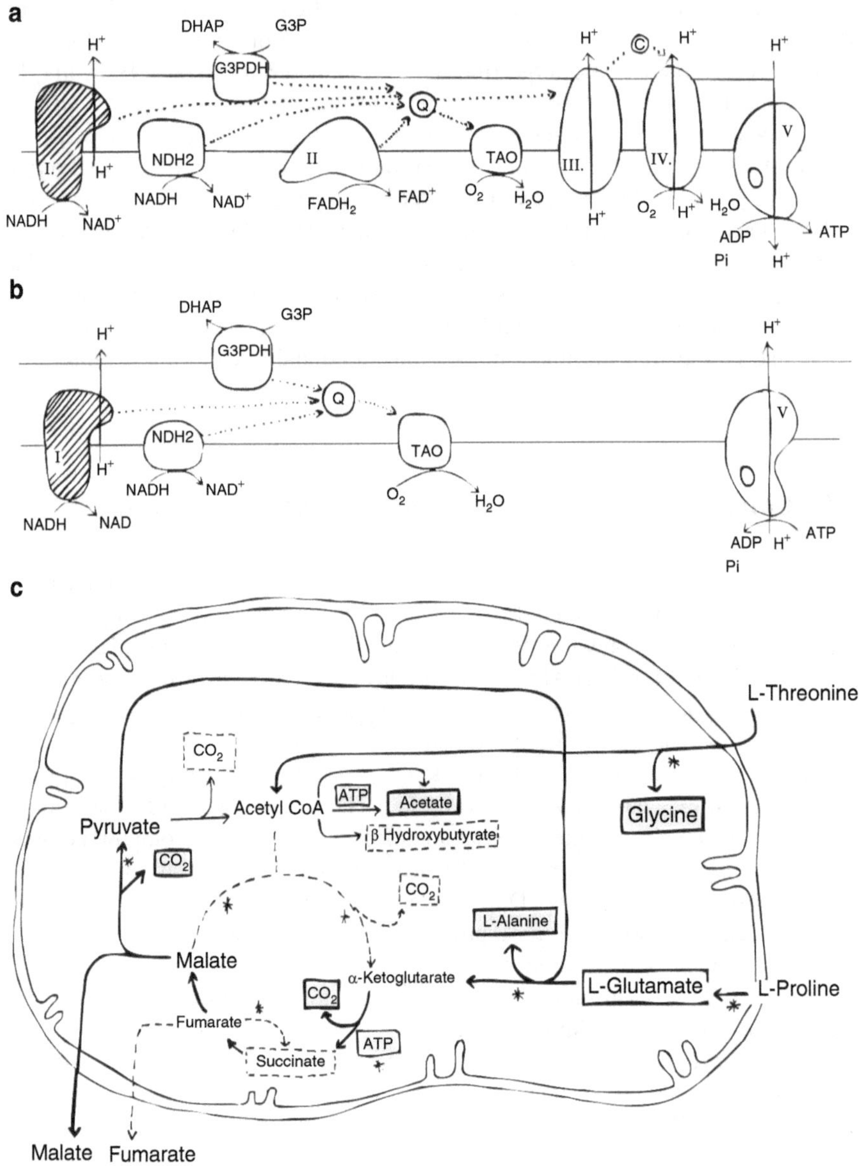

Fig. 6 *Respiratory chain and carbohydrate metabolism in the mitochondrion of T. brucei.* (**a**) *Procyclic stage from the tse-tse fly*. All canonical respiratory complexes (I–V) are physically present. Striped complex I emphasizes uncertainty of its biochemical function(s). Electrons are taken from complexes I and II and two single-peptide enzymes (alternative NADH dehydrogenase [NDH2] and glycerol-3-phosphate dehydrogenase [G3PDH]) and passed to ubiquinone [Q]. From here, about 25% flows to trypanosome alternative oxidase (TAO), the rest goes through cyanide sensitive pathway (complex III to cytochrome c) to complex IV. If not complex I, at least complexes III and IV generate proton-motive force that drives ATP synthesis by complex V. (**b**) *Bloodstream stage from mammalian blood.* Cytochromes (complex III, cytochrome c and complex

The prokaryotic homologue comprises three catalytic subunits, one of them having a Fe–S cluster and the others bearing heme. Eukaryotes added an additional six to eight more proteins to build their mitochondrial complex III (Attardi and Chomyn 1995). The *T. brucei* genome contains three subunits: Rieske Fe–S protein, cytochrome c1, and a putative cytochrome *c* hinge protein homologue. The two former proteins were also identified by mass spectrometry (Panigrahi et al. 2009). Since both of them, together with eight more proteins, were also isolated in the related parasite *L. tarentolae* (Horváth et al. 2000), their homologues are very likely to also be isolated in *T. brucei*. Moreover, experimental evidence is available for the Rieske Fe–S protein being a genuine component of complex III in *T. brucei* (Horváth et al. 2005).

Three subunits of complex IV (cytochrome *c* oxidase) are coded by maxicircle kDNA (Fig. 2a) and constitute the core shared between prokaryotes and eukaryotes. These proteins bear metal prosthetic groups (heme A and copper) that constitute three redox centers. Additional proteins were recruited during the evolution of eukaryotes, forming a complex of up to 13 subunits (Fontanesi et al. 2008). Seven nuclear-encoded subunits were identified in a mitoproteomic survey (Panigrahi et al. 2009).

Finally, complex V (F_1F_0-ATP synthase) requires only a single subunit coded by the organellar genome (Fig. 2a) (Hashimi et al. 2010). ATP synthase from *Escherichia coli* contains eight subunits with various degree of multimerization, while complex V from the yeast and bovine mitochondria contains additional ten subunits. The ATPase of *T. brucei* is more reminiscent of the bacterial homologue, as it is composed of just ten subunits (Zíková et al. 2009). However, since the apparent lack of hydrophobic nuclear-encoded subunits from these proteomic datasets may be due to technical reasons, other approaches are needed to verify the currently known composition of respiratory complexes in trypanosomatids.

Fig. 6 (continued) IV) are missing in this stage. Complex I is shown as we cannot unambiguously rule out its presence. Single-peptide alternative enzymes are present as in the procyclic stage. Notice that TAO is now the only electron sink. Complex V reverses its usual activity and burns ATP to produce proton gradient. (**c**) *"Krebs" cycle of the procyclic stage*. Proline is converted into glutamate that reacts with pyruvate to produce α-ketoglutarate. This follows standard Krebs cycle reactions to produce malate, which is either excreted or converted into pyruvate. Some of the pyruvate are used for production of acetyl coenzyme A. Most of pyruvate reacts with glutamate and closes the cycle. Acetyl coenzyme A is also produced from threonine. Through substrate phosphorylation, acetyl Coenzyme A gives rise to ATP and acetate, a major end-product. Note that there are two reactions leading to acetate; nevertheless, only one of them produces ATP. Part of acetyl coenzyme A is used for β-hydroxybutyrate, a minor end-product. Full lines and boxes depict main metabolic fluxes and end-products under glucose-low conditions; dashed lines and boxes represents minor or background reactions and end-products; asterisks indicate origins of reduced co-factors

6 Energy Metabolism of the *T. brucei* Mitochondrion

T. brucei faces two dramatically distinct environments during its life cycle. The mammalian bloodstream is an environment rich with glucose, where glycolytic ATP production is more than sufficient to support parasite growth. In contrast, the gut of the tse-tse fly is glucose-poor. Despite usage of amino acids extracted from this environment as a primary energy source, glucose is consumed at the same rate as in the bloodstream form (Cazzulo 1992). Amino acids enter the Krebs cycle through different intermediates. Reduced cofactors generated in this process are oxidized by the respiratory chain (Besteiro et al. 2005; van Weelden et al. 2005; Coustou et al. 2008).

6.1 Bloodstream Stage

In the stage that dwells in the bloodstream of vertebrate hosts, the mitochondrion is extremely suppressed. Pyruvate, the substrate of pyruvate dehydrogenase complex, is excreted as a main end-product of metabolism. The Krebs cycle thus has no substrate to use and its enzymes are absent, as are the cytochrome-containing respiratory complexes (Fig. 6b) (Hannaert et al. 2003). The biochemical presence of a genuine mitochondrial complex I remains the subject of an open debate (see Insect stage for details). Nevertheless, the presence of type II alternative NADH dehydrogenase, (Fang and Beattie 2003b) makes complex I expendable, at least for NAD regeneration. This alternative enzyme is represented by a single peptide, which is able to regenerate NAD, but cannot pump protons through the inner mitochondrial membrane (Fig. 6b).

The other electron source of respiratory chain of bloodstreams is mitochondrial FAD-dependent glycerol-3-phosphate dehydrogenase. This enzyme connects the mitochondrion with the glycosome via a glycerol-3-phosphate:dihydroxy-acetone phosphate shuttle, which helps to maintain a favorable redox state in the glycosomes. All gathered electrons are sent onto ubiquinone, from where they are subsequently transferred onto molecular oxygen at an enzyme called trypanosomal alternative oxidase (TAO) (Fig. 6b). This alternative oxidase catalyzes the same reaction as respiratory complex IV, but does not use cytochrome *c* as an electron mediator, and no protons are shuffled across the inner membrane (Hannaert et al. 2003).

Taken together, despite an electron flow, any enzymes or complexes that could use this flow to generate proton gradient seem to be absent in the respiratory chain of the bloodstream stage. Still, the proton gradient is indispensable for protein import (Schnaufer et al. 2005). Lacking any classical respiratory complexes to generate the proton gradient, a long lasting question was how bloodstreams cope with this challenge. This conundrum was solved when it was discovered that in the infectious bloodstream stage, complex V is able to reverse its action and maintain the proton gradient at the expense of ATP consumption (Schnaufer et al. 2005; Brown et al. 2006). Hence, in this pathogenic stage of *T. brucei*, mitochondrion is not the powerhouse of the cell. Quite the contrary, the organelle is as ATP consumer (Fig. 6b).

6.2 Procyclic Stage

In contrast, the procyclic stage from the tse-tse fly contains a mitochondrion that is physiologically and metabolically similar to that found in conventional eukaryotic model systems. All classical complexes are present and most of them appear to be active from a biochemical point of view (Fig. 6a). Although nowadays we have a clear evidence for the physical presence of complex I, or at least a part of it (Panigrahi et al. 2008), its biochemical status remains problematic. Alternative NADH dehydrogenase catalyzes the same reaction and rotenone, a specific inhibitor of complex I in other eukaryotes, was shown to be rather nonspecific in the trypanosomal respiratory chain, failing to distinguish between either dehydrogenase (Hernandez and Turrens 1998). On the other hand, acyl-carrier protein, which is a subunit of complex I, was shown to be actively involved in fatty acids synthesis (Guler et al. 2008). Also, we cannot exclude the possibility that the activity of complex I is strain-dependent. The UC strain of *L. tarentolae* and the 1S LdBob strain of *L. donovani* lost some of their kDNA minicircles, thus losing the capacity to edit some subunits of the respiratory chain, and ultimately resulting in biochemically inactive complexes (Thiemann et al. 1994; Neboháčová et al. 2009).

The activities of the discussed enzymes are highly environment-dependent. Under standard in vitro condition, the procyclic cells are cultivated in a glucose-rich SDM-79 medium. This situation results in conditions under which the mitochondrion seems to be "semiactive," in that all enzymes involved in mitochondrial energy metabolism are present but ATP is mainly derived from glycolysis. The supposed in vivo conditions within the insect midgut can be mimicked by removing glucose from the medium, in which the cells resort to catabolism of amino acids. It is worth noting that such conditions lead to the remarkably higher activities of the respiratory complexes (Coustou et al. 2008).

The glycerol-3-phosphate:dihydroxy-acetone phosphate reaction shuttle, alternative dehydrogenase and alternative oxidase are also present in the procyclic stage (Fig. 6a) (Fang and Beattie 2003b; Guerra et al. 2006). The presence of the first of these three enzymes is considered unambiguous as glycolysis works the same way as in bloodstreams with only a couple of enzymes relocated into the cytosol. However, the presence of alternative dehydrogenase further questions any vital biochemical role of complex I. Similarly, we still have to learn a lot about the role of alternative oxidase. So far the most feasible theory connects its presence with the potential flexibility to cope with a different availability of nutrients (Chaudhuri et al. 2006). It has been also proposed to be retained during the procyclic stage in order to reduce the level of reactive oxygen species (Fang and Beattie 2003a).

Mitochondrial energy generation is highly dependent on available carbon source (s). In vivo, with amino acids as a primary source, energy is usually obtained through the respiratory chain. Reduced cofactors are derived from reactions leading from amino acids to the intermediates of the Krebs cycle or acetyl-coenzyme A, and from reactions of this cycle itself (Fig. 6c). Proline is converted into glutamate, which enters the Krebs cycle through α-ketoglutarate and is metabolized in the canonical way until malate, which either leaves the mitochondrion and participates in gluconeogenesis, or is converted into pyruvate (Fig. 6c) (Coustou et al. 2008). The pyruvate is partly converted into acetyl-coenzyme A, while most of it is used for reaction with L-glutamate, thus completing the cycle (Fig. 6c) (Coustou et al. 2008).

When glucose is available, the *T. brucei* procyclics derive pyruvate mainly from glycolytic reactions. The ratio between the acetyl-coenzyme A and glutamate reactions is shifted in favor of the former pathway. Moreover, instead of being exported, malate is imported. Again, part of it is used for the production of pyruvate, while the rest goes in a reverse direction, producing succinate instead. Nevertheless, under both conditions a certain amount of ATP is produced by substrate phosphorylation. Acetyl-coenzyme A reacts with succinate producing succinyl-coenzyme A, which is converted back to succinate with concomitant production of ATP (van Weelden et al. 2005; Coustou et al. 2008) (Fig. 6c).

References

Abu-Elneel K, Robinson DR, Drew ME, Englund PT, Shlomai J (2001) Intramitochondrial localization of universal minicircle sequence-binding protein, a trypanosomatid protein that binds kinetoplast minicircle replication origin. J Cell Biol 153:725–733

Adler BK, Harris ME, Bertrand KI, Hajduk SL (1991) Modification of *Trypanosoma brucei* mitochondrial rRNA by posttranscriptional 3′ polyuridine tail formation. Mol Cell Biol 11:5878–5884

Alfonzo JD, Blanc V, Estévez AM, Rubio MA, Simpson L (1999) C to U editing of the anticodon of imported mitochondrial $tRNA^{Trp}$ allows decoding of the UGA stop codon in *Leishmania tarentolae*. EMBO J 24:7056–7062

Amaro RE, Schnaufer A, Interthal H, Hol W, Stuart KD, McCammon JA (2008) Discovery of drug-like inhibitors of an essential RNA-editing ligase in *Trypanosoma brucei*. Proc Natl Acad Sci USA 105:17278–17283

Ammerman ML, Fisk JC, Read LK (2008) gRNA/pre-mRNA annealing and RNA chaperone activities of RBP16. RNA 14:1069–1080

Aphasizhev R, Aphasizheva I, Simpson L (2003a) A tale of two TUTases. Proc Natl Acad Sci USA 100:10617–10622

Aphasizhev R, Aphasizheva I, Nelson RE, Simpson L (2003b) A 100-kD complex of two RNA-binding proteins from mitochondria of *Leishmania tarentolae* catalyzes RNA annealing and interacts with several RNA editing components. RNA 9:62–76

Attardi GM, Chomyn A (1995) Mitochondrial biogenesis and genetics. Academic, San Diego

Avliyakulov NK, Lukeš J, Ray DS (2004) Mitochondrial histone-like DNA-binding proteins are essential for normal cell growth and mitochondrial function in *Crithidia fasciculata*. Eukaryot Cell 3:518–526

Benne R, van den Burg J, Brakenhoff JP, Sloof P, Van Boom JH, Tromp MC (1986) Major transcript of the frameshifted coxII gene from trypanosome mitochondria contains four nucleotides that are not encoded in the DNA. Cell 46:819–826

Besteiro S, Barrett MP, Riviere L, Bringaud F (2005) Energy generation in insect stages of *Trypanosoma brucei*: metabolism in flux. Trends Parasitol 21:185–191

Blum B, Bakalara N, Simpson L (1990) A model for RNA editing in kinetoplastid mitochondria: "guide" RNA molecules transcribed from maxicircle DNA provide the edited information. Cell 60:189–198

Bouzaidi-Tiali N, Aeby E, Charriere F, Pusnik M, Schneider A (2007) Elongation factor 1a mediates the specificity of mitochondrial tRNA import in *T. brucei*. EMBO J 26:4302–4312

Bringaud F, Riviére L, Coustou V (2006) Energy metabolism of trypanosomatids: adaptation to available carbon sources. Mol Biochem Parasitol 149:1–9

Brown SV, Hosking P, Li J, Williams N (2006) ATP synthase is responsible for maintaining mitochondrial membrane potential in bloodstream form *Trypanosoma brucei*. Eukaryot Cell 5:45–53

Carnes J, Trotter JR, Ernst NL, Steinberg AG, Stuart K (2005) An essential RNase III insertion editing endonuclease in *Trypanosoma brucei*. Proc Natl Acad Sci USA 102:16614–16619

Carnes J, Trotter JR, Peltan A, Fleck M, Stuart K (2008) RNA editing in *Trypanosoma brucei* requires three different editosomes. Mol Cell Biol 28:122–130

Cazzulo JJ (1992) Aerobic fermentation of glucose by trypanosomatids. FASEB J 6:3153–3161

Chaudhuri M, Ott RD, Hill GC (2006) Trypanosome alternative oxidase: from molecule to function. Trends Parasitol 22:484–491

Chen JH, Rauch CA, White JH, Englund PT, Cozzarelli NR (1995) The topology of the kinetoplast DNA network. Cell 80:61–69

Coustou V, Biran M, Breton M, Guegan FR, Plazolles N, Nolan D, Barrett MP, Franconi JM, Bringaud F (2008) Glucose-induced remodeling of intermediary and energy metabolism in procyclic *Trypanosoma brucei*. J Biol Chem 283:16342–16354

Deng J, Schnaufer A, Salavati R, Stuart KD, Hol WG (2004) High resolution crystal structure of a key editosome enzyme from *Trypanosoma brucei*: RNA editing ligase 1. J Mol Biol 343: 601–613

Downey N, Hines JC, Sinha KM, Ray DS (2005) Mitochondrial DNA ligases of *Trypanosoma brucei*. Eukaryot Cell 4:765–774

Drew ME, Englund PT (2001) Intramitochondrial location and dynamics of *Crithidia fasciculata* kinetoplast minicircle replication intermediates. J Cell Biol 153:735–744

Dyková I, Fiala I, Lom J, Lukeš J (2003) *Perkinsiella* amoebae-like endosymbionts of *Neoparamoeba* spp., relatives of the kinetoplastid *Ichthyobodo*. Eur J Protistol 39:37–52

Englund PT (1979) Free minicircles of kinetoplast DNA in *Crihidia fasciculata*. J Biol Chem 254:4895–4900

Estevez AM, Simpson L (1999) Uridine insertion/deletion RNA editing in trypanosome mitochondria – a review. Gene 240:247–260

Etheridge RD, Aphasizheva I, Gershon PD, Aphasizhev R (2008) 3′ adenylation determines mRNA abundance and monitors completion of RNA editing in *T. brucei* mitochondria. EMBO J 27:1596–1608

Fang J, Beattie DS (2003a) Alternative oxidase present in procyclic *Trypanosoma brucei* may act to lower the mitochondrial production of superoxide. Arch Biochem Biophys 414:294–302

Fang J, Beattie DS (2003b) Identification of a gene encoding a 54 kDa alternative NADH dehydrogenase in *Trypanosoma brucei*. Mol Biochem Parasitol 127:73–77

Fisk JC, Ammerman ML, Presnyak V, Read LK (2008) TbRGG2, an essential RNA editing accessory factor in two *Trypanosoma brucei* life cycle stages. J Biol Chem 283: 23016–23025

Flegontov PN, Zhirenkina EN, Gerasimov ES, Ponirovsky EN, Strelkova MV, Kolesnikov AA (2009) Selective amplification of maxicircle classes during the life cycle of *Leishmania major*. Mol Biochem Parasitol 165:142–152

Fontanesi F, Soto IC, Barrientos A (2008) Cytochrome c oxidase biogenesis: new levels of regulation. IUBMB Life 60:557–568

Gabaldón T, Rainey D, Huynen MA (2005) Tracing the evolution of a large protein complex in the eukaryotes, NADH:ubiquinone oxidoreductase (complex I). J Mol Biol 348:857–870

Gluenz E, Shaw MK, Gull K (2007) Structural asymmetry and discrete nucleic acid subdomains in the *Trypanosoma brucei* kinetoplast. Mol Microbiol 64:1529–1539

Golas MM, Böhm C, Sander B, Effenberger K, Brecht M, Stark H, Göringer HU (2009) Snapshots of the RNA editing machine in trypanosomes captured at different assembly stages *in vivo*. EMBO J 28:766–778

Golden DE, Hajduk SL (2005) The 3′-untranslated region of cytochrome oxidase II mRNA functions in RNA editing of African trypanosomes exclusively as a cis guide RNA. RNA 11:29–37

Grams J, McManus MT, Hajduk SL (2000) Processing of polycistronic guide RNAs is associated with RNA editing complexes in *Trypanosoma brucei*. EMBO J 19:5525–5532

Grams J, Morris JC, Drew ME, Wang ZF, Englund PT, Hajduk SL (2002) A trypanosome mitochondrial RNA polymerase is required for transcription and replication. J Biol Chem 277:16952–16959

Guerra DG, Decottignies A, Bakker BM, Michels PAM (2006) The mitochondrial FAD-dependent glycerol-3-phosphate dehydrogenase of Trypanosomatidae and the glycosomal redox balance of insect stage of *Trypanosoma brucei* and *Leishmania* spp. Mol Biochem Parasitol 149:155–169

Guler JL, Kriegová E, Smith TK, Lukeš J, Englund PT (2008) Mitochondrial fatty acid synthesis is required for normal mitochondrial morphology and function in *Trypanosoma brucei*. Mol Microbiol 67:1125–1142

Hannaert V, Bringaud F, Opperdoes FR, Michels PAM (2003) Evolution of energy metabolism and its compartmentation in Kinetoplastida. Kinetoplastid Biol Dis 2:11–40

Hashimi H, Zíková A, Panigrahi AK, Stuart KD, Lukeš J (2008) TbRGG1, a component of a novel multi-protein complex involved in kinetoplastid RNA editing. RNA 14:970–980

Hashimi H, Čičová Z, Novotná L, Wen YZ, Lukeš J (2009) Kinetoplastid guide RNA biogenesis is dependant on subunits of the mitochondrial RNA binding complex and mitochondrial RNA polymerase. RNA 15:588–599

Hashimi H, Benkovičová V, Čermáková P, Lai D-H, Horváth A, Lukeš J (2010) The assembly of F_1F_O-ATP synthase is disrupted upon interference of RNA editing in *Trypanosoma brucei*. Int J Parasitol 40:45–54

Hernandez FR, Turrens JF (1998) Rotenone at high concenrations inhibits NADH-fumarate reductase and the mitochondrial respiratory chain of *Trypanosoma brucei* and *T. cruzi*. Mol Biochem Parasitol 93:135–137

Horváth A, Berry EA, Huang L, Maslov DA (2000) *Leishmania tarentolae*: a parallel isolation of cytochrome bc1 and cytochrome c oxidase. Exp Parasitol 96:160–167

Horváth A, Horáková E, Dunajčíková P, Verner Z, Pravdová E, Šlapetová I, Cuninková L, Lukeš J (2005) Down-regulation of the nuclear-encoded subunits of the complexes III and IV disrupts their respective complexes but not complex I in procyclic *Trypanosoma brucei*. Mol Microbiol 58:116–130

Jensen RE, Simpson L, Englund PT (2008) What happens when *Trypanosoma brucei* leaves Africa. Trends Parasitol 24:428–431

Kaneko T, Suzuki T, Kapushoc ST, Rubio MA, Ghazvini J, Watanabe K, Simpson L, Suzuki T (2003) Wobble modification differences and subcellular localization of tRNAs in *Leishmania tarentolae*: implication for tRNA sorting mechanism. EMBO J 22:657–667

Kao CY, Read LK (2005) Opposing effect of polyadenylation on the stability of edited and unedited mitochondrial RNAs in *Trypanosoma brucei*. Mol Cell Biol 25:1634–1644

Klingbeil MM, Motyka SA, Englund PT (2002) Multiple mitochondrial DNA polymerases in *Trypanosoma brucei*. Mol Cell 10:175–186

Koslowsky DJ, Yahampath G (1997) Mitochondrial mRNA 3′ cleavage/polyadenylation and RNA editing in *Trypanosoma brucei* are independent events. Mol Biochem Parasitol 90:81–94

Lai D-H, Hashimi H, Lun Z-R, Ayala FJ, Lukeš J (2008) Adaptation of *Trypanosoma brucei* to gradual loss of kinetoplast DNA: *T. equiperdum* and *T. evansi* are petite mutants of *T. brucei*. Proc Natl Acad Sci USA 105:1999–2004

Li Z, Lindsay ME, Motyka SA, Englund PT, Wang CC (2008) Identification of a bacterial-like HslVU protease in the mitochondria of *Trypanosoma brucei* and its role in mitochondrial DNA replication. PLoS Pathog 4:e1000048

Li F, Ge P, Hui W, Atanasov A, Rogers K, Guo Q, Osato D, Falick AM, Zhou H, Simpson L (2009) Structure of the core editing complex (L-complex) involved in uridine insertion/deletion editing in trypanosomatid mitochondria. Proc Natl Acad Sci USA 106:12306–12310

Lindsay ME, Gluenz E, Gull K, Englund PT (2008) A new function of *Trypanosoma brucei* mitochondrial topoisomerase II is to maintain kinetolpast DNA network topology. Mol Microbiol 70:1465–1476

Liu Y, Englund PT (2007) The rotational dynamics of kinetoplast DNA replication. Mol Microbiol 64:676–690

Liu B, Liu Y, Motyka SA, Agbo EEC, Englund PT (2005) Fellowship of the rings: the replication of kinetoplast DNA. Trends Parasitol 21:363–369

Liu Y, Molina K, Kalume D, Pandey A, Griffith JD, Englund PT (2006) Role of p38 in replication of *Trypanosoma brucei* kinetoplast DNA. Mol Cell Biol 26:5382–5393

Lukeš J, Votýpka J (2000) *Trypanosoma avium*: novel features of the kinetoplast structure. Exp Parasitol 96:178–181

Lukeš J, Guilbride DL, Votýpka J, Zíková A, Benne R, Englund PT (2002) The kinetoplast DNA network: evolution of an improbable structure. Eukaryot Cell 1:495–502

Lukeš J, Hashimi H, Zíková A (2005) Unexplained complexity of the mitochondrial genome and transcriptome in kinetoplastid flagellates. Curr Genet 48:277–299

Maslov DA, Simpson L (1992) The polarity of editing within a multiple gRNA-mediated domain is due to formation of anchors for upstream gRNAs by downstream editing. Cell 70:459–467

Maslov DA, Sharma MR, Butler E, Falick AM, Gingery M, Agrawal RK, Spremulli LL, Simpson L (2006) Isolation and characterization of mitochondrial ribosomes and ribosomal subunits from *Leishmania tarentolae*. Mol Biochem Parasitol 148:69–78

Mattiacio JL, Read LK (2008) Roles for TbDSS-1 in RNA surveillance and decay of maturation by-products from the 12S rRNA locus. Nucleic Acids Res 36:319–329

McManus MT, Adler BK, Pollard VW, Hajduk SL (2000) *Trypanosoma brucei* guide RNA poly (U) tail formation is stabilized by cognate mRNA. Mol Cell Biol 20:883–891

Militello KT, Read LK (2000) UTP-dependent and -independent pathways of mRNA turnover in *Trypanosoma brucei* mitochondria. Mol Cell Biol 20:2308–2316

Milman N, Motyka SA, Englund PT, Robinson D, Shlomai J (2007) Mitochondrial origin-binding protein UMSBP mediates DNA replication and segregation in trypanosomes. Proc Natl Acad Sci USA 104:19250–19255

Missel A, Souza AE, Nörskau G, Göringer HU (1997) Disruption of a gene encoding a novel mitochondrial DEAD-box protein in *Trypanosoma brucei* affects edited mRNAs. Mol Cell Biol 17:4895–4903

Morales J, Mogi T, Mineki S, Takshima E, Mineki R, Hirawake H, Sakamoto K, Omura S, Kita K (2009) Novel mitochondrial complex II isolated from *Trypanosoma cruzi* is composed of twelve peptides including a heterodimeric Ip subunit. J Biol Chem 284:7255–7263

Mukherjee S, Basu S, Home P, Dhar G, Adhya S (2007) Necessary and sufficient factors for the import of transfer RNA into the kinetoplast mitochondrion. EMBO Rep 8:589–595

Neboháčová M, Kim CE, Simpson L, Maslov DA (2009) RNA editing and mitochondrial activity in promastigotes and amastigotes of *Leishmania donovani*. Int J Parasitol 39:635–644

Ochsenreiter T, Anderson S, Wood ZA, Hajduk SL (2008a) Alternative RNA editing produces a novel protein involved in mitochondrial DNA maintenance in trypanosomes. Mol Cell Biol 28:5595–5604

Ochsenreiter T, Cipriano M, Hajduk SL (2008b) Alternative mRNA editing in trypanosomes is extensive and may contribute to mitochondrial protein diversity. PLoS One 3:e1566

Ogbadoyi EO, Robinson DR, Gull K (2003) A high-order transmembrane structural linkage is responsible for mitochondrial genome positioning and segregation by flagellar basal bodies in trypanosomes. Mol Biol Cell 14:1769–1779

Opperdoes FR, Michels PAM (2008) Complex I of Trypanosomatidae: does it exist? Trends Parasitol 24:310–317

Pagliarini DJ, Calvo SE, Chang B, Sheth SA, Vafai SB, Ong S-E, Walford GA, Sugiana C, Boneh A, Chen WK, Hill DE, Vidal M, Evans JG, Thornburn DR, Carr SA, Mootha VK (2008) A mitochondrial protein compendium elucidates complex I disease biology. Cell 134:112–123

Panigrahi AK, Zíková A, Halley RA, Acestor N, Ogata Y, Myler PJ, Stuart K (2008) Mitochondrial complexes in *Trypanosoma brucei*: a novel complex and a unique oxidoreductase complex. Mol Cell Proteomics 7:534–545

Panigrahi AK, Ogata Y, Zíková A, Anupama A, Dalley RA, Acestor N, Myler PJ, Stuart KD (2009) A comprehensive analysis of *Trypanosoma brucei* mitochondrial proteome. Proteomics 9:434–450

Paris Z, Rubio MAT, Lukeš J, Alfonzo JD (2009) Mitochondrial tRNA import in *Trypanosoma brucei* is independent of thiolation and the Rieske protein. RNA 15:1398–1406

Pelletier M, Read LK (2003) RBP16 is a multifunctional gene regulatory protein involved in editing and stabilization of specific mitochondrial mRNAs in *Trypanosoma brucei*. RNA 9:457–468

Penschow JL, Sleve DA, Ryan CM, Read LK (2004) TbDSS-1, an essential *Trypanosoma brucei* exoribonuclease homolog that has pleiotropic effects on mitochondrial RNA metabolism. Eukaryot Cell 3:1206–1216

Pusnik M, Small I, Read LK, Fabbro T, Schneider A (2007) Pentatricopeptide repeat proteins in *Trypanosoma brucei* function in mitochondrial ribosomes. Mol Cell Biol 27:6876–6888

Rubio MAT, Alfonzo J (2005) Editing and modification in trypanosomatids: the reshaping of non-coding RNAs. Top Curr Genet 12:71–86

Ryan CM, Read LK (2005) UTP-dependent turnover of *Trypanosoma brucei* mitochondrial mRNA requires UTP polymerization and involves the RET1 TUTase. RNA 11:1–11

Saxowsky TT, Choudhary G, Klingbeil MM, Englund PT (2003) *Trypanosoma brucei* has two distinct mitochondrial DNA polymerase beta enzymes. J Biol Chem 278:49095–49101

Schnaufer A, Panigrahi AK, Panicucci B, Igo RP Jr, Wirtz E, Salavati R, Stuart K (2001) An RNA ligase essential for RNA editing and survival of the bloodstream form of *Trypanosoma brucei*. Science 291:2159–2162

Schnaufer A, Domingo GJ, Stuart K (2002) Natural and induced dyskinetoplastic trypanosomatids: how to live without mitochondrial DNA. Int J Parasitol 32:1071–1084

Schnaufer A, Ernst NL, Palazzo SS, O'Rear J, Salavati R, Stuart K (2003) Separate insertion and deletion subcomplexes of the *Trypanosoma brucei* RNA editing complex. Mol Cell 12:307–319

Schnaufer A, Clark-Walker GD, Steinberg AG, Stuart K (2005) The F1-ATP synthase complex in bloodstram stage of trypanosomes has an unusual and essential function. EMBO J 24:4029–4040

Schneider A (2001) Unique aspects of mitochondrial biogenesi in trypanosomatids. Int J Parasitol 31:1403–1415

Schumacher MA, Karamooz E, Zíková A, Trantírek L, Lukeš J (2006) Crystal structures of *Trypanosoma brucei* MRP1/MRP2 guide-RNA-binding complex reveals RNA matchmaking mechanism. Cell 126:701–711

Sela D, Shlomai J (2009) Regulation of UMSBP activities through redox-sensitive protein domains. Nucleic Acids Res 37:279–288

Shapiro TA, Englund PT (1995) The structure and replication of kinetoplast DNA. Annu Rev Microbiol 49:117–143

Shlomai J (2004) The structure and replication of kinetoplast DNA. Curr Mol Med 4:623–647

Simpson L, Aphasizhev R, Gao G, Kang X (2004) Mitochondrial proteins and complexes in *Leishmania* and *Trypanosoma* involved in U-insertion/deletion RNA editing. RNA 10:159–170

Simpson AGB, Stevens JR, Lukeš J (2006) The evolution and diversity of kinetoplastid flagellates. Trends Parasitol 22:168–174

Simpson L, Aphasizhev R, Lukeš J, Cruz-Reyes J (2010) Guide to the nomenclature of kinetoplastid RNA editing: a proposal. Protist 161:2–6

Sinha KM, Hines JC, Ray DS (2006) Cell cycle-dependent localization and properties of a second mitochondrial DNA ligase in *Crithidia fasciculata*. Eukaryot Cell 5:54–61

Speijer D (2008) Evolutionary aspects of RNA editing. In: Goringer HU (ed) RNA editing. Springer, Berlin, pp 199–229

Stuart K, Allen TE, Heidmann S, Seiwert SD (1997) RNA editing in kinetoplastid protozoa. Microbiol Mol Biol Rev 61:105–120

Stuart K, Schnaufer A, Ernst NL, Panigrahi AK (2005) Complex management: RNA editing in trypanosomes. Trends Biochem Sci 30:97–105

Sturm NR, Simpson L (1990) Kinetoplast DNA minicircles encode guide RNAs for editing of cytochrome oxidase subunit III mRNA. Cell 61:879–884

Thiemann OH, Maslov DA, Simpson L (1994) Disruption of RNA editing in *Leishmania tarentolae* by the loss of minicircle-encoded guide RNA genes. EMBO J 13:5689–5700

Torri AF, Englund PT (1995) A DNA polymerace β in the mitochondrion of the trypanosomatid *Crithidia fasciculata*. J Biol Chem 270:3495–3497

Trotter JR, Ernst NL, Carnes J, Panicucci B, Stuart K (2005) A deletion site editing endonuclease in *Trypanosoma brucei*. Mol Cell 20:403–412

van Weelden SW, van Hellemond JJ, Opperdoes FR, Tielens AG (2005) New functions for parts of the Krebs cycle in procyclic *Trypanosoma brucei*, a cycle not operating as a cycle. J Biol Chem 280:12451–12460

Vanhamme L, Perez-Morga D, Marchal C, Speijer D, Lambert L, Geuskens M, Alexandre S, Ismaïli N, Göringer U, Benne R, Pays E (1998) *Trypanosoma brucei* TBRGG1, a mitochondrial oligo(U)-binding protien co-localizes with an *in vitro* RNA editing activity. J Biol Chem 273:21825–21833

Vondrušková E, van den Burg J, Zíková A, Ernst NL, Stuart K, Benne R, Lukeš J (2005) RNA interference analyses suggest a transcript-specific regulatory role for MRP1 and MRP2 in RNA editing and other RNA processing in *Trypanosoma brucei*. J Biol Chem 280:2429–2438

Wang Z, Englund PT (2001) RNA interference of a trypanosome topoisomerase II causes progressive loss of mitochondrial DNA. EMBO J 20:4674–4683

Weng J, Aphasizheva I, Etheridge RD, Huang L, Wang X, Falick AM, Aphasizhev R (2008) Guide RNA-binding complex from mitochondria of trypanosomatids. Mol Cell 32:1–12

Wohlgamuth-Benedum JM, Rubio MAT, Paris Z, Long S, Poliak P, Lukeš J, Alfonzo JD (2009) Thiolation controls cytoplasmic tRNA stability and acts as a negative determinant for tRNA editing in mitochondria. J Biol Chem 284:23947–23953

Woodward R, Gull K (1990) Timing of nuclear and kinetoplast DNA replication and early morphological events in the cell cycle of *Trypanosoma brucei*. J Cell Sci 95:49–57

Zhao Z, Lindsay ME, Roy Chowdhury A, Robinson DR, Englund PT (2008) p166, a link between the trypanosome mitochondrial DNA and flagellum, mediates genome segregation. EMBO J 27:143–154

Ziemann H (1898) Eine Methode der Doppelfarbung bei Flagellaten, Pilzen, Spirillen und Bakterien, sowie bei einigen Amoben. Zentralbl Bakt Parasitenkd Infekt 24:945–955

Zíková A, Panigrahi AK, Dalley RA, Acestor N, Anupama A, Ogata Y, Myler PJ, Stuart K (2008a) *Trypanosoma brucei* mitochondrial ribosomes: affinity purification and component identification by mass spectrometry. Mol Cell Proteomics 7:1286–1296

Zíková A, Kopečná J, Schumacher MA, Stuart KD, Trantírek L, Lukeš J (2008b) Structure and function of the native and recombinant mitochondrial MRP1/MRP2 complex from *Trypanosoma brucei*. Int J Parasitol 38:901–912

Zíková A, Schnaufer A, Dalley RA, Panigrahi AK, Stuart KD (2009) The F0F1-ATP synthase complex contains novel subunits and is essential for procyclic *Trypanosoma brucei*. PLoS Pathog 5:e1000436

The Apicoplast: An Ancient Algal Endosymbiont of Apicomplexa

Swati Agrawal, Sethu Nair, Lilach Sheiner, and Boris Striepen

Contents

Abstract The discovery of a chloroplast in the Apicomplexa came as a surprise as these are nonphotosynthetic parasites that historically had been the domain of zoologists. This organelle, the apicoplast is essential for parasite survival and its

S. Agrawal, S. Nair, L. Sheiner, and B. Striepen (✉)
Center for Tropical and Emerging Global Diseases & Department for Cellular Biology, University of Georgia, 500 D.W. Brooks Drive, Athens, GA 30602, USA
e-mail: striepen@cb.uga.edu

W. de Souza (ed.), *Structures and Organelles in Pathogenic Protists*,
Microbiology Monographs 17, DOI 10.1007/978-3-642-12863-9_11,

© Springer-Verlag Berlin Heidelberg 2010

metabolism is intensively pursued as the source of new targets for antiparasitic drugs, in particular new antimalarials. The apicoplast has a remarkable evolutionary history, and this history is reflected in its complex structure and cell biology. A cyanobacterium and two eukaryotes have contributed to the genesis of this organelle and their contributions can still be traced today. This chapter sets out by briefly summarizing the studies that led to the discovery of the apicoplast followed by an overview of our most current knowledge of the molecular mechanisms of apicoplast protein import, apicoplast division and replication and apicoplast metabolism.

1 The Surprising Photosynthetic Past of Apicomplexa

Apicomplexa are a phylum of unicellular eukaryotes that live as obligate intracellular parasites and maintain a complex life cycle that includes sexual and asexual reproduction. Over their long evolutionary history Apicomplexa have adapted to a tremendous variety of invertebrate and vertebrate hosts. Infection with many of these parasites, results in severe disease of the host. Members of the genus *Plasmodium* are the etiological agents of malaria, and *Toxoplasma gondii* causes neurological disease in immunosuppressed patients and upon congenital transmission. Parasites of the genera *Babesia* and *Theileria* are responsible for Texas and East Coast fever in cattle, and *Cryptosporidium*, *Cyclospora* and *Eimeria* cause gastro-intestinal diseases in humans and various domestic animals. Phylogenetic and morphological studies robustly position Apicomplexa within the Alveolata. In addition to Apicomplexa this group includes ciliates and dinoflagellates that share cortical alveoli as a defining morphological feature. These are flatted membranous cisternae that underlie the plasma membrane and form part of the complex pellicle (Gould et al. 2008). Following a hypothesis initially formulated by Cavalier-Smith, these organisms are thought to be part of a major branch of the eukaryotic tree of life, the chromalveolates, that includes single and multicellular organisms which at first sight show little resemblance and occupy a tremendous variety of ecological niches (Cavalier-Smith 2002). The origin of this superphylum is thought to lie in the symbiotic union of two single celled eukaryotes, a heterotroph and a red alga. While not without critics the chromalveolate hypothesis has steadily gained support over the years (Keeling 2009). One of the key elements of this support was the discovery of a plastid in Apicomplexa – the apicoplast. As indicated by their historical phylum name Sporozoa, Apicomplexa have long been the domain of zoologist, however, it appears now well established that Apicomplexa have a photosynthetic past in the ocean (Moore et al. 2008). This chapter briefly describes the discovery of the apicoplast and then outlines our current understanding of its evolution, cell biology, and metabolic function.

1.1 The Discovery of the Apicoplast: From Vacuole to Endosymbiont

The discovery of a plastid in Apicomplexa emerged from three observations made by researchers from different disciplines that initially did not appear to be related: a multimembranous structure seen on electron micrographs, a third density species of parasite DNA apparent by cesium chloride gradient centrifugation, and the puzzling sensitivity of eukaryotic parasites to antibiotics that target prokaryotic-type protein translation.

Ultrastructural studies reported the presence of a multimembrane bound organelle in a variety of apicomplexan parasites. This mysterious organelle received numerous descriptive names (reviewed in Siddall 1992), yet its identity and cellular function remained elusive. In an independent effort that was aimed at isolating mitochondrial DNA, a series of studies in different parasites described a circular DNA molecule distinct of nuclear chromosomal DNA (Borst et al. 1984; Dore et al. 1983; Gardner et al. 1988; Kilejian 1975; Williamson et al. 1985). The size of the molecule and investigator expectation initially let to its assignment as putative mitochondrial genome. However doubts in this interpretation emerged in 1987 when Vaidya and Arasu described another even smaller nonnuclear DNA molecule in *Plasmodium yoelii* (Vaidya and Arasu 1987).

Step by step the distinct nature of the two extrachromosomal DNA molecules was uncovered in *Plasmodium*: one a 6 kb repetitive linear molecule, which was assigned to the mitochondrion based on its genetic content, and the 35 kb circular molecule of unknown function. Phylogenetic evidence suggested that the two molecules were of different evolutionary origin (Feagin et al. 1992; Gardner et al. 1991a; Williamson et al. 1994), and fractionation data implied that they were likely located in different subcellular compartments (Wilson et al. 1992). Based on these insights Wilson and coworkers proposed the presence of an organelle corresponding to a residual plastid in apicomplexans (Gardner et al. 1994a). Features of the genome that supported this idea were an inverted tail to tail repeat of small and large subunit ribosomal RNA, which is reminiscent of red algal plastid genomes (Gardner et al. 1988, 1991b, 1993), and genes for elements of plastid-type transcription and translation machinery (Gardner et al. 1991a, 1994b; Preiser et al. 1995; Wilson et al. 1994, 1996). The notion of plastid protein translation also explained the observation that Apicomplexa are sensitive to certain antibiotics (see below).

As soon it was clear that apicomplexans harbor a plastid genome, the obvious question became: "where does it reside?" Maternal inheritance of the genome strongly suggested a nonnuclear residence (Creasey et al. 1994). The previously enigmatic multimembranous organelle seen in so many micrographs emerged as the favorite candidate. In situ hybridization experiments had the final word in confirming this hypothesis in *Plasmodium* and in *Toxoplasma* (Kohler et al. 1997; McFadden et al. 1996) and the apicoplast was formally identified and named. Apicoplasts have been described in many Apicomplexa with the marked exception of *Cyptosporidium* (Zhu et al. 2000).

1.2 The Apicoplast's Endosymbiotic Origin is the Source of Its Complex Biology

Plastids like mitochondria are thought to be the product of endosymbiosis. A eukaryotic cell engulfed a cyanobacterium, the metabolism and genomes of both partners became progressively intertwined ultimately resulting in a double membrane bound organelle that we now know as chloroplast (reviewed in Gray 1993; McFadden and van Dooren 2004). This early event gave rise to the common ancestor of three major groups of photosynthetic eukaryotes: glaucophytes, red algae and green algae. In several cases this was not the end of the journey though and plastids seem to have moved laterally from one lineage to another by secondary endosymbiosis. There is now robust evidence that plastids have been acquired secondarily several times during eukaryotes evolution giving rise to additional photosynthetic clades (Keeling 2009; Lane and Archibald 2008). These events occurred through the uptake of a single-celled alga carrying a primary plastid by a second eukaryote. As in the case of primary plastids and mitochondria, lateral gene transfer to the host nucleus resulted in subjugation of the symbiont into a dependent organelle (Howe and Purton 2007), in this case likely involving transfer from the plastid as well as the algal nuclear genome. A telltale of this evolutionary history is the additional membranes that surround secondary plastids which often are referred to as "complex" plastids (four in the case of the apicoplast see Fig. 1a and below for further detail). We know that this process occurred more than once as there are organisms with complex plastids that trace back to the green as well as the red algal lineage. Some of the most intriguing evidence for secondary endosymbiosis comes from the discovery and sequencing of nucleomorphs (Moore and Archibald 2009); these are remnant nuclei of the ancestral algal endosymbiont that are localized between the second and third membrane of the complex plastid.

Whereas most researchers agree that the apicoplast is the product of a secondary endosymbiotic event (with some exceptions (Kohler 2005)), the identity of the alga involved remains disputed between a red and a green lineage camp. Numerous genes and proteins have been subjected to phylogenetic analysis to trace the origin of the endosymbiont. These studies have focused on genes found in the apicoplast genome itself, nuclear genes that encode apicoplast proteins, and lastly nuclear genes that encode proteins that do not target to the apicoplast but might have been acquired from the endosymbiont's nucleus nonetheless. While the majority of studies pointed to a red algal origin a significant number of analyses favored an ancestor from the green algal lineages (see Feagin and Parsons (2007) for a comprehensive summary). These analyzes may be complicated by extensive gene loss in certain taxa and the potential lateral gene transfer between organellar genomes (Obornik et al. 2002). Fast and coworkers made an interesting breakthrough by noting that the plastid glyceraldehydes-3-phosphate dehydrogenase (GAPDH) from dinoflagelattes and apicomplexans are of eukaryotic origin, as opposed to the cyanobacterial-type targeted to the plastids of plants and green and red algae (Fast et al. 2001). Plastid-GAPDH phylogenies then grouped

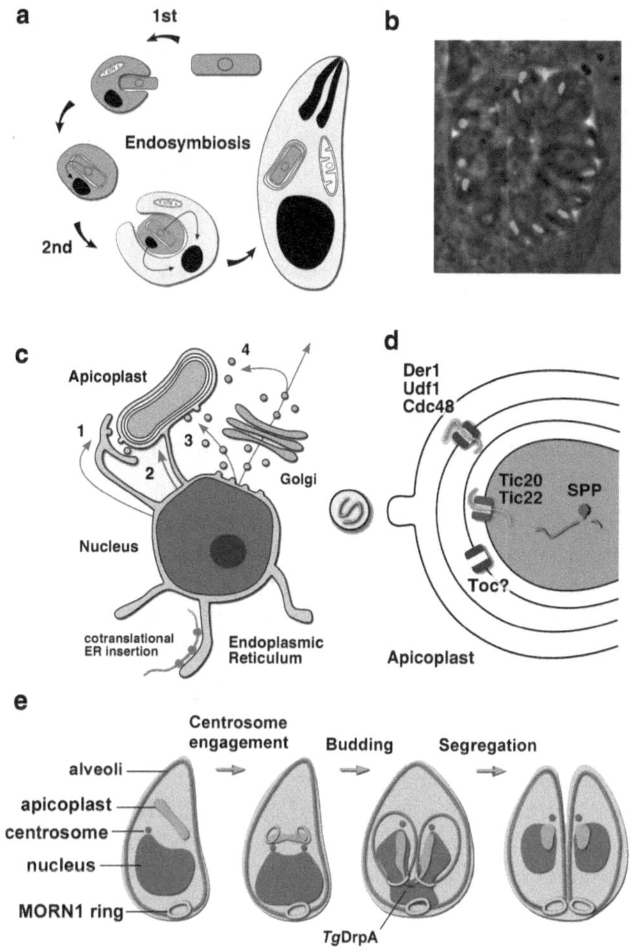

Fig. 1 (**a**) Schematic representation of the evolutionary origin of the apicoplast depicting primary and secondary endosymbiosis. (**b**) *Toxoplasma gondii* parasites, the apicoplast is visible in green due to expression of ACP–GFP. (**c**) Hypothetical routes of nuclear encoded apicoplast proteins along the secretory pathway in Apicomplexa. Vesicular trafficking directly from the ER to the apicoplast is a currently favored yet not fully validated hypothesis (from Vaishnava and Striepen 2006). (**d**) Simplified diagram of the four membranes of the apicoplast and the translocons that have been identified (modified from Agrawal et al. 2009). (**e**) Schematic representation of apicoplast division in *T. gondii* (modified from van Dooren et al. (2009)

apicomplexans as well as dinoflagellates tightly together with heterokonts and cryptomonads. This suggests that the apicoplast originated early in evolution, and that a single endosymbiosis gave rise to plastids in these taxa. Given that well established phylogenies, based on plastid genes as well as pigmentation strongly support a red algal origin of both, heterokont and cryptomonad plastids (Delwiche 1999), these findings meant that a red alga is also the likely origin of the apicoplast.

All together this picture sits well with the chromalveolate hypothesis, which groups these lineages together based on a common secondary endosymbiosis involving a red alga (Cavalier-Smith 1999, 2004; Keeling 2009). Additional support for red ancestry emerges from the details of apicoplast cell biology and metabolism detailed later in this chapter. This includes the association of its genome with histone-like HU proteins found only in Eubacteria and red algae (Kobayashi et al. 2002), the apparent red algal origin of elements of the apicoplast protein import machinery (Agrawal et al. 2009), and a cytoplasmic starch pathway that resembles red alga and is distinct from the chloroplast localized starch metabolism in the green lineage (Coppin et al. 2005).

1.3 The Apicoplast Genome is Mostly Concerned with Itself

The apicoplast genome is reduced in size and highly focused in the functions of the encoded proteins. Excluding seven hypothetical genes and two known genes, all encode components of the apicoplast transcription and translation machinery (Wilson et al. 1996). These include three subunits of a eubactrial-type RNA-polymerase, 17 ribosomal proteins, a complete set of tRNAs and the translation elongation factor Tu (Wilson et al. 1996). The apicoplast genome encoded subunits (rpoB, rpoC1 and rpoC2) together with a nuclear encoded rpoA homolog (Bahl et al. 2003) could compose a chloroplast-like RNA polymerase (Wilson and Williamson 1997) which is in agreement with the sensitivity of the parasite to rifampin, an inhibitor of eubacterial RNA polymerases (Dahl et al. 2006; McConkey et al. 1997). Antibiotics that typically only interfere with bacterial translation machinery have been known to show some efficacy in the treatment of malaria and toxoplasmosis for many years (Coatney and Greenberg 1952; Tabbara and O'Connor 1980). Subsequent studies in *Toxoplasma* connected this sensitivity to the sequence of the apicoplast ribosomal RNA (Beckers et al. 1995). Further experiments showed apicoplast specific effects following clindamycin treatment of *Toxoplasma* (Fichera and Roos 1997), and demonstrated association between resistance to this drug and mutations in the apicoplast ribosomal RNA genes (Camps et al. 2002). In addition to these reports, studies in *Plasmodium* linked the effects of thiostrepton and azithromycin to the apicoplast ribosomal RNA (Clough et al. 1997; Sidhu et al. 2007). As we will see below, most genes have been transferred to the nucleus. If the apicoplast genome largely encodes proteins concerned with itself, why has it not been lost? One idea is that actually it is in the process of loss but this process is not concluded yet in all Apicomplexa. Another one is that a small number of proteins has to be made in place because they cannot be transported (e.g., due to particular protein folding or sensitivity to changing redox environments). Potential candidates are two genes encoding a putative chaperone (ClpC) and an iron–sulfur cluster biogenesis protein (SufB).

2 The Long Journey from Nucleus to Apicoplast

As described above the proteome encoded by the apicoplast genome is small and restricted to housekeeping functions, including transcription and translation. However, about 500 proteins are predicted to populate the apicoplast stroma (Foth et al. 2003) and the biochemical pathways that have been localized to the organelle (see below). A closer look revealed that like mitochondria and chloroplasts, the biosynthetic and metabolic functions of the apicoplast are mediated in large part by nuclear encoded proteins. This is likely the result of massive gene transfer from endosymbiont to host. These proteins must now be translated in the cytosol from where they have to travel to their destination in order to perform their metabolic functions.

2.1 A Bipartite Leader Peptide is Required for Apicoplast Targeting

When compared to their cyanobacterial orthologs most nuclear encoded apicoplast proteins possess a pronounced N-terminal extension. Such extensions are known to mediate import of nuclear encoded chloroplast proteins in plants. Experimental work confirmed that the N-termini of several apicoplast proteins are necessary and sufficient to target a GFP reporter to the lumen of the organelle (Foth et al. 2003; Waller et al. 1998, 2000). Compared to their chloroplast counterparts the N-terminal extension of apicoplast proteins are longer and have a bipartite structure. Their first portion resembles the signal peptide present in secretory proteins (Waller et al. 1998), while the second portion (the transit peptide) appears to have features reminiscent of the transit peptides of chloroplast proteins. In fact, GFP engineered to carry the transit peptide of a *T. gondii* apicoplast ribosomal protein was shown to be imported efficiently into pea chloroplast (DeRocher et al. 2000). Similar bipartite leaders had been described for a variety of algae harboring secondary plastids and the initial trafficking in these systems appears to occur via the secretory pathway (Grossman et al. 1990; Sulli and Schwartzbach 1995). The bipartite model has been extensively tested using transgenic reporters in *Plasmodium* and *Toxoplasma* in studies that deleted each component of the leader. In the absence of the transit peptide, the signal sequence of apicoplast proteins targets a GFP reporter to the parasitophorous vacuole, the default secretory route in these parasites. Deletion of the signal peptide results in cytosolic (and in some cases mitochondrial) localization of the reporter (DeRocher et al. 2000; Harb et al. 2004; Waller et al. 2000). Transit peptides of both primary and secondary plastid proteins are enriched in hydrophilic and basic amino acids, have very low sequence conservation and vary in length from 50 to 200 amino acids (Bruce 2001; Claros et al. 1997). Similar to their choloroplast counterparts, apicoplast protein transit peptides have very few acidic and hydrophobic residues and possess a net positive charge. Detailed mapping of the transit peptide by point mutation and serial deletion revealed that a net positive charge at the N-terminus is essential for proper apicoplast targeting, but that their exact position can be varied (Foth et al. 2003; Harb et al. 2004; Tonkin et al. 2006a). Apicoplast transit peptides further appear to possess Hsp70 chaperone-binding sites suggesting that maintenance of preprotein in an unfolded state might be a prerequisite for successful import. Interestingly a stretch of 26 amino acids that separates Hsp70 binding sites in plant transit peptides is also present in apicoplast transit peptides suggesting a similar mechanism by which molecular chaperones facilitates the translocation of cargo proteins through membrane pores.

The bipartite leader is thought to guide preproteins to the apicoplast in a sequential fashion. The signal peptide is believed to result in the cotranslational insertion of the nascent peptide into the endoplasmic reticulum (ER) followed by its removal (full-length proteins with intact signal peptides have not been detected (Waller et al. 1998, 2000)). Cleavage of the signal peptide then would expose the transit peptide. The presence of the transit peptide and its processing has been observed both in *T. gondii* and *Plasmodium falciparum* (Vollmer et al. 2001; Waller et al. 1998, 2000; Yung et al. 2001). In Western blot analyses apicoplast proteins typically have a slower migrating precursor band, corresponding to the size of proteins with intact transit peptide, and a faster migrating mature band corresponding to the processed form. To understand the kinetics of transit peptide processing van Dooren and colleagues performed pulse-chase labeling with radioactive amino acids and immuno-precipitation of apicoplast-targeted proteins in *P. falciparum* (van Dooren et al. 2002). They demonstrated that processed apicoplast proteins appear after 45 min of labeling and complete processing might take as long as four hours. Processing of transit peptides in plants is mediated by a specific peptidase in the chloroplast stroma. A nuclear encoded stromal peptidase homolog with a bipartite targeting sequence is present in *T. gondii* and *P. falciparum* (He et al. 2001; van Dooren et al. 2002), however direct experimental evidence that this protein is capable of transit peptide cleavage, is still lacking.

2.2 Chloroplast Derived Membranes are Crossed Using Chloroplast Translocons

The apicoplast is surrounded by four membranes of divergent evolutionary origin. A number of intriguing models have been proposed to explain how proteins might cross these four membranes in Apicomplexa and other organisms bearing secondary plastids (Tonkin et al. 2008). These have invoked fusion and fission of membranes (Gibbs 1979), nonselective pores (Kroth and Strotmann 1999), or the activity of protein translocons derived from either the chloroplast envelope (van Dooren et al. 2008a) or the endoplasmic reticulum (Sommer et al. 2007). Until recently there was little evidence to distinguish and test the competing hypotheses. Two developments have dramatically accelerated research in this area. Full genome sequence information is now available for a growing number of apicomplexan and algal species, and increasingly sophisticated genetic tools can be used in species that were previously experimentally inaccessible. The model that emerges from these studies is that protein translocons are crucial to import and that these translocons are divergent in their origin reflecting the organisms that contributed the membrane they cross. We will describe our current knowledge of translocons as it emerged from the inside out.

The two innermost membranes are homologous to the membranes of the red algal chloroplast. Protein import across chloroplast membranes has been studied in great detail and is mediated by two multiprotein complexes: the translocon of the outer chloroplast membrane (Toc) and the translocon of the inner chloroplast membrane (Tic). These complexes and their protein components function in recognition and binding of the transit peptide, form the translocation channel, provide mechanical energy to drive translocation, and act as chaperones aiding in unfolding and refolding of cargo proteins (see Hormann et al. 2007; Jarvis 2008) for recent reviews of the extensive literature). A homolog of the plant import component Tic20 has been identified in a variety of Apicomplexa with the marked exception of plastid-less genus *Cryptosporidium* (van Dooren et al. 2008a). While the primary sequence conservation between these proteins and their plant counterparts is very low, they share the overall topology of a four-pass transmembrane protein. *Tg*Tic20 features a canonical bipartite signal that is processed. Biochemical and immunoelectron microscopic studies demonstrated that *Tg*Tic20 is a component of the apicoplast membranes. To further define its localization van Dooren et al. employed a split-GFP assay. In this experiment GFP is split into two segments that are fused to either the test protein or marker proteins of known subcellular localization (Cabantous and Waldo 2006). By themselves these fragments are nonfluorescent, however, they regain fluorescence by direct molecular interaction when the two fusion proteins are targeted to the same compartment. The analysis of *Tg*Tic20 split-GFP transgenics showed that the protein localizes to the innermost membrane of the organelle with both N and C-terminus projecting into the apicoplast stroma. Further functional insights emerged from a *Tg*Tic20 conditional mutant. Inducible knockdown of *Tg*Tic20 leads to impairment of protein import into the plastid, and subsequent defects in plastid biogenesis and parasite survival (van Dooren et al. 2008a). A homolog of a second putative member of the Tic complex, Tic22, has been identified in the *P. falciparum* and *T. gondii* apicoplast (Kalanon et al. 2009; van Dooren, Swati Agrawal and Boris Striepen unpublished). Tic22 in plants has been described as a small soluble protein that peripherally associates with the Tic and the Toc complex and interacts with preproteins in transit. It was proposed that Tic22 serves as an adaptor that facilitates preproteins translocation from the Toc translocon towards the Tic complex (Kouranov et al. 1998). Conditional knockout of Tic22 in *T. gondii* demonstrates that this protein is essential for apicoplast protein import and parasite survival (van Dooren, Swati Agrawal and Boris Striepen unpublished).

Extensive genome mining so far has failed to conclusively identify homologs of the proteins that make up the Toc complex in apicomplexan parasites (a putative

Toc34 has been described in *P. falciparum*) but this assignment has not been experimentally evaluated (Waller and McFadden 2005). It is possible that apicomplexan homologs of Toc proteins are too divergent to be identified. Alternatively a different translocon might have subsumed its function. While the details (in particular with respect to Toc) are still emerging it appears that transport across the membranes that are derived from the chloroplast employs conserved machinery that is homologous to the machinery that performed this task in the algal ancestor of the apicoplast.

2.3 Retooling an Endosymbiont Translocon from Export to Import

The third membrane, also known as periplastid membrane, is of particular interest as it represents the plasma membrane of the alga and as such the direct interface between host and endosymbiont. Establishing transport across this membrane was likely a prerequisite for gene transfer and a crucial event in the progressive adaptation of the endosymbiont to intracellular life. Key to the discovery of the mechanism that allows proteins to cross this membrane was the sequencing of the nucleomorph genome of *Guillardia theta* (the nucleomorph is the "fossil" remnant of the algal nucleus). Sommer and coworkers noted that this highly reduced genome encodes core elements of the endoplasmatic reticulum associated degradation (ERAD) system (Sommer et al. 2007). ERAD usually acts in ER homeostasis by retrieving misfolded secretory proteins from the ER and funneling them for degradation to the proteasome in the cytosol. The core components of the ERAD transport machinery are Der-1, the ATPase Cdc48 and its cofactor Ufd-1. Der-1 is a favored candidate for the proteinaceous pore in the ER membrane and has been shown to be essential for retrotranslocation of misfolded luminal proteins (Ye et al. 2004). Protein substrates destined to be degraded are polyubiquitinated and subsequently extracted from the pore by the Cdc48–Ufd-1–Npl4 complex (Ye et al. 2001). What could be the function of such a system in an endosymbiont that appears to have long lost its ER? In a bold stroke Sommer and colleagues formulated the hypothesis that an ERAD translocon had been retooled to import proteins into complex plastids (Sommer et al. 2007). Relocation of, an existing translocation machinery to a different membrane seems to be remarkably simple and elegant solution to engineer import into a newly acquired organelle.

The ERAD hypothesis has accumulated considerable support from a recent flurry of publications reporting the identification and plastid localization of ERAD components in cryptomonads, diatoms, and Apicomplexa (Agrawal et al. 2009; Kalanon et al. 2009; Sommer et al. 2007; Spork et al. 2009). In our own work we have demonstrated that the *T. gondii* genome encodes multiple homologs of Der1, Cdc48 and Ufd-1. Immunofluorescence analysis of parasite cell lines expressing epitope tagged forms of these proteins reveal that while one complete set of components is associated with the ER and likely performs their classical role in ERAD, at least one homolog of each of these components localizes to the outer membranes of the apicoplast. Split GFP assays performed in the diatom *Phaeodactylum tricornitum*, suggest that Der-1 is indeed associated with the third membrane as predicted by the Sommer hypothesis (Hempel et al. 2009). Furthermore phylogenetic analysis of the two *T. gondii* Cdc48 proteins demonstrates that they are of divergent evolutionary origins. The apicoplast localized Cdc48 forms a well-supported clade with its red algal lineage counterparts (including the protein encoded on the *G. theta* nucleomorph) while the cytoplasmic protein branches with proteins that reflect the current view of vertical evolution for Apicomplexa (Agrawal et al. 2009). Genetic ablation of $\mathrm{Der1}_{\mathrm{Ap}}$ in *T. gondii* results in swift and complete ablation of apicoplast protein import as measured using a variety of biochemical assays (Agrawal et al. 2009; van Dooren et al. 2008a) demonstrating a direct role of $\mathrm{Der1}_{\mathrm{Ap}}$ and the endosymbiont derived ERAD system in apicoplast protein import.

ERAD mediated protein retrotranslocation across the ER membrane coincides with polyubiquitination or the cargo protein. This modification is critical for the subsequent degradation of the protein but also appears to have a potential role in the translocation step. The amino terminus of Ufd-1 is known to bind polyubiquitin. Deletion of yeast Ufd-1 amino terminus results in disruption of translocation of proteins across the ER membrane (Park et al. 2005; Walters 2005). Ubiquitylation is brought about by the sequential action of three enzymes, ubiquitin activating enzyme (E1), ubiquitin conjugating enzyme (E2) and ubiquitin ligase (E3). Genome mining and emerging experimental studies point to the likely presence of plastid-specific ubiquitylation factors in cryptomonads, Apicomplexa and diatoms (Hempel et al. 2009; Spork et al. 2009; Swati Agrawal, Giel van Dooren and Boris Striepen unpublished). At this point the mechanistic role that ubiquitylation might play in the apicoplast is not understood. It is possible that ubiquitylation provides the necessary signal for movement of preproteins across the membrane. However, it remains to be experimentally demonstrated whether apicoplast targeted preproteins are ubiquitylated and whether the organelle possesses all the other necessary components for ubiquitylation.

2.4 Still a Black Box: How Do Apicoplast Proteins Find the Apicoplast?

While we now have some understanding of how proteins translocate across the inner membranes of the apicoplast, we know little about their way from the ER to the outermost compartment of the organelle. This part of the journey shows some diversity when comparing the targeting to different complex plastids. In heterokonts and cryptophytes the outermost compartment of the plastid is directly continuous with the ER and the nuclear envelope (Gibbs 1979; Gould et al. 2006a;

Kilian and Kroth 2005; Wastl and Maier 2000). This setup makes plastid targeting straightforward. The signal peptide provides access to the ER, and once in ER proteins have direct access to the periplastid membrane and its translocon (Gould et al. 2006a, b). Targeting to complex plastids that are surrounded by three membranes in the euglenophytes and dinoflagellates appears to proceed from the ER to the Golgi and then the plastid. The hallmark of his route is sensitivity to the drug brefeldin A and the presence of a second hydrophobic segment at the end of the transit peptide that acts as a stop transfer signal thus exposing a potential sorting motif on the cytoplasmic face of vesicles (Durnford and Gray 2006; Nassoury et al. 2003; Patron et al. 2005; Sulli and Schwartzbach 1995, 1996). In the case of Apicomplexa, brefeldin A treatment or low temperature incubation (both known to block Golgi trafficking) do not affect the steady state distribution of apicoplast targeted GFP reporters (DeRocher et al. 2005; Tonkin et al. 2006b). This would argue for direct trafficking from the ER to the apicoplast side-stepping the Golgi. Electron tomographic studies have found no evidence for a permanent connection between ER and the apicoplast; however, they noted sites of close apposition between the membranes of both organelles, which may reflect functional interaction (Tomova et al. 2006, 2009). Evidence for a vesicular step at this point is mostly circumstantial. A number of groups noted what appear to be vesicles carrying apicoplast proteins. These can be more apparent when markers are overexpressed, or when import is blocked due to loss of the import machinery or loss of the apicoplast (DeRocher et al. 2008; Karnataki et al. 2007a; van Dooren et al. 2008b, 2009). If these structures truly represent transport vesicles en route remains to be studied in greater detail. If apicoplast proteins are packed into vesicles for transport from the ER, what directs these vesicles to the apicoplast? Identifying specific marker proteins for such transport vesicles would be a major step towards a deeper mechanistic understanding.

2.5 Additional Signals and Mechanisms may be Involved in the Trafficking of Apicoplast Membrane Proteins

While the bulk of our understanding of apicoplast protein import is based on the study of luminal proteins, insights into proteins that target to the membranous compartments that surround the organelle are emerging. Some of these proteins, like the recently identified ERAD and Tic components appear to have canonical bipartite leaders and undergo processing at the N-terminus (Agrawal et al. 2009; Kalanon et al. 2009; Spork et al. 2009; van Dooren et al. 2008b). However, other membrane proteins lack identifiable leader peptides. Most of these proteins feature transmembrane domains that potentially could serve as a signal anchor. One example is FtsH1 a putative zinc metalloprotease that localizes to several apicoplast membranes and appears to undergo complex processing at the N-terminus and C-terminus (Karnataki et al. 2007b, 2009). Similarly, the thioredoxin-like protein

Atrx-1 is targeted to the outer membranes of the apicoplast in the absence of a canonical bipartite signal (DeRocher et al. 2008). At this moment it not clear how exactly these differences in transit peptide structure relate to differential targeting. As another example, in *P. falciparum* two phosphate translocators have been identified, one has a leader and is targeted to the innermost membrane and a second protein that lacks a leader and is found in the outer membranes (Mullin et al. 2006). In *T. gondii* a single translocator appears to traffic to all membranes, and it does so without a leader (Fleige et al. 2007; Karnataki et al. 2007a). The question as to how different proteins are restricted to different apicoplast subcompartments is still wide open. Mechanistic models could invoke defined positive forward signals, negative signals that block proteins from being substrates for certain translocons (and thus restrict them to outer compartments), or invoke import to the lumen and subsequent reexport to outer compartments. Such re-export is seen for elements of the outer chloroplast membrane in plants (Tranel et al. 1995; Tranel and Keegstra 1996). The fact that many proteins are processed by what is believed to be a luminal signal peptidase might provide some support for the latter. Clearly more work is needed to unravel the considerable complexity of the system.

3 Replicating and Dividing the Apicoplast

As the apicoplast posses its own gnome, de novo formation is not possible, and replication and partition must occur in each cell division cycle to ensure inheritance to all daughter cells. Similar to apicoplast protein import, the machinery that replicates the organelle is a phylogenetic mosaic with the most ancient components at its core and more recent additions acting at the periphery.

3.1 The Apicoplast Genome is Replicated by Prokaryotic-Type Machinery

Apicoplast genome replication has been studied in *Plasmodium* and *Toxoplasma*. In both species the genome is present in multiple copies per cell but there is an important difference in their structure. In *Plasmodium* most apicoplast genome molecules are circular, while in *T. gondii* they are organized in linear tandem arrays (Williamson et al. 2001, 2002). Branch point tracking through two dimensional gel electrophoresis, indicated that in *Plasmodium* replication initiates at two sites within the inverted ribosomal RNA repeats and proceeds, via a D-loop intermediate, towards a circular replicate (Singh et al. 2005; Williamson et al. 2002). Williamson and coworkers also reported an additional, rolling circle, mechanism that starts at sites found outside the inverted repeats (Williamson et al. 2002). Rolling circle is the mechanism for replication proposed for the linear genome found in the *T. gondii* apicoplast (Williamson et al. 2001). The apicoplast genome itself does not seem to encode proteins involved in its replication, and those are therefore presumably imported. Indeed, several nuclear-encoded homologs of such components were identified: Prex (*p*lastid-DNA *r*eplication *e*nzyme comple*x*), a multidomain protein showing DNA-helicase, primase, and polymerase activities were shown to target the apicoplast (Seow et al. 2005). Evidence for the potential activity of gyrase, a strand relaxation topoisomerase typical for bacteria and plastids, initially came from pharmacological experiments with ciprofloxacin in *Plasmodium* and *Toxoplasma* (Fichera and Roos 1997; Weissig et al. 1997). Indeed, both the A and B subunits of gyrase were later identified in *Plasmodium* and their activity and interaction with each other have been confirmed (Ahmed and Sharma 2008; Dar et al. 2007; Raghu Ram et al. 2007). DNA replication is a complex process and additional components of the machinery undoubtedly remain to be identified. Our efforts to systematically identify all apicoplast proteins in *T. gondii* by experimental localization have recently yielded several candidate genes. These include a DNA ligase, the gyrase B subunit, a helicase, and two hypothetical proteins with domains reminiscent of DNA-repair components (Lilach Sheiner and Boris Striepen unpublished).

3.2 Organelle Division and Segregation: Fission with a Twist

Research into the mechanisms of plant chloroplast division has shown significant similarities between chloroplasts and their cyanobacterial ancestors. The most conserved element in this process is FtsZ (Armbrust et al. 2004; Fraunholz et al. 1998; Miyagishima 2005; Osteryoung and Nunnari 2003). FtsZ is a GTPase, structurally related to tubulin, and localizes to the division ring in bacteria and chloroplasts (Miyagishima 2005; Stokes et al. 2000). The site of assembly of the FtsZ ring is controlled by homologs of the bacterial MinD and MinE proteins which have been found in the nuclear genome of plants and certain algae (Colletti et al. 2000; Itoh et al. 2001). Chloroplast division also has eukaryotic elements, most importantly a constrictive ring formed by the dynamin-related protein ARC5. The position of the FtsZ ring is ingeniously transduced and coordinated from the stroma to the outer membrane through two interacting membrane proteins (Glynn et al. 2008). While there is some variation with respect to particular elements the overall mechanism and the central role of FtsZ seems conserved among all plastids including secondary plastids. This is to the marked exception of Apicomplexa which lack FtsZ and, any of its associated other factors (Vaishnava and Striepen 2006). How does the apicoplast divide in the absence of the conserved machinery?

An important initial observation was that in Apicomplexa plastid development is tightly associated with the development of the nucleus. While extracellular forms of various species carry small ovoid organelles that resemble each other they show significant morphological diversity in intracellular stages (Stanway et al. 2009; Striepen et al. 2000; Vaishnava and Striepen 2006; van Dooren et al. 2005;

Waller et al. 2000). In species that replicate through large polyploid intermediates through schizogongy or endopolygeny (see Striepen et al (2007) for detail) the apicoplast grows into a large tubular or branched reticulate structure. These structures then fragment during the final budding process that produces new infectious cells. Imaging experiments in *Toxoplasma* and *Sarcocystis* demonstrated that apicoplast division coincides with nuclear division and that this coordination is due to physical association between the apicoplast and the centrosome of the intranuclear spindle (Striepen et al. 2000; Vaishnava et al. 2005). This has been studied in greatest detail in *Toxoplasma*. Early in M-phase the apicoplast associates with the recently divided centrosomes and is suspended between be two poles of the spindle. As mitosis and budding proceeds both the nucleus and the apicoplast appear U-shaped, and this U elongates until fission occurs concurrently with buddy, leaving each new organelle in a separate new cell (Striepen et al. 2000; van Dooren et al. 2009). Centrosome association provides a satisfactory model to explain faithful segregation into daughter cells yet does not fully explain fission. Ultrastructural studies have reported the observation of plastid constrictions that may be indicative of a division ring (Ferguson et al. 2005; Matsuzaki et al. 2001) but did not address the composition of these putative rings. A first candidate emerged with the description of a constrictive cytokinetic ring marked by the recently described repeat protein MORN1 (Gubbels et al. 2006). The position of the MORN1 rings coincides with apicoplast constrictions lending support to a pull and cut model in which centrosomes and cytokinetic ring cooperate in apicoplast division (Vaishnava and Striepen 2006; van Dooren et al. 2009). Most recently another player was discovered, the dynamin-related protein DrpA. Genetic and cell biological studies using dominant negative mutants demonstrated that DrpA is essential for the final apicoplast fission step. In these mutants plastids segregate and pinch, however they remain continuous tubules connecting recently divided daughter cells (van Dooren et al. 2009). The assembly of dynamin-related proteins into mulitmeric active fission complexes is thought to require an initial constriction of the target area (Legesse-Miller et al. 2003). A new unified model shown in Fig. 1e takes this into account and proposes that centrosome and MORN1 rings generate constrictions that are then the site of DrpA activity. Interestingly, TgDrpA is phylogenetically distinct from the ARC5 dynamin involved in chloroplast division (van Dooren et al. 2009). This suggests that dynamins have been recruited independently multiple times to aid in the division of endosymbiont organelles.

4 What are the Metabolic Functions of the Apicoplast?

With the discovery of *Chromera* it now appears well established that Apicomplexa are derived from a photosynthetic ancestor (Moore et al. 2008). In all likelihood the acquisition of photosynthesis was the driving benefit in the relationship between the ancestor of apicomplexans and the red algal endosymbiont. An obvious question is why Apicomplexa maintained the apicoplast, through a dramatic change of ecological niche, from a life in the ocean to a life as an obligate intracellular parasite? Chloroplasts in plants and algae not only harvest the energy of light but are also home to several anabolic pathways that use precursors generated through photosynthesis. A plausible reason for the continued presence of the apicoplast is metabolic dependence. We dub this the "addiction to free candy" hypothesis. Under this hypothesis, Apicomplexa evolved to rely on the anabolic capabilities of the algal endosymbiont. It might have been energetically more favorable to rely on endosymbiont pathways that were directly tied to primary production through photosynthesis. In this adaptation process, redundant host pathways might have been lost thus generating dependence that persisted after the loss of photosynthesis. The most important insights into apicoplast metabolism arose from mining the genomes of Apicomplexa. This included directed searches for the genes of enzymes known to be chloroplast localized in plants and algae (Jomaa et al. 1999; Waller et al. 1998) as well as broader screens that attempted to identify all proteins that carry a potential bipartite leader (Foth et al. 2003; Ralph et al. 2004b). The metabolic map built in this effort now attributes three main functions to the apicoplast: fatty acid, isoprenoid and heme biosynthesis. These pathways trace their evolutionary history back to the initial primary endosymbiosis and are of cyanobacterial origin. As detailed below there are significant differences between these apicoplast pathways and those used by mammals to make equivalent metabolites. These differences have made the apicoplast one of the prime targets for the developments of new drugs for Apicomplexa. Pharmacological studies suggest that the pathways are essential (at least in certain life cycle stages) and that inhibitors of key enzymes involved in these pathways can successfully arrest the growth of the parasites in vitro and in vivo (Goodman et al. 2007; Wiesner et al. 2008). This assertion has been reinforced (and in part modified) by recent genetic studies that have targeted specific enzymes. The following section will review studies on the three main pathways in detail. We will also highlight new insights into how the apicoplast anabolic pathways are supplied with energy, carbon and reduction power through import of metabolites from the cytoplasm of the parasite.

4.1 Type II Fatty Acid Biosynthesis (FASII): A Slippery Target

In most organisms de novo synthesis of fatty acids is achieved using one of two types of fatty acid synthetases (FAS). FASI is found typically in animals and fungi and combines all required enzymatic activities on a single large polypeptide. This megasynthase architecture is shared with the related polyketide synthases (Smith et al. 2003). In contrast, in the FASII system, each enzyme is expressed as an independent protein. This system is found in many Eubacteria and the plastids of plants. The key feature of any fatty acid biosynthesis is the sequential extension of an alkalonic chain, two carbons at a time, by a series of decarboxylative condensation reactions (Smith et al. 2003). The central molecule in the process is a small protein called acyl-carrier protein (ACP). A primer substrate is generated by the transfer of a malonyl group from malonyl-CoA to ACP by FabD. The decarboxylative condensation of this primer with acetyl-CoA by FabH yields a two carbon extension of the acyl chain that is subsequently reduced, dehydrated, and reduced again by the sequential actions of FabG, FabZ and FabI. This cycle is repeated and progressive sequential decarboxylative extension using malonyl-ACP as donor yields even chained fatty acid molecules. All enzymes of the *P. falciparum* FASII pathway have now been expressed heterologously and the recombinant proteins have been studied in considerable biochemical and structural detail. Particular effort, including extensive medicinal chemistry, has been placed on FabI the target of the antibiotic triclosan and FabH the target of thiolactamycin (see Goodman and McFadden (2007), Mazumdar and Striepen (2007) for specific reference).

Acetyl-CoA forms the primary carbon source of fatty acid biosynthesis. Acetyl-CoA is generated in the apicoplast lumen from pyruvate by the pyruvate dehydrogenase complex (PDH). Pyruvate, in turn, is synthesized from imported phosphoenol pyruvate (PEP) by the enzyme pyruvate kinase. An apicoplast localized PDH complex as well as pyruvate kinase has been reported in both *Plasmodium* and *Toxoplasma* (Fleige et al. 2007; Foth et al. 2005; Maeda et al. 2009), and it is important to note that this is the sole PDH in the cell. Subsequent conversion of acetyl-CoA to malonyl-CoA is the first committed step in the FASII pathway. This conversion is mediated by acetyl-CoA carboxylase (ACCase) a large polypeptide carrying a biotin prosthetic group. The genomes of both *Toxoplasma* and *Plasmodium* encode a putative ACCase that was shown to localize to the apicoplast (Gardner et al. 2002; Jelenska et al. 2001; Zuther et al. 1999) and a biotinylated protein of suitable size has been documented in the organelle in *T. gondii* (van Dooren et al. 2008a). *Toxoplasma* also has a cytosolic FASI system which is likely served by a second and cytosolic ACCase (Mazumdar and Striepen 2007).

Bulk production of fatty acids supplying parasite lipid synthesis would appear to be the most obvious function of the FASII pathway. Initial pharmacological studies in *Plasmodium* lent support to this hypothesis. Following metabolic labeling with ^{14}C-acetate, radiolabeled fatty acids were apparent in extracts from *Plasmodium* infected red blood cells. This synthetic activity was linked to the apicoplast based on its sensitivity to triclosan, an inhibitor of the FASII enzyme FabI/enoyl-reductase (Surolia and Surolia 2001). Furthermore, triclosan inhibited the growth of *Plasmodium* and *Toxoplasma* in culture supporting the critical importance of the pathway (McLeod et al. 2001; Surolia and Surolia 2001). A conditional knockout study in *Toxoplasma* targeting ACP has shown that the pathway is essential for parasite growth in culture and in infected animals. However, and somewhat surprisingly, mutants show no difference in their ability to synthesize ^{14}C-actetate labeled fatty acids (Mazumdar et al. 2006). More recently, genetic studies in *Plasmodium* have targeted the enzymes FabI and FabB/F and showed that FASII, while essential for liver cell development, is not required for the blood stage (Vaughan et al. 2009; Yu et al. 2008). Furthermore, triclosan will kill parasites in which its presumptive enzyme target FabI has been genetically ablated (Yu et al. 2008) or is naturally absent as in *Theileria* (Lizundia et al. 2009). This indicates strong off-target effects of triclosan. From a drug development perspective FASII appears to be a strong target in *Toxoplasma* and *Plasmodium* liver stages but not for the treatment of blood stage malaria.

A number of genetic and pharmacological studies have shown that FASII is essential for the lipoylation of the apicoplast PDH-E2 subunit (Crawford et al. 2006; Cronan et al. 2005; Mazumdar and Striepen 2007; Mazumdar et al. 2006; Thomsen-Zieger et al. 2003; Wrenger and Muller 2004). Lipoic acid is an essential prosthetic group and its synthesis depends on a FASII derived precursor (octanoic acid-ACP) and two plastid localized enzymes, LipA and B that constitute the de novo synthesis pathway. In contrast, lipoylated mitochondrial enzymes depend on a salvage system using LplA. Both pathways are independent and it appears that the mitochondrial enzyme uses lipoic acid derived from the host cell rather than the apicoplast (Crawford et al. 2006; Mazumdar et al. 2006). *Toxoplasma* is fully dependent on FASII for the lipoylation of PDH. In contrast, in *Plasmodium*, LipB, the initial step of de novo lipoylation of apicoplast PDH was found to be not essential in blood stages (Gunther et al. 2007). Careful measurements suggested that despite the absence of LipB, PDH still showed a significant amount of lipoylation, and that this might be due to the activity of a dually targeted second LplA protein. While there is now excellent experimental support for the hypothesis that FASII provides essential precursors for PDH lipoylation, this does not provide a fully satisfactory answer to the question as to what the essential role of FASII might be. At the moment the main function of PDH appears to be to supply FASII with acetyl-CoA generating a circular argument. There are potential solutions to this conundrum that await experimental validation (1) PDH activity might be required for processes other than FASII, (2) fatty acid synthesis might be the main function of FASII but this activity is poorly measured by acetate labeling experiments or (3) FASII might be required for the synthesis of specialized lipids that are not needed in large quantities, but are still essential for parasite survival. Deeper biochemical and metabolomic analysis of the various mutants might provide a resolution to this puzzle.

4.2 Prokaryotic Isoprenoid Biosynthesis: The Pathway Found in All Apicoplasts

Isoprenoids function in many aspects of cell metabolism as well as membrane structure and function (Moreno and Li 2008; Wanke et al. 2001). Despite their enormous structural and functional diversity (sterols, cholesterol, retinoids, carotenoids, ubiquinones and prenylated proteins), all isoprenoids are derived from a simple five carbon precursor: isopentenyl pyrophosphate (IPP) and its allyl isomer dimethylallyl pyrophosphate (DMAPP). Until recently, IPP and DMAPP were thought to be synthesized exclusively via the mevalonate pathway, which had been extensively characterized in mammals and yeast. Yet unexpectedly the existence of an alternate pathway was discovered through isotope incorporation studies in a number of bacteria and plant species (Arigoni et al. 1997; Rohmer et al. 1993). This pathway is known by its key intermediates either as MEP (2-C-methyl-D-erythritol-4-phosphate) or DOXP (1-deoxy-D-xylulose-5-phosphate) pathway. We know now that many eukaryotes, archaebacteria and certain eubacteria rely on the classical mevalonate pathway, while most eubacteria and plastids use the MEP pathway (Rohdich et al. 2001). The pathway is initiated by the condensation of pyruvate and glyceraldehyde-3-phosphate to yield DOXP catalyzed by the enzyme DOXP synthase. DOXP is then rearranged and reduced to MEP by the enzyme DOXP reductoisomerase. The activity of three additional enzymes results in the formation of a cyclic diphosphate that is transformed to yield either IPP or DMAPP in the final step (see Eisenreich et al. (2004) for detailed review of the enzymology). Genomic and experimental analyses have shown that apicomplexan parasites harbor the DOXP pathway in the apicoplast but lack a cytoplasmic mevalonate pathway (Clastre et al. 2007; Coppens et al. 2000; Jomaa et al. 1999; Ralph et al. 2004b). In a seminal paper Jomaa and colleagues showed evidence for two enzymes of the DOXP pathway in the genome of *P. falciparum*. They demonstrated apicoplast localization of the protein product when expressed in *T. gondii* and also showed that *Plasmodium* is highly susceptible to fosmidomycin, an antibiotic that inhibits DOXP reductoisomerase (Jomaa et al. 1999). Using radiolabeling and HPLC analyses, Cassera and colleagues then demonstrated the presence of diagnostic intermediates of the DOXP pathway in extracts derived from *Plasmodium* infected red blood cells and such labeling could be abolished by pretreatment with fosmidomycin (Cassera et al. 2004). Subsequent studies have now identified all enzymes of the pathway itself as well as enzymes to furnish the required substrates (Fleige et al. 2007; Wiesner and Jomaa 2007). While some eubacteria encode an isomerase that mediates the inter conversions of IPP and DMAPP, in apicomplexan parasites the final enzyme LytB, is generating both isomers (Wiesner and Jomaa 2007). The last two enzymes of the pathway contain a 4Fe–4S cluster that acts in single electron transfers. Interestingly, LytB has been shown to interact with and accept electrons from the apicoplast localized ferredoxin/ferredoxinNADP-reductase system, a hold out of the electron transfer system associated with the chloroplast photosystem I (Rohrich et al. 2005).

In *Plasmodium* the DOXP pathway appears to be essential based on the sensitivity of the parasites to fosmidomycin in culture. In combination with clyndamycin, fosmidomycin has also shown promising efficacy in the treatment of clinical malaria (Borrmann et al. 2006). The DOXP pathway is conserved in all Apicomplexa that have apicoplasts making it an attractive target for drug development and a candidate for the raison d'être of the apicoplast. Surprisingly however, fosmidomycin kills *Plasmodium* and *Babesia*, yet has no effect on *Theileria*, *Toxoplasma* or *Eimeria* (Clastre et al. 2007; Lizundia et al. 2009; Sivakumar et al. 2008). It is conceivable that different host cell environments (i.e., red blood cells versus nucleated cells) provide different levels of access to isoprenoids and are responsible for this difference. This could make the pathway dispensable in some parasites. To test this we recently constructed conditional mutants in LytB and DOXP reductoisomerase in *T. gondii*. The studies demonstrated that the DOXP pathway is essential for *T. gondii* and, that fosmidomycin resistance is caused by differences in drug uptake, potentially at the level of the apicoplast membranes (Sethu Nair and Boris Striepen unpublished). Fosmidomycin is a phosphorylated compound that likely depends on transporters to cross membranes. Our observations suggest that drugs targeting the DOXP pathway that does not depend on specific transporters such as fosmidomycin might be highly active in other Apicomplexa. Furthermore, fosmidomycin could be vulnerable to the rapid development of drug resistance in malaria by import transporter mutation as seen in a variety of bacterial infections against fosfomycin, which is structurally related and shares resistance mechanisms e.g., (Castaneda-Garcia et al. 2009). Malaria parasites resistant to fosmidomycin have recently been generated in the laboratory. Resistance in these strains was associated with amplification of the target and a drug efflux pump, however, the level of this resistance was more modest than that observed in *T. gondii* (Dharia et al. 2009).

4.3 Heme Biosynthesis has Many Homes and Remains to be Fully Characterized

Heme is a porphyrin, a complex molecule made up of a four modified pyrroline rings that coordinate iron in the center of the molecule. Heme forms the prosthetic group of numerous important proteins and enzymes acting in catalysis, electron transfer and oxygen transport. In plants tetrapyrrole biosynthesis is plastid localized and provides both heme and chlorophyll (Heinemann et al. 2008). With the loss of photosynthesis the apicoplast has also lost the ability to synthesize chlorophyll, however, some apicomplexan parasites have maintained the ability to synthesize heme (Ralph et al. 2004a). De novo synthesis was detected in *P. falciparum* before the discovery of the apicoplast (Surolia and Padmanaban 1992). The different parts of this complex pathway are localized to three different compartments (mitochondrion, apicoplast, and cytoplasm). The pathway is initiated in the mitochondrion of the parasite where glycine and succinyl-CoA are converted to D-aminolevalonic acid (ALA) by delta aminolevolonate synthase (ALAS) (Varadharajan et al. 2002). The next step is the conversion of ALA to porphobilinogen by δ-aminolevulinate dehydratase (ALAD or HemB), which is an apicoplast localized enzyme. This obviously implies that ALA generated in the mitochondrion is transported to the lumen of the apicoplast, which could be one of the reasons for the observed close proximity of the two organelles (Rao et al. 2008). An apicoplast localized HemC (porphobilinogen deaminase) has been identified in *Plasmodium* which has been shown to catalyze the two next steps to yield uroporphynogenIII (Nagaraj et al. 2009a). The hydrophilic uroporphynogenIII is converted to hydrophobic coproporphinogenIII by HemE and in *Plasmodium* this step is clearly localized to the apicoplast (Sato et al. 2004). The last three steps of the pathway are catalyzed by the enzymes HemF, HemG, and HemH to yield functional heme. HemF and G lack obvious targeting signals and might be cytosolic. *P. falciparum* HemH (ferrochelatase) can complement the respective *E. coli* mutant (Sato and Wilson 2003) and has recently been shown to target to the mitochondrion (Nagaraj et al. 2009b). Interestingly, there have also been reports that *Plasmodium* might import enzymes from its host cell including ALAD and ferrochelatase (Bonday et al. 1997, 2000; Varadharajan et al. 2004). Such a mechanism could generate redundancy and limit the potential value of heme biosynthesis as a target for antimalarials. How important this concern is remains to be established.

4.4 Apicoplast Transporters: Feeding a Chloroplast in the Dark

Establishing metabolic exchange between host and endosymbiont through proteins that allow for metabolite transport across the membranes of the endosymbiont was likely an early and important step in endosymbiosis (Cavalier-Smith 2000; Weber et al. 2006). In higher plants, carbon skeletons generated through photosynthesis and Calvin cycle are exported in the form of triose phosphates by the triose phosphate/phosphate translocator (TPT) (Flügge et al. 2003). TPT is a member of a larger family of plastid phosphate translocators (Knappe et al. 2003). These translocators act as antiport systems exchanging inorganic phosphate for phosphorylated C3, C5 or C6 compounds. The TPT is involved in the export of carbon; in contrast, the other subfamilies import metabolites into plastids, namely phosphoenol pyruvate (PPT; Fischer et al. 1997), glucose-6-phosphate (GPT; Kammerer et al. 1998) and xylulose-5-phophate (Eicks et al. 2002). Proteins homologous to chloroplast phosphate translocators have been identified in Apicomplexa and have been shown to localize to the membranes of the apicoplast (Fleige et al. 2007; Karnataki et al. 2007a; Mullin et al. 2006). As the apicoplast has lost its photosynthetic capacity its anabolic pathways must now be supplied from the cytoplasm. To test this idea the *T. gondii* apicoplast phosphate translocator *Tg*APT was expressed in yeast and reconstituted into artificial liposomes (Brooks et al. 2009). Using this system, transport substrates could be tested in a controlled biochemical environment. Unique among phosphate translocators *Tg*APT combines the activities of chloroplast PPT and TPT by robustly exchanging phosphate against PEP, glyceraldehyde-3- phosphate and triose phosphate thus providing crucial substrates for at least two apicoplast pathways (FASII and DOXP). A conditional knockout mutant in *T. gondii* revealed that parasites die very quickly once the protein is lost. Interestingly, the phenotype of *Tg*APT loss is more severe than that of loss of FASII reinforcing a potential crucial role of the DOXP pathway in *Toxoplasma* (Brooks et al. 2009). Obviously, the apicoplast has additional metabolic needs that in all likelihood are met by a number of additional transporters (nucleotides for RNA and DNA synthesis, or iron for iron–sulfur cluster assembly are only two examples). Studies on such transporters might take leads from extensive work on plant chloroplasts and such transporter may be important as targets for drugs or as indirect determinants of drug sensitivity or resistance.

5 Conclusions

The apicoplast is a joint venture of three organisms (the alveolate host, the red alga and the cyanobacterium) and this is apparent in many aspects of its unique biology. Despite millions of years of cohabitation the boundaries between them as demarcated by the various membranes, are still important divides. The most obvious indication of this divided governance is that the proteins that execute and control apicoplast biology (e.g., those involved in biogenesis and replication) in the different subdomains of the organelle are derived from the organism that contributed this domain. Why did the complex multimembrane structure persist? One of several hypotheses to test is that the organelle has locked itself into a biochemical compartmentalization that is not easily undone. Understanding the structural and functional differences between the different domains will be the key to answering this question. A prime interest in this organelle stems from its potential value as a drug target. The first phase of this pursuit shows that rigorous genetic evaluation of targets is an absolute requirement. The fortune of some targets rose and those of others fell in these analyses. While the apicoplast has not yet delivered a pharmacological silver bullet, this promise is still unbroken. Significant advances were made dissecting the apicoplast metabolism and these will undoubtedly guide and accelerate the drug development process.

Acknowledgments Research in our laboratory is funded by grants from the National Institutes of Health to Boris Striepen, to Swati Agrawal who is the recipient of a predoctoral fellowship from the American Heart Association, to Lilach Sheiner who is supported by a postdoctoral fellowship from the Swiss National Science Fund, and to thank Giel van Dooren for many contributions.

References

Agrawal S, van Dooren GG, Beatty WL, Striepen B (2009) Genetic evidence that an endosymbiont-derived ERAD system functions in import of apicoplast proteins. J Biol Chem 284 (48):33683–33691

Ahmed A, Sharma YD (2008) Ribozyme cleavage of Plasmodium falciparum gyrase A gene transcript affects the parasite growth. Parasitol Res 103(4):751–763

Arigoni D, Sagner S, Latzel C, Eisenreich W, Bacher A, Zenk MH (1997) Terpenoid biosynthesis from 1-deoxy-D-xylulose in higher plants by intramolecular skeletal rearrangement. Proc Natl Acad Sci USA 94(20):10600–10605

Armbrust EV, Berges JA, Bowler C, Green BR, Martinez D, Putnam NH, Zhou S, Allen AE, Apt KE, Bechner M, Brzezinski MA, Chaal BK, Chiovitti A, Davis AK, Demarest MS, Detter JC, Glavina T, Goodstein D, Hadi MZ, Hellsten U, Hildebrand M, Jenkins BD, Jurka J, Kapitonov VV, Kroger N, Lau WW, Lane TW, Larimer FW, Lippmeier JC, Lucas S, Medina M, Montsant A, Obornik M, Parker MS, Palenik B, Pazour GJ, Richardson PM, Rynearson TA, Saito MA, Schwartz DC, Thamatrakoln K, Valentin K, Vardi A, Wilkerson FP, Rokhsar DS (2004) The genome of the diatom Thalassiosira pseudonana: ecology, evolution, and metabolism. Science 306(5693):79–86

Bahl A, Brunk B, Crabtree J, Fraunholz MJ, Gajria B, Grant GR, Ginsburg H, Gupta D, Kissinger JC, Labo P, Li L, Mailman MD, Milgram AJ, Pearson DS, Roos DS, Schug J, Stoeckert CJ Jr,

Whetzel P (2003) PlasmoDB: the Plasmodium genome resource A database integrating experimental and computational data. Nucleic Acids Res 31(1):212–215

Beckers CJ, Roos DS, Donald RG, Luft BJ, Schwab JC, Cao Y, Joiner KA (1995) Inhibition of cytoplasmic and organellar protein synthesis in *Toxoplasma gondii*. Implications for the target of macrolide antibiotics. J Clin Invest 95(1):367–376

Bonday ZQ, Dhanasekaran S, Rangarajan PN, Padmanaban G (2000) Import of host delta-aminolevulinate dehydratase into the malarial parasite: identification of a new drug target. Nat Med 6(8):898–903

Bonday ZQ, Taketani S, Gupta PD, Padmanaban G (1997) Heme biosynthesis by the malarial parasite Import of delta-aminolevulinate dehydrase from the host red cell. J Biol Chem 272 (35):21839–21846

Borrmann S, Lundgren I, Oyakhirome S, Impouma B, Matsiegui PB, Adegnika AA, Issifou S, Kun JF, Hutchinson D, Wiesner J, Jomaa H, Kremsner PG (2006) Fosmidomycin plus clindamycin for treatment of pediatric patients aged 1 to 14 years with Plasmodium falciparum malaria. Antimicrob Agents Chemother 50(8):2713–2718

Borst P, Overdulve JP, Weijers PJ, Fase-Fowler F, Van den Berg M (1984) DNA circles with cruciforms from Isospora (Toxoplasma) gondii. Biochim Biophys Acta 781(1–2):100–111

Brooks CF, Johnsen H, van Dooren GG, Muthalagi Liu SS M, Bohne W, Fischer K, Striepen B (2010) The phosphate translocator is the source of carbon and energy for the *Toxoplasma* apicoplast and essential for parasite survival. Cell Host & Microbe 7:63–73

Bruce BD (2001) The paradox of plastid transit peptides: conservation of function despite divergence in primary structure. Biochim Biophys Acta 1541(1–2):2–21

Cabantous S, Waldo GS (2006) In vivo and in vitro protein solubility assays using split GFP. Nat Methods 3(10):845–854

Camps M, Arrizabalaga G, Boothroyd J (2002) An rRNA mutation identifies the apicoplast as the target for clindamycin in Toxoplasma gondii. Mol Microbiol 43(5):1309–1318

Cassera MB, Gozzo FC, D'Alexandri FL, Merino EF, del Portillo HA, Peres VJ, Almeida IC, Eberlin MN, Wunderlich G, Wiesner J, Jomaa H, Kimura EA, Katzin AM (2004) The methylerythritol phosphate pathway is functionally active in all intraerythrocytic stages of Plasmodium falciparum. J Biol Chem 279(50):51749–51759

Castaneda-Garcia A, Rodriguez-Rojas A, Guelfo JR, Blazquez J (2009) Glycerol-3-phosphate permease GlpT is the only fosfomycin transporter in Pseudomonas aeruginosa. J Bacteriol 191(22):6968–6974

Cavalier-Smith T (1999) Principles of protein and lipid targeting in secondary symbiogenesis: euglenoid, dinoflagellate, and sporozoan plastid origins and the eukaryote family tree. J Eukaryot Microbiol 46(4):347–366

Cavalier-Smith T (2000) Membrane heredity and early chloroplast evolution. Trends Plant Sci 5(4):174–182

Cavalier-Smith T (2002) The phagotrophic origin of eukaryotes and phylogenetic classification of Protozoa. Int J Syst Evol Microbiol 52(Pt 2):297–354

Cavalier-Smith T (2004) Only six kingdoms of life. Proc Biol Sci 271(1545):1251–1262

Claros MG, Brunak S, von Heijne G (1997) Prediction of N-terminal protein sorting signals. Curr Opin Struct Biol 7(3):394–398

Clastre M, Goubard A, Prel A, Mincheva Z, Viaud-Massuart MC, Bout D, Rideau M, Velge-Roussel F, Laurent F (2007) The methylerythritol phosphate pathway for isoprenoid biosynthesis in coccidia: presence and sensitivity to fosmidomycin. Exp Parasitol 116(4):375–384

Clough B, Strath M, Preiser P, Denny P, Wilson IR (1997) Thiostrepton binds to malarial plastid rRNA. FEBS Lett 406(1–2):123–125

Coatney GR, Greenberg J (1952) The use of antibiotics in the treatment of malaria. Ann N Y Acad Sci 55(6):1075–1081

Colletti KS, Tattersall EA, Pyke KA, Froelich JE, Stokes KD, Osteryoung KW (2000) A homologue of the bacterial cell division site-determining factor MinD mediates placement of the chloroplast division apparatus. Curr Biol 10(9):507–516

Coppens I, Sinai AP, Joiner KA (2000) Toxoplasma gondii exploits host low-density lipoprotein receptor-mediated endocytosis for cholesterol acquisition. J Cell Biol 149(1):167–180

Coppin A, Varre JS, Lienard L, Dauvillee D, Guerardel Y, Soyer-Gobillard MO, Buleon A, Ball S, Tomavo S (2005) Evolution of plant-like crystalline storage polysaccharide in the protozoan parasite Toxoplasma gondii argues for a red alga ancestry. J Mol Evol 60(2):257–267

Crawford MJ, Thomsen-Zieger N, Ray M, Schachtner J, Roos DS, Seeber F (2006) Toxoplasma gondii scavenges host-derived lipoic acid despite its de novo synthesis in the apicoplast. EMBO J 25(13):3214–3222

Creasey A, Mendis K, Carlton J, Williamson D, Wilson I, Carter R (1994) Maternal inheritance of extrachromosomal DNA in malaria parasites. Mol Biochem Parasitol 65(1):95–98

Cronan JE, Zhao X, Jiang Y (2005) Function, attachment and synthesis of lipoic acid in Escherichia coli. Adv Microb Physiol 50:103–146

Dahl EL, Shock JL, Shenai BR, Gut J, DeRisi JL, Rosenthal PJ (2006) Tetracyclines specifically target the apicoplast of the malaria parasite Plasmodium falciparum. Antimicrob Agents Chemother 50(9):3124–3131

Delwiche CF (1999) Tracing the Thread of Plastid Diversity through the Tapestry of Life. Am Nat 154:S164–S177

Dar MA, Sharma A, Mondal N, Dhar SK (2007) Molecular cloning of apicoplast-targeted Plasmodium falciparum DNA gyrase genes: unique intrinsic ATPase activity and ATP-independent dimerization of PfGyrB subunit. Eukaryot Cell 6(3):398–412

DeRocher A, Gilbert B, Feagin JE, Parsons M (2005) Dissection of brefeldin A-sensitive and -insensitive steps in apicoplast protein targeting. J Cell Sci 118(Pt 3):565–574

DeRocher A, Hagen CB, Froehlich JE, Feagin JE, Parsons M (2000) Analysis of targeting sequences demonstrates that trafficking to the Toxoplasma gondii plastid branches off the secretory system. J Cell Sci 113(Pt 22):3969–3977

DeRocher AE, Coppens I, Karnataki A, Gilbert LA, Rome ME, Feagin JE, Bradley PJ, Parsons M (2008) A thioredoxin family protein of the apicoplast periphery identifies abundant candidate transport vesicles in Toxoplasma gondii. Eukaryot Cell 7(9):1518–1529

Dharia NV, Sidhu AB, Cassera MB, Westenberger SJ, Bopp SE, Eastman RT, Plouffe D, Batalov S, Park DJ, Volkman SK, Wirth DF, Zhou Y, Fidock DA, Winzeler EA (2009) Use of high-density tiling microarrays to identify mutations globally and elucidate mechanisms of drug resistance in Plasmodium falciparum. Genome Biol 10(2):R21

Dore E, Frontali C, Forte T, Fratarcangeli S (1983) Further studies and electron microscopic characterization of Plasmodium berghei DNA. Mol Biochem Parasitol 8(4):339–352

Durnford DG, Gray MW (2006) Analysis of Euglena gracilis plastid-targeted proteins reveals different classes of transit sequences. Eukaryot Cell 5(12):2079–2091

Eicks M, Maurino V, Knappe S, Flugge UI, Fischer K (2002) The plastidic pentose phosphate translocator represents a link between the cytosolic and the plastidic pentose phosphate pathways in plants. Plant Physiol 128(2):512–522

Eisenreich W, Bacher A, Arigoni D, Rohdich F (2004) Biosynthesis of isoprenoids via the non-mevalonate pathway. Cell Mol Life Sci 61(12):1401–1426

Fast NM, Kissinger JC, Roos DS, Keeling PJ (2001) Nuclear-encoded, plastid-targeted genes suggest a single common origin for apicomplexan and dinoflagellate plastids. Mol Biol Evol 18(3):418–426

Feagin JE, Parsons M (2007) The apicoplast and mitochondrion of *Toxoplasma gondii*. In: Weiss LM and Kim K (Eds) *Toxoplasma gondii* the model apicomplexan: perspectives and methods. Elsevier, London, pp. 207–244

Feagin JE, Werner E, Gardner MJ, Williamson DH, Wilson RJ (1992) Homologies between the contiguous and fragmented rRNAs of the two Plasmodium falciparum extrachromosomal DNAs are limited to core sequences. Nucleic Acids Res 20(4):879–887

Ferguson DJ, Henriquez FL, Kirisits MJ, Muench SP, Prigge ST, Rice DW, Roberts CW, McLeod RL (2005) Maternal inheritance and stage-specific variation of the apicoplast in

Toxoplasma gondii during development in the intermediate and definitive host. Eukaryot Cell 4(4):814–826

Fichera ME, Roos DS (1997) A plastid organelle as a drug target in apicomplexan parasites. Nature 390(6658):407–409

Fischer K, Kammerer B, Gutensohn M, Arbinger B, Weber A, Hausler RE, Flugge UI (1997) A new class of plastidic phosphate translocators: a putative link between primary and secondary metabolism by the phosphoenolpyruvate/phosphate antiporter. Plant Cell 9(3): 453–462

Fleige T, Fischer K, Ferguson DJ, Gross U, Bohne W (2007) Carbohydrate metabolism in the Toxoplasma gondii apicoplast: localization of three glycolytic isoenzymes, the single pyruvate dehydrogenase complex, and a plastid phosphate translocator. Eukaryot Cell 6(6):984–996

Flügge UI, Häusler RE, Ludewig F, Fischer K (2003) Functional genomics of phosphate antiport systems. Physiol Plant 118:475–482

Foth BJ, Ralph SA, Tonkin CJ, Struck NS, Fraunholz M, Roos DS, Cowman AF, McFadden GI (2003) Dissecting apicoplast targeting in the malaria parasite Plasmodium falciparum. Science 299(5607):705–708

Foth BJ, Stimmler LM, Handman E, Crabb BS, Hodder AN, McFadden GI (2005) The malaria parasite Plasmodium falciparum has only one pyruvate dehydrogenase complex, which is located in the apicoplast. Mol Microbiol 55(1):39–53

Fraunholz MJ, Moerschel E, Maier UG (1998) The chloroplast division protein FtsZ is encoded by a nucleomorph gene in cryptomonads. Mol Gen Genet 260(2–3):207–211

Gardner MJ, Bates PA, Ling IT, Moore DJ, McCready S, Gunasekera MB, Wilson RJ, Williamson DH (1988) Mitochondrial DNA of the human malarial parasite Plasmodium falciparum. Mol Biochem Parasitol 31(1):11–17

Gardner MJ, Feagin JE, Moore DJ, Rangachari K, Williamson DH, Wilson RJ (1993) Sequence and organization of large subunit rRNA genes from the extrachromosomal 35 kb circular DNA of the malaria parasite Plasmodium falciparum. Nucleic Acids Res 21(5):1067–1071

Gardner MJ, Feagin JE, Moore DJ, Spencer DF, Gray MW, Williamson DH, Wilson RJ (1991a) Organisation and expression of small subunit ribosomal RNA genes encoded by a 35-kilobase circular DNA in Plasmodium falciparum. Mol Biochem Parasitol 48(1):77–88

Gardner MJ, Goldman N, Barnett P, Moore PW, Rangachari K, Strath M, Whyte A, Williamson DH, Wilson RJ (1994a) Phylogenetic analysis of the rpoB gene from the plastid-like DNA of Plasmodium falciparum. Mol Biochem Parasitol 66(2):221–231

Gardner MJ, Hall N, Fung E, White O, Berriman M, Hyman RW, Carlton JM, Pain A, Nelson KE, Bowman S, Paulsen IT, James K, Eisen JA, Rutherford K, Salzberg SL, Craig A, Kyes S, Chan MS, Nene V, Shallom SJ, Suh B, Peterson J, Angiuoli S, Pertea M, Allen J, Selengut J, Haft D, Mather MW, Vaidya AB, Martin DM, Fairlamb AH, Fraunholz MJ, Roos DS, Ralph SA, McFadden GI, Cummings LM, Subramanian GM, Mungall C, Venter JC, Carucci DJ, Hoffman SL, Newbold C, Davis RW, Fraser CM, Barrell B (2002) Genome sequence of the human malaria parasite Plasmodium falciparum. Nature 419(6906):498–511

Gardner MJ, Preiser P, Rangachari K, Moore D, Feagin JE, Williamson DH, Wilson RJ (1994b) Nine duplicated tRNA genes on the plastid-like DNA of the malaria parasite Plasmodium falciparum. Gene 144(2):307–308

Gardner MJ, Williamson DH, Wilson RJ (1991b) A circular DNA in malaria parasites encodes an RNA polymerase like that of prokaryotes and chloroplasts. Mol Biochem Parasitol 44(1): 115–123

Gibbs SP (1979) The route of entry of cytoplasmically synthesized proteins into chloroplasts of algae possessing chloroplast ER. J Cell Sci 35:253–266

Glynn JM, Froehlich JE, Osteryoung KW (2008) Arabidopsis ARC6 coordinates the division machineries of the inner and outer chloroplast membranes through interaction with PDV2 in the intermembrane space. Plant Cell 20(9):2460–2470

Goodman CD, McFadden GI (2007) Fatty acid biosynthesis as a drug target in apicomplexan parasites. Curr Drug Targets 8(1):15–30

Goodman CD, Su V, McFadden GI (2007) The effects of anti-bacterials on the malaria parasite Plasmodium falciparum. Mol Biochem Parasitol 152(2):181–191

Gould SB, Sommer MS, Hadfi K, Zauner S, Kroth PG, Maier UG (2006a) Protein targeting into the complex plastid of cryptophytes. J Mol Evol 62(6):674–681

Gould SB, Sommer MS, Kroth PG, Gile GH, Keeling PJ, Maier UG (2006b) Nucleus-to-nucleus gene transfer and protein retargeting into a remnant cytoplasm of cryptophytes and diatoms. Mol Biol Evol 23(12):2413–2422

Gould SB, Tham WH, Cowman AF, McFadden GI, Waller RF (2008) Alveolins, a new family of cortical proteins that define the protist infrakingdom Alveolata. Mol Biol Evol 25(6): 1219–1230

Gray MW (1993) Origin and evolution of organelle genomes. Curr Opin Genet Dev 3(6): 884–890

Grossman A, Manodori A, Snyder D (1990) Light-harvesting proteins of diatoms: their relationship to the chlorophyll a/b binding proteins of higher plants and their mode of transport into plastids. Mol Gen Genet 224(1):91–100

Gubbels MJ, Vaishnava S, Boot N, Dubremetz JF, Striepen B (2006) A MORN-repeat protein is a dynamic component of the Toxoplasma gondii cell division apparatus. J Cell Sci 119: 2236–2245

Gunther S, Wallace L, Patzewitz EM, McMillan PJ, Storm J, Wrenger C, Bissett R, Smith TK, Muller S (2007) Apicoplast lipoic acid protein ligase B is not essential for Plasmodium falciparum. PLoS Pathog 3(12):e189

Harb OS, Chatterjee B, Fraunholz MJ, Crawford MJ, Nishi M, Roos DS (2004) Multiple functionally redundant signals mediate targeting to the apicoplast in the apicomplexan parasite Toxoplasma gondii. Eukaryot Cell 3(3):663–674

He CY, Striepen B, Pletcher CH, Murray JM, Roos DS (2001) Targeting and processing of nuclear-encoded apicoplast proteins in plastid segregation mutants of Toxoplasma gondii. J Biol Chem 276(30):28436–28442

Heinemann IU, Jahn M, Jahn D (2008) The biochemistry of heme biosynthesis. Arch Biochem Biophys 474(2):238–251

Hempel F, Bullmann L, Lau J, Zauner S, Maier UG (2009) ERAD-derived preprotein transport across the second outermost plastid membrane of diatoms. Mol Biol Evol 26(8):1781–1790

Hormann F, Soll J, Bolter B (2007) The chloroplast protein import machinery: a review. Methods Mol Biol 390:179–193

Howe CJ, Purton S (2007) The little genome of apicomplexan plastids: its raison d'etre and a possible explanation for the 'delayed death' phenomenon. Protist 158(2):121–133

Itoh R, Fujiwara M, Nagata N, Yoshida S (2001) A chloroplast protein homologous to the eubacterial topological specificity factor minE plays a role in chloroplast division. Plant Physiol 127(4):1644–1655

Jarvis P (2008) Targeting of nucleus-encoded proteins to chloroplasts in plants. New Phytol 179(2):257–285

Jelenska J, Crawford MJ, Harb OS, Zuther E, Haselkorn R, Roos DS, Gornicki P (2001) Subcellular localization of acetyl-CoA carboxylase in the apicomplexan parasite Toxoplasma gondii. Proc Natl Acad Sci USA 98(5):2723–2728

Jomaa H, Wiesner J, Sanderbrand S, Altincicek B, Weidemeyer C, Hintz M, Turbachova I, Eberl M, Zeidler J, Lichtenthaler HK, Soldati D, Beck E (1999) Inhibitors of the nonmevalonate pathway of isoprenoid biosynthesis as antimalarial drugs. Science 285(5433):1573–1576

Kalanon M, Tonkin CJ, McFadden GI (2009) Characterization of two putative protein translocation components in the apicoplast of Plasmodium falciparum. Eukaryot Cell 8(8): 1146–1154

Kammerer B, Fischer K, Hilpert B, Schubert S, Gutensohn M, Weber A, Flugge UI (1998) Molecular characterization of a carbon transporter in plastids from heterotrophic tissues: the glucose 6-phosphate/phosphate antiporter. Plant Cell 10(1):105–117

Karnataki A, Derocher A, Coppens I, Nash C, Feagin JE, Parsons M (2007a) Cell cycle-regulated vesicular trafficking of Toxoplasma APT1, a protein localized to multiple apicoplast membranes. Mol Microbiol 63(6):1653–1668

Karnataki A, Derocher AE, Coppens I, Feagin JE, Parsons M (2007b) A membrane protease is targeted to the relict plastid of toxoplasma via an internal signal sequence. Traffic 8(11):1543–1553
Karnataki A, DeRocher AE, Feagin JE, Parsons M (2009) Sequential processing of the Toxoplasma apicoplast membrane protein FtsH1 in topologically distinct domains during intracellular trafficking. Mol Biochem Parasitol 166(2):126–133
Keeling PJ (2009) Chromalveolates and the evolution of plastids by secondary endosymbiosis. J Eukaryot Microbiol 56(1):1–8
Kilejian A (1975) Circular mitochondrial DNA from the avian malarial parasite Plasmodium lophurae. Biochim Biophys Acta 390(3):276–284
Kilian O, Kroth PG (2005) Identification and characterization of a new conserved motif within the presequence of proteins targeted into complex diatom plastids. Plant J 41(2):175–183
Knappe S, Flugge UI, Fischer K (2003) Analysis of the plastidic phosphate translocator gene family in Arabidopsis and identification of new phosphate translocator-homologous transporters, classified by their putative substrate-binding site. Plant Physiol 131(3):1178–1190
Kobayashi T, Takahara M, Miyagishima SY, Kuroiwa H, Sasaki N, Ohta N, Matsuzaki M, Kuroiwa T (2002) Detection and localization of a chloroplast-encoded HU-like protein that organizes chloroplast nucleoids. Plant Cell 14(7):1579–1589
Kohler S (2005) Multi-membrane-bound structures of Apicomplexa: I. the architecture of the Toxoplasma gondii apicoplast. Parasitol Res 96(4):258–272
Kohler S, Delwiche CF, Denny PW, Tilney LG, Webster P, Wilson RJ, Palmer JD, Roos DS (1997) A plastid of probable green algal origin in Apicomplexan parasites. Science 275(5305): 1485–1489
Kouranov A, Chen X, Fuks B, Schnell DJ (1998) Tic20 and Tic22 are new components of the protein import apparatus at the chloroplast inner envelope membrane. J Cell Biol 143(4): 991–1002
Kroth P, Strotmann H (1999) Diatom plastids: secondary endocytobiosis, plastid genome and protein import. Physiol Plant 107:136–141
Lane CE, Archibald JM (2008) The eukaryotic tree of life: endosymbiosis takes its TOL. Trends Ecol Evol 23(5):268–275
Legesse-Miller A, Massol RH, Kirchhausen T (2003) Constriction and Dnm1p recruitment are distinct processes in mitochondrial fission. Mol Biol Cell 14(5):1953–1963
Lizundia R, Werling D, Langsley G, Ralph SA (2009) Theileria apicoplast as a target for chemotherapy. Antimicrob Agents Chemother 53(3):1213–1217
Louis MW, Kami K (eds) (2007) The model apicomplexan: perspectives and methods, vols. 1, 9. Elsevier, London, pp 207–244
Maeda T, Saito T, Harb OS, Roos DS, Takeo S, Suzuki H, Tsuboi T, Takeuchi T, Asai T (2009) Pyruvate kinase type-II isozyme in Plasmodium falciparum localizes to the apicoplast. Parasitol Int 58(1):101–105
Matsuzaki M, Kikuchi T, Kita K, Kojima S, Kuroiwa T (2001) Large amounts of apicoplast nucleoid DNA and its segregation in Toxoplasma gondii. Protoplasma 218(3–4):180–191
Mazumdar J, Striepen B (2007) Make it or take it fatty acid metabolism of apicomplexan parasites. Eukaryot Cell 6:1727–1735
Mazumdar J, Wilson E, Masarek K, Hunter C, Striepen B (2006) Apicoplast fatty acid synthesis is essential for organelle biogenesis and parasite survival in Toxoplasma gondii. Proc Natl Acad Sci USA 103:13192–13197
McConkey GA, Rogers MJ, McCutchan TF (1997) Inhibition of Plasmodium falciparum protein synthesis Targeting the plastid-like organelle with thiostrepton. J Biol Chem 272(4): 2046–2049
McFadden GI, Reith ME, Munholland J, Lang-Unnasch N (1996) Plastid in human parasites. Nature 381(6582):482
McFadden GI, van Dooren GG (2004) Evolution: red algal genome affirms a common origin of all plastids. Curr Biol 14(13):R514–R516
McLeod R, Muench SP, Rafferty JB, Kyle DE, Mui EJ, Kirisits MJ, Mack DG, Roberts CW, Samuel BU, Lyons RE, Dorris M, Milhous WK, Rice DW (2001) Triclosan inhibits the growth

of Plasmodium falciparum and Toxoplasma gondii by inhibition of apicomplexan Fab I. Int J Parasitol 31(2):109–113

Miyagishima SY (2005) Origin and evolution of the chloroplast division machinery. J Plant Res 118(5):295–306

Moore CE, Archibald JM (2009) Nucleomorph genomes. Annu Rev Genet 43:251–254

Moore RB, Obornik M, Janouskovec J, Chrudimsky T, Vancova M, Green DH, Wright SW, Davies NW, Bolch CJ, Heimann K, Slapeta J, Hoegh-Guldberg O, Logsdon JM, Carter DA (2008) A photosynthetic alveolate closely related to apicomplexan parasites. Nature 451(7181):959–963

Moreno SN, Li ZH (2008) Anti-infectives targeting the isoprenoid pathway of Toxoplasma gondii. Expert Opin Ther Targets 12(3):253–263

Mullin KA, Lim L, Ralph SA, Spurck TP, Handman E, McFadden GI (2006) Membrane transporters in the relict plastid of malaria parasites. Proc Natl Acad Sci USA 103:9572–9577

Nagaraj VA, Arumugam R, Chandra NR, Prasad D, Rangarajan PN, Padmanaban G (2009a) Localisation of Plasmodium falciparum uroporphyrinogen III decarboxylase of the heme-biosynthetic pathway in the apicoplast and characterisation of its catalytic properties. Int J Parasitol 39(5):559–568

Nagaraj VA, Prasad D, Rangarajan PN, Padmanaban G (2009b) Mitochondrial localization of functional ferrochelatase from Plasmodium falciparum. Mol Biochem Parasitol 168(1): 109–112

Nassoury N, Cappadocia M, Morse D (2003) Plastid ultrastructure defines the protein import pathway in dinoflagellates. J Cell Sci 116(Pt 14):2867–2874

Obornik M, Van de Peer Y, Hypsa V, Frickey T, Slapeta JR, Meyer A, Lukes J (2002) Phylogenetic analyses suggest lateral gene transfer from the mitochondrion to the apicoplast. Gene 285(1–2):109–118

Osteryoung KW, Nunnari J (2003) The division of endosymbiotic organelles. Science 302(5651):1698–1704

Park S, Isaacson R, Kim HT, Silver PA, Wagner G (2005) Ufd1 exhibits the AAA-ATPase fold with two distinct ubiquitin interaction sites. Structure 13(7):995–1005

Patron NJ, Waller RF, Archibald JM, Keeling PJ (2005) Complex protein targeting to dinoflagellate plastids. J Mol Biol 348(4):1015–1024

Preiser P, Williamson DH, Wilson RJ (1995) tRNA genes transcribed from the plastid-like DNA of Plasmodium falciparum. Nucleic Acids Res 23(21):4329–4336

Raghu Ram EV, Kumar A, Biswas S, Chaubey S, Siddiqi MI, Habib S (2007) Nuclear gyrB encodes a functional subunit of the Plasmodium falciparum gyrase that is involved in apicoplast DNA replication. Mol Biochem Parasitol 154(1):30–39

Ralph SA, Foth BJ, Hall N, McFadden GI (2004a) Evolutionary pressures on apicoplast transit peptides. Mol Biol Evol 21(12):2183–2194

Ralph SA, van Dooren GG, Waller RF, Crawford MJ, Fraunholz MJ, Foth BJ, Tonkin CJ, Roos DS, McFadden GI (2004b) Tropical infectious diseases: metabolic maps and functions of the Plasmodium falciparum apicoplast. Nat Rev Microbiol 2(3):203–216

Rao A, Yeleswarapu SJ, Srinivasan R, Bulusu G (2008) Localization of heme biosynthesis pathway enzymes in Plasmodium falciparum. Indian J Biochem Biophys 45(6):365–373

Rohdich F, Kis K, Bacher A, Eisenreich W (2001) The non-mevalonate pathway of isoprenoids: genes, enzymes and intermediates. Curr Opin Chem Biol 5(5):535–540

Rohmer M, Knani M, Simonin P, Sutter B, Sahm H (1993) Isoprenoid biosynthesis in bacteria: a novel pathway for the early steps leading to isopentenyl diphosphate. Biochem J 295(Pt 2):517–524

Rohrich RC, Englert N, Troschke K, Reichenberg A, Hintz M, Seeber F, Balconi E, Aliverti A, Zanetti G, Kohler U, Pfeiffer M, Beck E, Jomaa H, Wiesner J (2005) Reconstitution of an apicoplast-localised electron transfer pathway involved in the isoprenoid biosynthesis of Plasmodium falciparum. FEBS Lett 579(28):6433–6438

Sato S, Clough B, Coates L, Wilson RJ (2004) Enzymes for heme biosynthesis are found in both the mitochondrion and plastid of the malaria parasite Plasmodium falciparum. Protist 155(1):117–125

Sato S, Wilson RJ (2003) Proteobacteria-like ferrochelatase in the malaria parasite. Curr Genet 42(5):292–300

Seow F, Sato S, Janssen CS, Riehle MO, Mukhopadhyay A, Phillips RS, Wilson RJ, Barrett MP (2005) The plastidic DNA replication enzyme complex of Plasmodium falciparum. Mol Biochem Parasitol 141(2):145–153

Siddall ME (1992) Hohlzylinders. Parasitol Today 8(3):90–91

Sidhu AB, Sun Q, Nkrumah LJ, Dunne MW, Sacchettini JC, Fidock DA (2007) In vitro efficacy, resistance selection, and structural modeling studies implicate the malarial parasite apicoplast as the target of azithromycin. J Biol Chem 282(4):2494–2504

Singh D, Kumar A, Raghu Ram EV, Habib S (2005) Multiple replication origins within the inverted repeat region of the Plasmodium falciparum apicoplast genome are differentially activated. Mol Biochem Parasitol 139(1):99–106

Sivakumar T, Aboulaila MRA, Khukhuu A, Iseki H, Alhassan A, Yokoyama N, Igarashi I (2008) In vitro inhibitory effect of fosmidomycin on the asexual growth of Babesia bovis and Babesia bigemina. J Protozool Res 18:71–78

Smith S, Witkowski A, Joshi AK (2003) Structural and functional organization of the animal fatty acid synthase. Prog Lipid Res 42(4):289–317

Sommer MS, Gould SB, Lehmann P, Gruber A, Przyborski JM, Maier UG (2007) Der1-mediated preprotein import into the periplastid compartment of chromalveolates? Mol Biol Evol 24:918–928

Spork S, Hiss JA, Mandel K, Sommer M, Kooij TW, Chu T, Schneider G, Maier UG, Przyborski JM (2009) An unusual ERAD-like complex is targeted to the apicoplast of Plasmodium falciparum. Eukaryot Cell 8(8):1134–1145

Stanway RR, Witt T, Zobiak B, Aepfelbacher M, Heussler VT (2009) GFP-targeting allows visualization of the apicoplast throughout the life cycle of live malaria parasites. Biol Cell 101(7):415–430

Stokes KD, McAndrew RS, Figueroa R, Vitha S, Osteryoung KW (2000) Chloroplast division and morphology are differentially affected by overexpression of FtsZ1 and FtsZ2 genes in Arabidopsis. Plant Physiol 124(4):1668–1677

Striepen B, Crawford MJ, Shaw MK, Tilney LG, Seeber F, Roos DS (2000) The plastid of Toxoplasma gondii is divided by association with the centrosomes. J Cell Biol 151(7): 1423–1434

Striepen B, Jordan CN, Reiff S, van Dooren GG (2007) Building the perfect parasite: cell division in apicomplexa. PLoS Pathog 3(6):e78

Sulli C, Schwartzbach SD (1995) The polyprotein precursor to the Euglena light-harvesting chlorophyll a/b-binding protein is transported to the Golgi apparatus prior to chloroplast import and polyprotein processing. J Biol Chem 270(22):13084–13090

Sulli C, Schwartzbach SD (1996) A soluble protein is imported into Euglena chloroplasts as a membrane- bound precursor. Plant Cell 8(1):43–53

Surolia N, Padmanaban G (1992) de novo biosynthesis of heme offers a new chemotherapeutic target in the human malarial parasite. Biochem Biophys Res Commun 187(2):744–750

Surolia N, Surolia A (2001) Triclosan offers protection against blood stages of malaria by inhibiting enoyl-ACP reductase of Plasmodium falciparum. Nat Med 7(2):167–173

Tabbara KF, O'Connor GR (1980) Treatment of ocular toxoplasmosis with clindamycin and sulfadiazine. Ophthalmology 87(2):129–134

Thomsen-Zieger N, Schachtner J, Seeber F (2003) Apicomplexan parasites contain a single lipoic acid synthase located in the plastid. FEBS Lett 547(1–3):80–86

Tomova C, Geerts WJ, Muller-Reichert T, Entzeroth R, Humbel BM (2006) New comprehension of the apicoplast of sarcocystis by transmission electron tomography. Biol Cell 98(9): 535–545

Tomova C, Humbel BM, Geerts WJ, Entzeroth R, Holthuis JC, Verkleij AJ (2009) Membrane contact sites between apicoplast and ER in Toxoplasma gondii revealed by electron tomography. Traffic 10(10):1471–1480

Tonkin CJ, Kalanon M, McFadden GI (2008) Protein targeting to the malaria parasite plastid. Traffic 9(2):166–175

Tonkin CJ, Roos DS, McFadden GI (2006a) N-terminal positively charged amino acids, but not their exact position, are important for apicoplast transit peptide fidelity in Toxoplasma gondii. Mol Biochem Parasitol 150(2):192–200

Tonkin CJ, Struck NS, Mullin KA, Stimmler LM, McFadden GI (2006b) Evidence for Golgi-independent transport from the early secretory pathway to the plastid in malaria parasites. Mol Microbiol 61(3):614–630

Tranel PJ, Froehlich J, Goyal A, Keegstra K (1995) A component of the chloroplastic protein import apparatus is targeted to the outer envelope membrane via a novel pathway. EMBO J 14(11):2436–2446

Tranel PJ, Keegstra K (1996) A novel, bipartite transit peptide targets OEP75 to the outer membrane of the chloroplastic envelope. Plant Cell 8(11):2093–2104

Vaidya AB, Arasu P (1987) Tandemly arranged gene clusters of malarial parasites that are highly conserved and transcribed. Mol Biochem Parasitol 22(2–3):249–257

Vaishnava S, Morrison DP, Gaji RY, Murray JM, Entzeroth R, Howe DK, Striepen B (2005) Plastid segregation and cell division in the apicomplexan parasite Sarcocystis neurona. J Cell Sci 118(Pt 15):3397–3407

Vaishnava S, Striepen B (2006) The cell biology of secondary endosymbiosis–how parasites build, divide and segregate the apicoplast. Mol Microbiol 61(6):1380–1387

van Dooren GG, Marti M, Tonkin CJ, Stimmler LM, Cowman AF, McFadden GI (2005) Development of the endoplasmic reticulum, mitochondrion and apicoplast during the asexual life cycle of Plasmodium falciparum. Mol Microbiol 57(2):405–419

van Dooren GG, Reiff SB, Tomova C, Meissner M, Humbel BM, Striepen B (2009) A novel dynamin-related protein has been recruited for apicoplast fission in Toxoplasma gondii. Curr Biol 19(4):267–276

van Dooren GG, Su V, D'Ombrain MC, McFadden GI (2002) Processing of an apicoplast leader sequence in Plasmodium falciparum and the identification of a putative leader cleavage enzyme. J Biol Chem 277(26):23612–23619

van Dooren GG, Tomova C, Agrawal S, Humbel BM, Striepen B (2008a) Toxoplasma gondii Tic20 is essential for apicoplast protein import. Proc Natl Acad Sci USA 105(36): 13574–13579

van Dooren GG, Tomova C, Agrawal S, Humbel BM, Striepen B (2008b) Toxoplasma gondii Tic20 is essential for apicoplast protein import. Proc Natl Acad Sci USA 105:13574–13579

Varadharajan S, Dhanasekaran S, Bonday ZQ, Rangarajan PN, Padmanaban G (2002) Involvement of delta-aminolaevulinate synthase encoded by the parasite gene in de novo haem synthesis by Plasmodium falciparum. Biochem J 367(Pt 2):321–327

Varadharajan S, Sagar BK, Rangarajan PN, Padmanaban G (2004) Localization of ferrochelatase in Plasmodium falciparum. Biochem J 384(Pt 2):429–436

Vaughan AM, O'Neill MT, Tarun AS, Camargo N, Phuong TM, Aly AS, Cowman AF, Kappe SH (2009) Type II fatty acid synthesis is essential only for malaria parasite late liver stage development. Cell Microbiol 11(3):506–520

Vollmer M, Thomsen N, Wiek S, Seeber F (2001) Apicomplexan parasites possess distinct nuclear-encoded, but apicoplast-localized, plant-type ferredoxin-NADP+ reductase and ferredoxin. J Biol Chem 276(8):5483–5490

Waller RF, Keeling PJ, Donald RG, Striepen B, Handman E, Lang-Unnasch N, Cowman AF, Besra GS, Roos DS, McFadden GI (1998) Nuclear-encoded proteins target to the plastid in *Toxoplasma gondii* and *Plasmodium falciparum*. Proc Natl Acad Sci USA 95(21): 12352–12357

Waller RF, McFadden GI (2005) The apicoplast: a review of the derived plastid of apicomplexan parasites. Curr Issues Mol Biol 7(1):57–79

Waller RF, Reed MB, Cowman AF, McFadden GI (2000) Protein trafficking to the plastid of Plasmodium falciparum is via the secretory pathway. EMBO J 19:1794–1802

Walters KJ (2005) Ufd1 exhibits dual ubiquitin binding modes. Structure 13(7):943–944

Wanke M, Skorupinska-Tudek K, Swiezewska E (2001) Isoprenoid biosynthesis via 1-deoxy-D-xylulose 5-phosphate/2-C-methyl-D-erythritol 4-phosphate (DOXP/MEP) pathway. Acta Biochim Pol 48(3):663–672

Wastl J, Maier UG (2000) Transport of proteins into cryptomonads complex plastids. J Biol Chem 275(30):23194–23198

Weber AP, Linka M, Bhattacharya D (2006) Single, ancient origin of a plastid metabolite translocator family in Plantae from an endomembrane-derived ancestor. Eukaryot Cell 5(3): 609–612

Weissig V, Vetro-Widenhouse TS, Rowe TC (1997) Topoisomerase II inhibitors induce cleavage of nuclear and 35-kb plastid DNAs in the malarial parasite Plasmodium falciparum. DNA Cell Biol 16(12):1483–1492

Wiesner J, Jomaa H (2007) Isoprenoid biosynthesis of the apicoplast as drug target. Curr Drug Targets 8(1):3–13

Wiesner J, Reichenberg A, Heinrich S, Schlitzer M, Jomaa H (2008) The plastid-like organelle of apicomplexan parasites as drug target. Curr Pharm Des 14(9):855–871

Williamson DH, Denny PW, Moore PW, Sato S, McCready S, Wilson RJ (2001) The in vivo conformation of the plastid DNA of Toxoplasma gondii: implications for replication. J Mol Biol 306(2):159–168

Williamson DH, Gardner MJ, Preiser P, Moore DJ, Rangachari K, Wilson RJ (1994) The evolutionary origin of the 35 kb circular DNA of Plasmodium falciparum: new evidence supports a possible rhodophyte ancestry. Mol Gen Genet 243(2):249–252

Williamson DH, Preiser PR, Moore PW, McCready S, Strath M, Wilson RJ (2002) The plastid DNA of the malaria parasite Plasmodium falciparum is replicated by two mechanisms. Mol Microbiol 45(2):533–542

Williamson DH, Wilson RJ, Bates PA, McCready S, Perler F, Qiang BU (1985) Nuclear and mitochondrial DNA of the primate malarial parasite Plasmodium knowlesi. Mol Biochem Parasitol 14(2):199–209

Wilson RJ, Denny PW, Preiser PR, Rangachari K, Roberts K, Roy A, Whyte A, Strath M, Moore DJ, Moore PW, Williamson DH (1996) Complete gene map of the plastid-like DNA of the malaria parasite Plasmodium falciparum. J Mol Biol 261(2):155–172

Wilson RJ, Fry M, Gardner MJ, Feagin JE, Williamson DH (1992) Subcellular fractionation of the two organelle DNAs of malaria parasites. Curr Genet 21(4–5):405–408

Wilson RJ, Williamson DH (1997) Extrachromosomal DNA in the Apicomplexa. Microbiol Mol Biol Rev 61(1):1–16

Wilson RJ, Williamson DH, Preiser P (1994) Malaria and other Apicomplexans: the "plant" connection. Infect Agents Dis 3(1):29–37

Wrenger C, Muller S (2004) The human malaria parasite Plasmodium falciparum has distinct organelle-specific lipoylation pathways. Mol Microbiol 53(1):103–113

Ye Y, Meyer HH, Rapoport TA (2001) The AAA ATPase Cdc48/p97 and its partners transport proteins from the ER into the cytosol. Nature 414(6864):652–656

Ye Y, Shibata Y, Yun C, Ron D, Rapoport TA (2004) A membrane protein complex mediates retro-translocation from the ER lumen into the cytosol. Nature 429(6994):841–847

Yu M, Kumar TR, Nkrumah LJ, Coppi A, Retzlaff S, Li CD, Kelly BJ, Moura PA, Lakshmanan V, Freundlich JS, Valderramos JC, Vilcheze C, Siedner M, Tsai JH, Falkard B, Sidhu AB, Purcell LA, Gratraud P, Kremer L, Waters AP, Schiehser G, Jacobus DP, Janse CJ, Ager A, Jacobs WR Jr, Sacchettini JC, Heussler V, Sinnis P, Fidock DA (2008) The fatty acid biosynthesis enzyme FabI plays a key role in the development of liver-stage malarial parasites. Cell Host Microbe 4(6):567–578

Yung S, Unnasch TR, Lang-Unnasch N (2001) Analysis of apicoplast targeting and transit peptide processing in Toxoplasma gondii by deletional and insertional mutagenesis. Mol Biochem Parasitol 118(1):11–21

Zhu G, Marchewka MJ, Keithly JS (2000) Cryptosporidium parvum appears to lack a plastid genome. Microbiology 146:315–321

Zuther E, Johnson JJ, Haselkorn R, McLeod R, Gornicki P (1999) Growth of Toxoplasma gondii is inhibited by aryloxyphenoxypropionate herbicides targeting acetyl-CoA carboxylase. Proc Natl Acad Sci USA 96(23):13387–13392

The Glycosome of Trypanosomatids

Fred R. Opperdoes

Contents

Abstract Glycosomes are the microbodies of the Kinetoplastida. They belong to the family of peroxisomes present in almost all eukaryotic cells. Glycosomes share the same machinery for their biosynthesis and several metabolic pathways with the microbodies of other organisms. However, glycosomes contain in addition the enzymes of glycolysis and glycerol metabolism, gluconeogenesis, purine salvage, and pyrimidine biosynthesis, traits normally not encountered in other microbodies. The unique aspects of the trypanosomatid glycosome render this organelle and its constituents interesting targets for the development of new antitrypanosome drugs.

F.R. Opperdoes
Research Unit for Tropical Diseases, de Duve Institute and Biochemistry Unit, Université catholique de Louvain, Avenue Hipprocrate 75, 1200, Brussels, Belgium
e-mail: fred.opperdoes@uclouvain.be

W. de Souza (ed.), *Structures and Organelles in Pathogenic Protists*,
Microbiology Monographs 17, DOI 10.1007/978-3-642-12863-9_12,
© Springer-Verlag Berlin Heidelberg 2010

1 Introduction

Peroxisomes are the single membrane bounded cytoplasmic organelles, also called microbodies, which range from 0.1 to 1 μm in diameter. They lack DNA and normally contain a number of enzymes involved in oxidative functions, resulting in the production of reactive oxygen species (ROS) such as superoxide anions and hydroxyl radicals. Characteristic enzymes of ROS-producing peroxisomes are the flavine oxidases xanthine oxidase, D-amino acid oxidase and acyl-CoA oxidase. Other peroxisomal enzymes such as superoxide dismutase (SOD) and catalase inactivate ROS to molecular oxygen and water. Thus, catalase is considered a typical marker enzyme of peroxisomes. In addition to ROS metabolism, peroxisomes share a number of enzymes involved in fatty-acid beta-oxidation and the biosynthesis of ether lipids (see Michels et al. 2005 for a review on the functions of peroxisomes). Although the presence of catalase has been reported for certain trypanosomatids, the peroxisomes of the pathogenic members, such as *Trypanosoma* and *Leishmania*, all lack catalase and the above flavine oxidases. However, the microbodies of all Trypanosomatidae share the unique property that they contain a number of enzymes of the glycolytic and the gluconeogenic pathway, For this reason the name glycosome was coined for this organelle (Opperdoes and Borst 1977). Later research carried out in several laboratories have not only revealed the presence of many more enzymes and pathways that glycosomes share with the peroxisomes of other organisms but also have consolidated the unique nature of these organelles by the detection of numerous enzymes and metabolic pathways that turn out to be exclusively associated with the glycosomes of the trypanosomatids. In this chapter the author will give an account of these findings and make a comparison between the glycosomes of Trypanosomatidae and the peroxisomes of other eukaryotes.

2 Discovery of the Glycosome

Bloodstream-form *Trypanosoma brucei,* the causative agent of African sleeping sickness, lacks a functional mitochondrion and cytochrome system. It's rate of oxygen consumption is 50 times that of mammalian cells (Von Brandt 1973) and its respiration is insensitive to the classical inhibitors of the respiratory chain such as cyanide, azide, and antimycin, while it is inhibited by aromatic hydroxamic acids. (Evans and Brown 1973). In the sixties investigators reported the presence of an enzyme called glycerol-3-phosphate oxidase, or GPO, responsible for this cyanide-insensitive respiration (Grant and Sargent 1960). GPO is a constituent of the glycerophosphate cycle, which together with the NAD-dependent glycerol-3-phosphate dehydrogenase is responsible for the reoxidation of glycolytically produced NADH. This cycle was shown to be of paramount importance for the proper functioning of the glycolytic pathway of the trypanosome (Grant and Sargent 1960). Based on the reduction of tetrazolium salts into insoluble formazan deposits by glycerol 3-phosphate, several investigators have suggested that the GPO enzyme is located in the microbodies of *T. brucei* (Vickerman 1965; Ryley 1962; Bayne et al. 1969) and the name "glycerophosphate oxidase body" or GPO body was thus proposed (Bayne et al. 1969; Müller 1975; Hill 1976). What made GPO intriguing was that this oxidase is absent from humans, and can be inhibited by aromatic hydroxamic acids. For these reasons this enzyme system seemed an interesting drug target and this triggered more detailed research on this peculiar organelle. Subsequent subcellular fractionation experiments using markers enzymes for the different subcellular organelles of the trypanosome, however, quickly revealed that, contrary to the general belief, GPO was not at all located in microbodies but in the mitochondria (Opperdoes et al. 1977a, b). However, the other member of the glycerophosphate cycle, NAD-dependent glycerol-3-phosphate dehydrogenase, which in most other organisms is a cytosolic enzyme, turned out to be located within the microbodies of *T. brucei* and this novel location within the cell was held responsible for the observed association of formazan granules with these organelles (Opperdoes et al. 1977b). Such an unexpected localization inside organelles of an otherwise soluble NAD-dependent enzyme involved in the reoxidation of glycolytically produced NADH, then triggered further research in the localization of the other enzymes of the glycolytic pathway in the trypanosome and this resulted in the notion that in total, seven enzymes of the glycolytic pathway and two enzymes of glycerol metabolism all resided inside this peculiar peroxisome, henceforward called glycosomes (Opperdoes and Borst 1977).

3 Glycosomes in Other Members of the Kinetoplastida

The glycosome is not unique to the bloodstream-form stage of *T. brucei*. Also the peroxisomes of its insect stage, which utilizes amino acids rather than glucose, as well as the representatives of all other major trypanosomatid genera, such as *T. cruzi*, *Leishmania*, *Crithidia* and *Phytomonas*, were shown to have peroxisomes with glycolytic enzymes (Opperdoes et al. 1988). Glycosomes have also been demonstrated in nontrypanosomatid members of the order of the Kinetoplastida. In the fish bodonine parasite *Trypanoplasma borreli* several of the glycolytic enzymes were shown to be associated with glycosomes (Opperdoes et al. 1988), while there is indirect evidence that at least one free-living bodonid, *Parabodo caudatus*, has glycosomes as well, because a typical glycosomal bifunctional pyrimidine biosynthetic enzyme orotidine-5′-monophosphate decarboxylase/orotate phosphoribosyltransferase (OPRT/ODC, see below) carries a peroxisomal targeting signal (Makiuchi et al. 2007). Euglenids, bodonids and trypanosomatids all belong to the same group of the Euglenozoa, but so far no evidence has been found for a similar organization of the glycolytic pathway in *Euglena* spp. (Opperdoes et al. 1988). Therefore, it is likely that only the order of the Kinetoplastida have glycosomes.

4 Morpholgy and Properties of the Organelle

Glycosomes in *T. brucei* are abundant (Fig. 1a). Their reported number ranges from 80 to 340 per cell. (Opperdoes et al. 1984; Tetley and Vickerman 1991). They are remarkably homogeneous with respect to their size and appearance, with an average diameter of 0.27 μm. The organelle is surrounded by a single membrane containing phospholipids, only phosphatidyl choline, and phosphatidyl ethanolamine in a ratio of 2:1 (Opperdoes et al. 1984). There is no evidence for the presence of DNA in the organelle. Their protein content has been estimated to amount to 150 mg/ml which may explain why often crystalloid inclusions inside glycosomes are observed (Fig. 1b). In mammalian peroxisomes these crystalloids have been shown to consist mainly of urate oxidase, but such an enzyme is not present in trypanosomes.

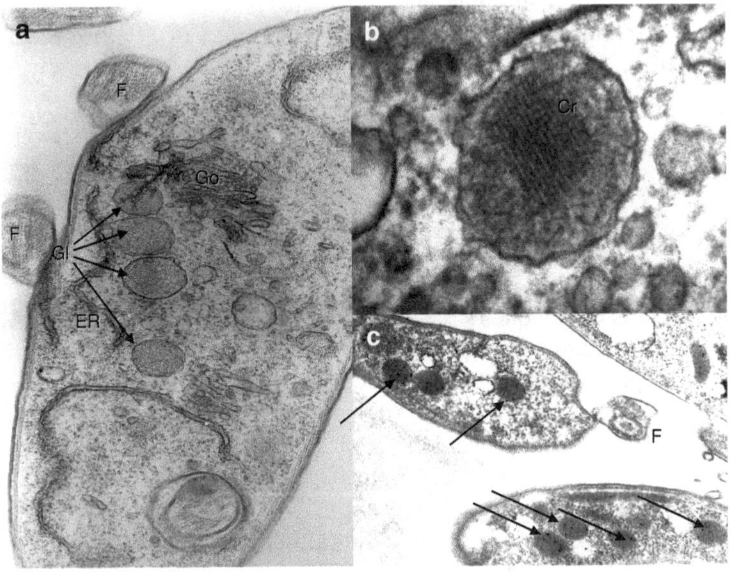

Fig. 1 Glycosomes in *T. brucei* bloodstream forms. A cluster of glycosomes is visible in close proximity of the endoplasmic reticulum and Golgi apparatus (Courtesy I. Coppens). (**b**) Crystalloid inclusion in a glycosome. (**c**) Immunolocalization of the glycolytic enzyme phosphoglycerate kinase inside glycosomes (Courtesy M. veenhuis). *ER* endoplasmic reticulum ; *F* flagellum ; *Gl* glycosome ; *Go* golgi; *Cr* crystalloid. *Arrows* point at colloidal-gold labeled antiphosphoglycerate kinase antibodies

In *T. brucei* glycosomes the crystalloid inclusions are thought to consist of either hexokinase or phosphofructokinase, or both (Misset et al. 1986). The organelles can be isolated from cellular extracts by a combination of differential and sucrose gradient centrifugation by making use of their homogeneous buoyant density in sucrose of 1.23 g/ml (Opperdoes et al. 1984).

5 Biogenesis of Glycosomes

Peroxisomes and glycosomes are related organelles. Their proteins are synthesized in the cytosol and are imported posttranslationally (Lazarow and Fujiki 1985) as fully folded proteins, or even as oligomeric complexes (McNew and Goodman 1994; Titorenko et al. 2002; Walton et al. 1995). This import is typical for peroxisomes, since in other organelles, such as mitochondria and the endoplasmic reticulum, proteins are imported posttranslationally as unfolded polypeptides. The protein import machinery of peroxisomes is relatively well conserved and many of its constituent, called peroxins (acronym PEX), which mediate the various steps in different organisms are homologous (Eckert and Erdmann 2003; Heiland and Erdmann 2005; Moyersoen et al. 2004). Most of the counterparts of the glycosomal import machinery have been identified in trypanosomes as well (Jardim et al. 2002; Moyersoen et al. 2003; Choe et al. 2003; Madrid et al. 2004; Madrid and Jardim 2005; Krazy and Michels 2006; Galland et al. 2007; Pilar et al. 2008; Cyr et al. 2008; Verplaetse et al. 2009). Import of the majority of glycosomal proteins occurs via recognition of either a C- or N-terminal import signal (Brocard and Hartig 2006; Petriv et al. 2004) present in the polypeptide by receptor proteins known as PEX5 and PEX7, respectively. However, because of the fact that proteins can enter the organelle as multimeric complexes, not all their constituents require a targeting signal for entry. This import process is also known as piggybacking. Interaction of the loaded receptors with other peroxins associated with the glycosomal membrane then leads to the interiorization of the cargo protein followed by the subsequent return of the PEX5 and PEX7 receptors to the cytosol. At present well over ten peroxins have been identified and characterized in trypanosomatids (Jardim et al. 2002 ; Moyersoen et al. 2003; Choe et al. 2003; Madrid et al. 2004; Madrid and Jardim 2005; Krazy and Michels 2006; Galland et al. 2007; Cyr et al. 2008; Pilar et al. 2008; Verplaetse et al. 2009). Biogenesis of glycosomes in trypanosomatids is extensively studied because it offers interesting possibilities as a drug target. In trypanosomes glycolysis, the major energy generating pathway, is present inside glycosomes, while in mammals it is cytosolic, all glycolytic enzymes need to be imported and the peroxins involved in this process have a poor degree of sequence similarity with their human counterparts opening the possibility for specific targeting. Moreover, the correct localization of the glycolytic enzymes is of vital importance, since misrouting has been shown to lead to impaired growth or death of the parasite (Blattner et al. 1998; Bakker et al. 2000; Helfert et al. 2001; Moyersoen et al. 2003, 2004; Krazy and Michels 2006; Galland et al. 2007).

6 Metabolic Pathways Associated with Glycosomes

6.1 Glycolysis

African trypanosomes shuttle between the mammalian bloodstream and the midgut of the tse-tse fly, requiring drastic adaptations of the parasite's metabolism to adapt to these highly different environments in these two hosts. The insect midgut is rich in amino acids, notably proline, threonine and glutamic acid and an active mitochondrial metabolism is required for the metabolism of these carbon sources. The mammalian bloodstream is rich in glucose and the bloodstream trypanosome adapts to this situation by suppressing its mitochondrial metabolism and by up-regulating the enzymes of its glycolytic pathway (Hart et al. 1984; Vertommen et al. 2008). As mentioned already most of the enzymes of glycolysis are located within the glycosomes, which are abundantly present in this life-cycle stage. Ninety percent of their protein is made up by glycolytic enzymes (Misset et al. 1986; Hart et al. 1984). The initial seven enzymes of the glycolyic pathway, from hexokinase to phosphoglycerate kinase, are present in glycosomes, while the last three enzymes of the pathway, phosphoglycerate mutase, enolase and pyruvate kinase are present in the cytosol (Fig. 2).

In bloodstream-forms of *T. brucei*, glucose is almost completely converted into pyruvate, which is excreted by the parasite (Fig. 2). Because of the complete repression of mitochondrial respiratory chain and TCA-cycle enzymes glycolysis is the only ATP supplying process. The organization of the pathway over two compartments (glycosome and cytosol) is such that net ATP consumption and synthesis within the organelle are balanced. For every glucose molecule two molecules of ATP are required to generate fructose 1,6-bisphosphate and two molecules of ATP are produced by phosphoglycerate kinase-catalyzed substrate level phosphorylation. Net ATP synthesis is achieved in the cytosol at the step catalyzed by pyruvate kinase. The redox state within glycosomes is also balanced: NADH produced in the glyceraldehyde-3-phosphate dehydrogenase reaction is reoxidized through electron transfer to oxygen by the glycerol-phosphate cycle (see above) comprising a glycosomal NAD-glycerol-3-phosphate dehydrogenase and a mitochondrial FAD-glycerol-3-phosphate oxidase system (Michels et al. 2006). Insect stages of the trypanosomatids, including those of *T. brucei,* do not produce pyruvate but instead excrete significant amounts of succinate as the result of the presence in glycosomes of additional enzymes that allow the conversion of phospho*enol*pyruvate to oxaloacetate followed by its reduction to fumatate and succinate by glycosomal NADH (Fig. 2).

6.2 Gluconeogenesis

Insect stage trypanosomes feed mainly on amino acids due to their limited access to glucose in the insect midgut. On the other hand sugar residues required by these

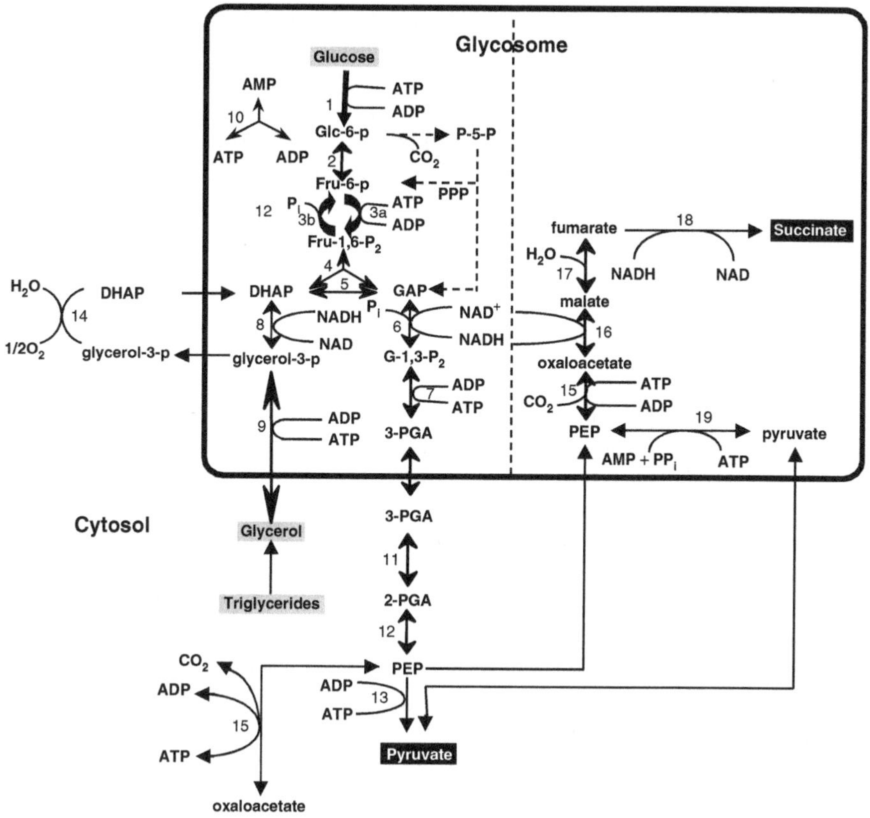

Fig. 2 Compartmentation of the glycolysis and gluconeogenesis and associated metabolic reactions in *T. brucei*. *Boxed* metabolites are nutrients (*in gray*) or end-products (*in black*) of metabolism. PPP, pentose-phosphate pathway. Enzymes: 1, hexokinase; 2, phosphoglucose isomerase; 3a, phosphofructokinase; 3b, fructose-1,6-bisphosphatase ; 4, fructosebisphosphate aldolase; 5, triosephosphate isomerase; 6, glyceraldehyde-3-phosphate dehydrogenase; 7, phosphoglycerate kinase; 8, glycerol-3-phosphate dehydrogenase; 9 glycerol kinase; 10, adenylate kinase; 11, phosphoglycerate mutase; 12, enolase; 13, pyruvate kinase; 14, glycerophosphate oxidase; 15, phosphoenolpyruvate carboxykinase; 16, malate dehydrogenase; 17, fumarate hydratase; 18, NADH-dependent fumarate reductase; 19, pyruvate phosphate dikinase. The reactions depicted in the *right-hand side* of the glycosome are operational in the insect stage only

stages for protein glycosylation, glycophospholipid-anchor formation and polysaccharides thus have to be formed de novo via the gluconeogenic pathway. The enzymes essential for gluconeogenesis, i.e., phospho*enol*pyruvate carboxykinase and fructose-1,6-bisphosphatase, are equally present inside glycosomes and thus there is the need for a tight regulation of both the glycolytic and gluconeogenic fluxes within the organelle. In most eukaryotes this regulation takes place at the level of the enzymes phosphofructokinase and fructose-1,6-bisphosphatase by the allosteric regulator fructose 2,6 bisphosphate. However, in trypanosomes these two enzymes are insensitive to this regulator. Moreover, fructose 2,6-bisphosphate is synthesized in the cytosol, rather than in the glycosomes and, contrary to the situation encountered in all other organisms, in trypanosomes fructose 2,6-bisphosphate functions as an allosteric regulator of the cytosolic enzyme pyruvate kinase (Van Schaftingen et al. 1985). How the fluxes of glycolysis and gluconeogenesis are regulated within the glycosome is still an open question (Opperdoes and Michels 2008).

6.3 *Pentosephosphate Shunt*

In addition to glycolysis and gluconeogenesis involving C6 sugars, glycosomes are also involved in the metabolism of C5 and C4 sugars such as ribose and erythrose. These sugars are formed from glucose-6-phosphate via the pentose-phosphate shunt. The enzymes glucose-6-phosphate dehydrogenase and 6-phosphogluconate dehydrogenase have been localized to glycosomes (Vertommen et al. 2008) and are involved in the formation of ribulose-5-phosphate and the production of NADPH necessary for the detoxification of ROS produced inside glycosomes. For several other enzymes of the same pathway either peroxisomal targeting signals (Opperdoes and Szikora 2006) have been identified, or the enzymes have been detected in purified glycosomal fractions (Vertommen et al. 2008). The major product of the pathway, ribose-5-phosphate, is required for the synthesis of pyrimidine and purine ribonucleotides and many enzymes involved in nucleotide metabolism have been found to be associated with these organelles, as well (see below). Also the presence of ribokinase, xylulokinase and ribulokinase in *Leishmania* glycosomes (Berriman et al. 2005; Opperdoes and Szikora 2006) suggests that in addition to glucose, glycosomes may also be capable of phophorylating C5 sugars.

6.4 *Lipid Metabolism*

6.4.1 Fatty Acid Oxidation

In lower eukaryotes, peroxisomes are the exclusive site of beta-oxidation of fatty acids. It is not clear whether glycosomes are also involved in the beta-oxidation of fatty acids as is the case in yeast and fungi. For *T. brucei* several reports have made mention of the possible presence of beta-oxidation enzymes in glycosomes (Opperdoes and Szikora 2006; Wiemer et al. 1996) but a key enzyme of the beta-oxidation pathway, acyl coA oxidase, was never demonstrated. Also in the case of *Leishmania* spp. there is evidence for the presence of beta-oxidation enzymes associated with glycosomes (Hart and Opperdoes 1984), but the functioning of such a pathway in these organisms has not been demonstrated either.

6.4.2 Ether-Lipid Synthesis

Glycosomes contain the first two enzymes of the ether-lipid biosynthetic pathway: DHAP acyltransferase and alkyl DHAP synthase (Lux et al. 2000; Zomer et al. 1999; Opperdoes 1984; Heise and Opperdoes 1997; Opperdoes and Szikora 2006; Vertommen et al. 2008). Both enzymes carry a peroxisome targeting signal and have been demonstrated in glycosomal fractions. For the third enzyme, acyl G3P: NADPH oxidoreductase, the corresponding gene could, so far, not be identified.

6.4.3 Sterol Synthesis

The isoprenoid biosynthetic pathway generates sterols, dolichols, coenzyme Q, heme and prenylated proteins. In trypanosomatids sterol biosynthesis is distributed over multiple intracellular compartments. The production of hydroxymethylglutaryl coenzyme A from acetyl coenzyme A and the generation of mevalonate occur mainly in the mitochondrion while later steps of the pathway are almost exclusively located in glycosomes (Carrero-Lérida et al. 2009). In agreement with this several of the enzymes carry a glycosome targeting signal: i.e., mevalonate kinase, 5-diphospho-mevalonate decarboxylase, isopentenyl- diphosphate-isomerase and squalene synthase/farnesyl transferase (Opperdoes and Szikora 2006). The final steps in the formation of sterols probably take place in the ER.

6.5 Nucleotide Metabolism

Like most other parasites trypanosomatids are unable to synthesize their purines de novo and thus they are totally dependent on the scavenging of nucleosides from the host and their subsequent interconversion by the enzymes of the purine salvage pathway. Interestingly most, or all, of the enzymes required for the salvage of purines and their subsequent conversion into purine nucleotides are associated with glycosomes, as a recent proteomics analysis of highly purified glycosomes has revealed (Vertommen et al. 2008) (Fig. 3). Contrary to purines, pyrimidine nucleotides are synthesized by trypanosomatids de novo, but the subcellular compartmentalization of this pathway differs dramatically from that found in other eukaryotes. The first three enzymes of the pathway are cytosolic, as in other eukaryotes, the fourth enzyme, dihydroorotate dehydrogenase, is a cytosolic enzyme, while in other eukaryotes this enzyme is associated with the mitochondrial respiratory chain, and the last two enzymes (OPRT/ODC), as well as the pyrimidine salvage enzymes uracil phosphoribosyltransferase and a uridine-specific nucleoside hydrolase, are all present inside glycosomes (Hammond et al. 1981; Gao et al. 1999; Takashima et al. 2002; Annoura et al. 2005; Vertommen et al. 2008). Thus based on their enzyme content it is predicted that glycosomes synthesize both AMP and GMP from either their free bases (i.e., adenine and guanine) or from hypoxanthine as well as from the

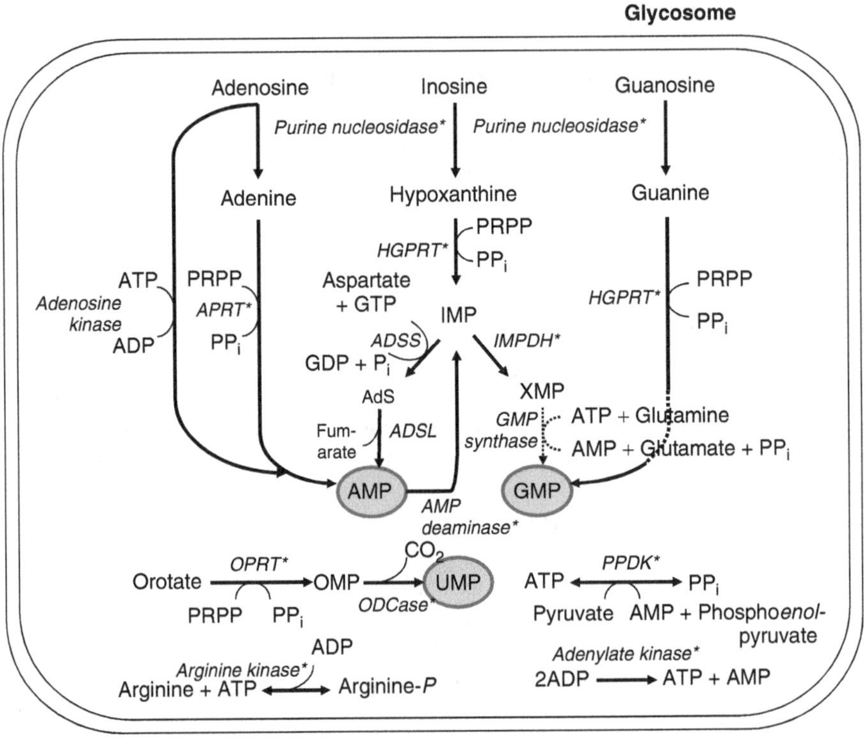

Fig. 3 Nucleotide metabolism in *T. brucei* glycosomes. An *asterisk* indicates an enzyme for which a peroxisome targeting signal (PTS) has been identified. *Thick arrows* represent enzymes which have been identified by 3D-LC–MS/MS (Vertommen et al. 2008). For the enzyme GMP synthase, there is no evidence for its presence inside glycosomes. Although a PTS-containing adenylate kinase and AMP deaminase are encoded in the *T. brucei* genome, their peptides were not detected. However, adenylate kinase has been reported to be present in glycosomes previously (Opperdoes et al. 1981). Enzyme abbreviations used are: *ADSL* adenylosuccinate lyase; *ADSS* adenylosuccinate synthetase; *APRT* adenine phosphoribosyltransferase; *HGPRT* hypoxanthine guanine phosphoribosyltransferase; *OPRT* orotate phosphoribosyltransferase; *ODC* orotidine-monophosphate decarboxylase; *IMPDH* inosine-monophosphate dehydrogenase; *INGNH* inosine adenine guanosine nucleoside hydrolase; *PPDK* pyruvate phosphate dikinase; *PRS* phosphoribosylpyrophosphate synthetase; *UPRT* uracil phosphoribosyltransferase. Modified from Vertommen et al. (2008)

corresponding nucleosides (i.e., adenosine, inosine and guanosine) following their hydrolysis by a glycosomal purine-specific nucleosidase. Glycosomal UMP can be synthesized from either uracil or uridine, or de novo from cytosolically produced orotate. To prevent nucleotide-monophosphate synthesis to come to a halt by mass action, the accumulation of pyrophosphate (PPi) inside glycosomes needs to be prevented. In other eukaryotes these reactions take place in the cytosol and PPi is hydrolyzed into inorganic phosphate by a cytosolic pyrophosphatase. However, glycosomes are devoid of any pyrophosphatase activity (Opperdoes, unpublished).

This probably explains why glycosomes do not only contain enzymes involved in the production of inorganic pyrophosphate, but also enzymes involved in its degradation (Fig. 4). The need for PPi inactivation is probably also the reason why glycosomes contain pyruvate phosphate dikinase, an enzyme that converts all glycosomal PPi to ATP at the expense of phospho*enol*pyruvate and AMP (Bringaud et al. 1998).

7 Conclusions

Glycosomes are the unique organelles of the Kinetoplastida. Morphologically they very much resemble the microbodies of other organisms and biochemical studies have shown that they belong to the same family of peroxisomes. Glycosomes use a protein-import machinery similar to that described for other peroxisomes and they also share some enzymes and metabolic pathways (such as catalase in the case of some nonpathogenic trypanosomatids and ether-lipid biosynthetic enzymes) with other peroxisomes. However, glycosomes are unique by the fact that they contain the enzymes of glycolysis and glycerol metabolism, gluconeogenesis, purine salvage and pyrimidine biosynthesis, a situation not encoutered in the other members of the peroxisome family. The unique aspects of the trypanosomatid glycosomes have attracted a lot of attention, because not only the individual enzymes of the glycolytic pathway, but also glycosome integrity, has been shown to be essential for the proper functioning and the survival of the trypanosome. Glycosomes therefore constitute excellent targets for the development of new antitrypanosome drugs.

References

Annoura T, Nara T, Makiuchi T, Hashimoto T, Aoki T (2005) The origin of dihydroorotate dehydrogenase genes of kinetoplastids, with special reference to their biological significance and adaptation to anaerobic, parasitic conditions. J Mol Evol 60:113–127

Bakker BM, Mensonides FI, Teusink B, van Hoek P, Michels PA, Westerhoff HV (2000) Compartmentation protects trypanosomes from the dangerous design of glycolysis. Proc Natl Acad Sci USA 97:2087–2092

Bayne RA, Muse KE, Roberts JF (1969) Isolation of bodies containing the cyanide-insensitive glycerophosphate oxidase of Trypanosoma equiperdum. Comp Biochem Physiol 30:1049–1054

Berriman M, Ghedin E, Hertz-Fowler C, Blandin G, Renauld H, Bartholomeu DC, Lennard NJ, Caler E, Hamlin NE, Haas B, Böhme U, Hannick L, Aslett MA, Shallom J, Marcello L, Hou L, Wickstead B, Alsmark UC, Arrowsmith C, Atkin RJ, Barron AJ, Bringaud F, Brooks K, Carrington M, Cherevach I, Chillingworth TJ, Churcher C, Clark LN, Corton CH, Cronin A, Davies RM, Doggett J, Djikeng A, Feldblyum T, Field MC, Fraser A, Goodhead I, Hance Z,

Harper D, Harris BR, Hauser H, Hostetler J, Ivens A, Jagels K, Johnson D, Johnson J, Jones K, Kerhornou AX, Koo H, Larke N, Landfear S, Larkin C, Leech V, Line A, Lord A, Macleod A, Mooney PJ, Moule S, Martin DM, Morgan GW, Mungall K, Norbertczak H, Ormond D, Pai G, Peacock CS, Peterson J, Quail MA, Rabbinowitsch E, Rajandream MA, Reitter C, Salzberg SL, Sanders M, Schobel S, Sharp S, Simmonds M, Simpson AJ, Tallon L, Turner CM, Tait A, Tivey AR, Van Aken S, Walker D, Wanless D, Wang S, White B, White O, Whitehead S, Woodward J, Wortman J, Adams MD, Embley TM, Gull K, Ullu E, Barry JD, Fairlamb AH, Opperdoes F, Barrell BG, Donelson JE, Hall N, Fraser CM, Melville SE, El-Sayed NM (2005) The genome of the African trypanosome Trypanosoma brucei. Science 309:416–422

Blattner J, Helfert S, Michels P, Clayton C (1998) Compartmentation of phosphoglycerate kinase in Trypanosoma brucei plays a critical role in parasite energy metabolism. Proc Natl Acad Sci USA 95:11596–11600

Bringaud F, Baltz D, Baltz T (1998) Functional and molecular characterization of a glycosomal PPi-dependent enzyme in trypanosomatids: pyruvate, phosphate dikinase. Proc Natl Acad Sci USA 95:7963–7968

Brocard C, Hartig A (2006) Peroxisome targeting signal 1: is it really a simple tripeptide? Biochim Biophys Acta 1763:1565–1573

Carrero-Lérida J, Pérez-Moreno G, Castillo-Acosta VM, Ruiz-Pérez LM, González-Pacanowska D (2009) Intracellular location of the early steps of the isoprenoid biosynthetic pathway in the trypanosomatids Leishmania major and Trypanosoma brucei. Int J Parasitol 39:307–314

Choe J, Moyersoen J, Roach C, Carter TL, Fan E, Michels PA, Hol WG (2003) Analysis of the sequence motifs responsible for the interactions of peroxins 14 and 5, which are involved in glycosome biogenesis in Trypanosoma brucei. Biochemistry 42:10915–10922

Cyr N, Madrid KP, Strasser R, Aurousseau M, Finn R, Ausio J, Jardim A (2008) Leishmania donovani peroxin 14 undergoes a marked conformational change following association with peroxin 5. J Biol Chem 283(46):31488–31499

Eckert JH, Erdmann R (2003) Peroxisome biogenesis. Rev Physiol Biochem Pharmacol 147:75–121

Evans DA, Brown RC (1973) m-Chlorobenzhydroxyamic acid–an inhibitor of cyanide-insensitive respiration in Trypanosoma brucei. J Protozool 20:157–160

Galland N, Demeure F, Hannaert V, Verplaetse E, Vertommen D, Van der Smissen P, Courtoy PJ, Michels PA (2007) Characterization of the role of the receptors PEX5 and PEX7 in the import of proteins into glycosomes of Trypanosoma brucei. Biochim Biophys Acta 1773:521–535

Gao G, Nara T, Nakajima-Shimada J, Aoki T (1999) Novel organization and sequences of five genes encoding all six enzymes for de novo pyrimidine biosynthesis in Trypanosoma cruzi. J Mol Biol 285:149–161

Grant PT, Sargent JR (1960) Properties of L-alpha-glycerophosphate oxidase and its role in the respiration of Trypanosoma rhodesiense. Biochem J 76:229–237

Hammond DJ, Gutteridge WE, Opperdoes FR (1981) A novel location for two enzymes of de novo pyrimidine biosynthesis in trypanosomes and Leishmania. FEBS Lett 128:27–29

Hart DT, Misset O, Edwards SW, Opperdoes FR (1984) A comparison of the glycosomes (microbodies) isolated from Trypanosoma brucei bloodstream form and cultured procyclic trypomastigotes. Mol Biochem Parasitol 12:25–35

Hart DT, Opperdoes FR (1984) The occurrence of glycosomes (microbodies) in the promastigote stage of four major Leishmania species. Mol Biochem Parasitol 13:159–172

Heiland I, Erdmann R (2005) Biogenesis of peroxisomes Topogenesis of the peroxisomal membrane and matrix proteins. FEBS J 272:2362–2372

Heise N, Opperdoes FR (1997) The dihydroxyacetonephosphate pathway for biosynthesis of ether lipids in Leishmania mexicana promastigotes. Mol Biochem Parasitol 89:61–72

Helfert S, Estévez AM, Bakker B, Michels P, Clayton C (2001) Roles of triosephosphate isomerase and aerobic metabolism in Trypanosoma brucei. Biochem J 357:117–125

Hill GC (1976) Electron transport systems in kinetoplastida. Biochim Biophys Acta 456: 149–193

Jardim A, Rager N, Liu W, Ullman B (2002) Peroxisomal targeting protein 14 (PEX14) from Leishmania donovani. Molecular, biochemical, and immunocytochemical characterization. Mol Biochem Parasitol 124:51–62

Krazy H, Michels PA (2006) Identification and characterization of three peroxins–PEX6, PEX10 and PEX12–involved in glycosome biogenesis in Trypanosoma brucei. Biochim Biophys Acta 1763:6–17

Lazarow PB, Fujiki Y (1985) Biogenesis of peroxisomes. Annu Rev Cell Biol 1:489–530

Lux H, Heise N, Klenner T, Hart D, Opperdoes FR (2000) Ether–lipid (alkyl-phospholipid) metabolism and the mechanism of action of ether–lipidanalogues in Leishmania. Mol Biochem Parasitol 111:1–14

Madrid KP, De Crescenzo G, Wang S, Jardim A (2004) Modulation of the Leishmania donovani peroxin 5 quaternary structure by peroxisomal targeting signal 1 ligands. Mol Cell Biol 24:7331–7344

Madrid KP, Jardim A (2005) Peroxin 5-peroxin 14 association in the protozoan Leishmania donovani involves a novel protein-protein interaction motif. Biochem J 391:105–114

Makiuchi T, Nara T, Annoura T, Hashimoto T, Aoki T (2007) Occurrence of multiple, independent gene fusion events for the fifth and sixth enzymes of pyrimidine biosynthesis in different eukaryotic groups. Gene 394:78–86

McNew JA, Goodman JM (1994) An oligomeric protein is imported into peroxisomes in vivo. J Cell Biol 127:1245–1257

Michels PA, Bringaud F, Herman M, Hannaert V (2006) Metabolic functions of glycosomes in trypanosomatids. Biochim Biophys Acta 1763:1463–1477

Michels PA, Moyersoen J, Krazy H, Galland N, Herman M, Hannaert V (2005) Peroxisomes, glyoxysomes and glycosomes (review). Mol Membr Biol 22:133–145

Misset O, Bos OJ, Opperdoes FR (1986) Glycolytic enzymes of Trypanosoma brucei. Simultaneous purification, intraglycosomal concentrations and physical properties. Eur J Biochem 157:441–453

Moyersoen J, Choe J, Fan E, Hol WG, Michels PA (2004) Biogenesis of peroxisomes and glycosomes: trypanosomatid glycosome assembly is a promising new drug target. FEMS Microbiol Rev 28:603–643

Moyersoen J, Choe J, Kumar A, Voncken FG, Hol WG, Michels PA (2003) Characterization of Trypanosoma brucei PEX14 and its role in the import of glycosomal matrix proteins. Eur J Biochem 270:2059–2067

Müller M (1975) Biochemistry of protozoan microbodies: peroxisomes, alpha-glycerophosphate oxidase bodies, hydrogenosomes. Annu Rev Microbiol 29:467–483

Opperdoes FR, Michels PAM (2008) The metabolic repertoire of Leishmania and implications for drug discovery. In: Myler P, Fasel N (eds) Leishmania after the genome. Academic, Norfolk, pp 123–158

Opperdoes FR (1984) Localization of the initial steps in alkoxyphospholipid biosynthesis in glycosomes (microbodies) of Trypanosoma brucei. FEBS Lett 169:35–39

Opperdoes FR, Baudhuin P, Coppens I, De Roe C, Edwards SW, Weijers PJ, Misset O (1984) Purification, morphometric analysis, and characterization of the glycosomes (microbodies) of the protozoan hemoflagellate Trypanosoma brucei. J Cell Biol 98:1178–1184

Opperdoes FR, Borst P (1977) Localization of nine glycolytic enzymes in a microbody-like organelle in Trypanosoma brucei: the glycosome. FEBS Lett 80:360–364

Opperdoes FR, Borst P, Bakker S, Leene W (1977a) Localization of glycerol-3-phosphate oxidase in the mitochondrion and particulate NAD+-linked glycerol-3-phosphate dehydrogenase in the microbodies of the bloodstream form to Trypanosoma brucei. Eur J Biochem 76:29–39

Opperdoes FR, Borst P, Spits H (1977b) Particle-bound enzymes in the bloodstream form of Trypanosoma brucei. Eur J Biochem 76:21–28

Opperdoes FR, Markos A, Steiger RF (1981) Localization of malate dehydrogenase, adenylate kinase and glycolytic enzymes in glycosomes and the threonine pathway in the mitochondrion of cultured procyclic trypomastigotes of Trypanosoma brucei. Mol Biochem Parasitol 4:291–309

Opperdoes FR, Nohynkova E, Van Schaftingen E, Lambeir AM, Veenhuis M, Van Roy J (1988) Demonstration of glycosomes (microbodies) in the Bodonid flagellate Trypanoplasma borelli (Protozoa, Kinetoplastida). Mol Biochem Parasitol 30:155–163
Opperdoes FR, Szikora JP (2006) In silico prediction of the glycosomal enzymes of Leishmania major and trypanosomes. Mol Biochem Parasitol 147:193–206
Petriv OI, Tang L, Titorenko VI, Rachubinski RA (2004) A new definition for the consensus sequence of the peroxisome targeting signal type 2. J Mol Biol 341:119–134
Pilar AV, Madrid KP, Jardim A (2008) Interaction of Leishmania PTS2 receptor peroxin 7 with the glycosomal protein import machinery. Mol Biochem Parasitol 158(1):72–81
Ryley JF (1962) Studies on the metabolism of the protozoa. 9. Comparative metabolism of blood-stream and culture forms of Trypanosoma rhodesiense. Biochem J 85:211–223
Takashima E, Inaoka DK, Osanai A, Nara T, Odaka M, Aoki T, Inaka K, Harada S, Kita K (2002) Characterization of the dihydroorotate dehydrogenase as a soluble fumarate reductase in Trypanosoma cruzi. Mol Biochem Parasitol 122:189–200
Tetley L, Vickerman K (1991) The glycosomes of trypanosomes: number and distribution as revealed by electron spectroscopic imaging and 3-D reconstruction. J Microsc 162:83–90
Titorenko VI, Nicaud JM, Wang H, Chan H, Rachubinski RA (2002) Acyl-CoA oxidase is imported as a heteropentameric, cofactor-containing complex into peroxisomes of Yarrowia lipolytica. J Cell Biol 156:481–494
van Schaftingen E, Opperdoes FR, Hers HG (1985) Stimulation of Trypanosoma brucei pyruvate kinase by fructose 2, 6-bisphosphate. Eur J Biochem 153:403–406
Verplaetse E, Rigden DJ, Michels PA (2009) Identification, characterization and essentiality of the unusual peroxin 13 from Trypanosoma brucei. Biochim Biophys Acta 1793:516–527
Vertommen D, Van Roy J, Szikora JP, Rider MH, Michels PA, Opperdoes FR (2008) Differential expression of glycosomal and mitochondrial proteins in the two major life-cycle stages of Trypanosoma brucei. Mol Biochem Parasitol 158:189–201
Vickerman K (1965) Polymorphism and mitochondrial activity in sleeping sickness trypanosomes. Nature 208(5012):762–766
Von Brandt T (1973) Biochemistry of parasites, 2nd edn. Academic, New York
Walton PA, Hill PE, Subramani S (1995) Import of stably folded proteins into peroxisomes. Mol Biol Cell 6:675–683
Wiemer EA, IJlst L, van Roy J, Wanders RJ, Opperdoes FR (1996) Identification of 2-enoyl coenzyme A hydratase and NADP(+)-dependent 3-hydroxyacyl-CoA dehydrogenase activity in glycosomes of procyclic Trypanosoma brucei. Mol Biochem Parasitol 82:107–111
Zomer AW, Michels PA, Opperdoes FR (1999) Molecular characterisation of Trypanosoma brucei alkyl dihydroxyacetone-phosphate synthase. Mol Biochem Parasitol 104:55–66

Calcium Homeostasis and Acidocalcisomes in *Trypanosoma cruzi*

Paul Ulrich, Roxana Cintrón, and Roberto Docampo

Contents

Abstract Calcium ion (Ca^{2+}) is an important second messenger in *Trypanosoma cruzi* and is essential for invasion of host cells by this parasite. A number of transporters and channels in the plasma membrane, endoplasmic reticulum, and mitochondria regulate cytosolic calcium concentration. Additionally, the *T. cruzi* genome contains a wide variety of signaling and regulatory proteins that bind calcium as well as many putative calcium-binding proteins that await further characterization. In *T. cruzi*, acidic organelles known as acidocalcisomes are the primary reservoir of intracellular calcium and mediate polyphosphate metabolism, osmoregulation, and calcium and pH homeostasis.

P. Ulrich, R. Cintrón, and R. Docampo (✉)
Center for Tropical and Emerging Global Diseases and Department of Cellular Biology, University of Georgia, Athens, GA 30602, USA
e-mail: rdocampo@uga.edu

W. de Souza (ed.), *Structures and Organelles in Pathogenic Protists*,
Microbiology Monographs 17, DOI 10.1007/978-3-642-12863-9_13,
© Springer-Verlag Berlin Heidelberg 2010

Abbreviations

AQP	Aquaporin
cADPR	Cyclic ADP ribose
Ca^{2+}	Calcium ion
$[Ca^{2+}]_i$	Cytosolic Ca^{2+} concentration
CaM	Calmodulin
CaMK	Ca^{2+}/calmodulin dependent kinase
CICR	Calcium induced calcium release
Cn	Calcineurin
FCaBP	Flagellar calcium binding protein
$InsP_3$	Inositol 1,4,5-trisphosphate
$InsP_3R$	$InsP_3$ receptor
NAADP	Nicotinic acid adenine dinucleotide phosphate
PIP_2	Phosphatidylinositol 4,5-bisphosphate
PI-PLC	Phosphatidylinositol phospholipase C
PMCA	Plasma membrane Ca^{2+}-ATPase
poly P	Polyphosphate
RyR	Ryanodine receptor
SERCA	Sarcoplasmic-endoplasmic reticulum Ca^{2+}-ATPase
V-H^+-ATPase	Vacuolar proton ATPase
V-H^+-PPase	Vacuolar proton pyrophosphatase
VTC	Vacuolar transporter chaperone

1 Introduction

All cells use calcium as a second messenger to control cellular functions. Cells maintain free cytosolic Ca^{2+} concentration $[Ca^{2+}]_i$ at very low levels ($\sim 10^{-7}$ M) relative to the concentration in the extracellular medium ($\sim 10^{-3}$ M). This strong ion gradient allows cells to respond rapidly to stimuli by coupling changes in Ca^{2+} with the activity of Ca^{2+}-dependent and Ca^{2+}-controlled proteins. Free calcium in cells represents only a small fraction of total cellular calcium because the bulk of this ion is sequestered inside organelles or bound to proteins, polyphosphate, membranes, or other cellular constituents (Irvine 1986).

Eukaryotic cells control intracellular Ca^{2+} with a variety of Ca^{2+} transporting systems, several of which have been demonstrated in *Trypanosoma cruzi*. The plasma membrane regulates Ca^{2+} influx through channels and actively extrudes Ca^{2+} via a Ca^{2+}/Na^+ exchanger and a Ca^{2+}-ATPase (PMCA) (Carafoli 1987). The endoplasmic reticulum and the nuclear membrane also possess a Ca^{2+}-ATPase (SERCA) for influx and a channel for efflux. In contrast to the plasma membrane and the endoplasmic reticulum, mitochondria do not possess Ca^{2+}-ATPases. The cation moves into mitochondria down an electrochemical gradient through a uniport mechanism whose molecular nature remains unidentified. Efflux from mitochondria appears driven by electroneutral exchange of matrix Ca^{2+} with external Na^+ or H^+ (Nicholls et al. 1984).

In addition to these conserved homeostatic mechanisms, calcium plays specific roles in the interactions between *T. cruzi* and its hosts. Calcium is critical for *T. cruzi* invasion of host cells.

2 Cytosolic Ca^{2+} Concentration and the Role of the Plasma Membrane

Intracellular Ca^{2+} in epimastigotes, amastigotes, and trypomastigotes is 50, 20, and 20 nM, respectively, when measured with fura 2-loaded cells in the absence of extracellular Ca^{2+} (presence of excess Ca^{2+} chelator EGTA) (Docampo et al. 1995). These concentrations are in the range observed in many studies with eukaryotic cells (Grynkiewicz et al. 1985) (Fig. 1). However, relatively little is known about the proteins responsible for calcium movement across the plasma membrane of *T. cruzi*.

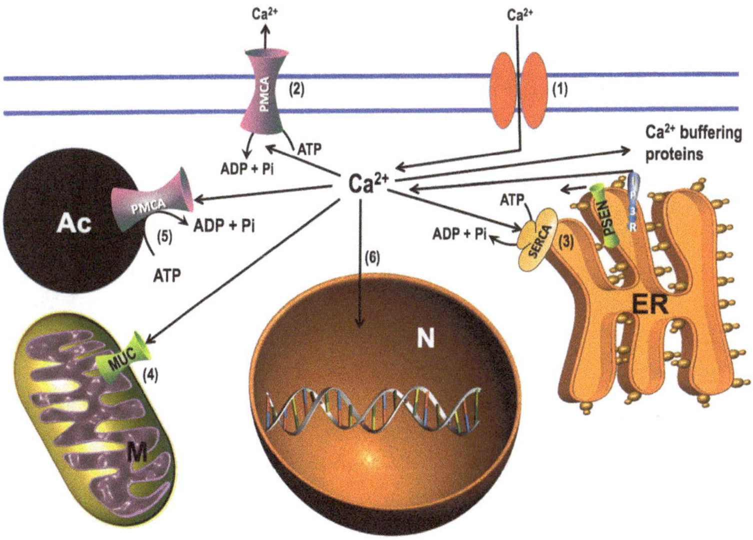

Fig. 1 *Schematic representation of the distribution of Ca^{2+} in T. cruzi.* Ca^{2+} entry is probably through a Ca^{2+} channel (1). Once inside the cell, Ca^{2+} can be translocated back to the extracellular environment by the action of the PMCA Ca^{2+}-ATPase (2). In addition, Ca^{2+} will interact with Ca^{2+}-binding proteins or become sequestered by the endoplasmic reticulum through a SERCA Ca^{2+}-ATPase (3), by the mitochondrion through a uniporter (4), by acidocalcisomes through a PMCA Ca^{2+}-ATPase (5), or by the nucleus through the nuclear pores (6). Further details are discussed in the text. *ER* endoplasmic reticulum; *M* mitochondrion; *N* nucleus; *Ac* acidocalcisome; *SERCA* sarcoplasmic-endoplasmic reticulum Ca^{2+}-ATPase; *PMCA* plasma membrane Ca^{2+}-ATPase; *PSEN* presenilin; *IP3R* inositol 1,4,5-trisphosphate receptor; *MUC* mitochondrial uniporter channel. Drawing adapted from a SABiosciences pathway map

Table 1 Calcium channels and pumps identified in *Trypanosoma cruzi* (CL strain)

Type	GenBank ID Number	GeneDB ID number	Expression
PMCA-Ca^{2+}-ATPase	EAN95492.1[a]	Tc00.1047053508543.90[b]	Yes (Tca1)[a]
	EAN94362.1	Tc00.1047053509647.150	No
SERCA-Ca^{2+}-ATPase	EAN92377.1[c]	Tc00.1047053509770.70[d]	Yes (TcSCA1)[a]
	EAN96035.1	Tc00.1047053506241.70[e]	No
TRP channel	EAN97848.1	Tc00.1047053504105.130 (H)	No
$InsP_3R$-type channel	EAN89926.1	Tc00.1047053509461.90[f]	No
Presenilin	EAN98414.1	Tc00.1047053508277.50[g]	No

H homozygous
[a]Similar to AAC38969.1 from Y strain (Lu et al. 1998)
[b]Allele of Tc00.1047053506401.170 (EAN99420.1)
[c]Similar to AAD08694.1 from Y strain (Furuya et al. 2001)
[d]Allele of Tc00.1047053503563.10 (EAN83220.1)
[e]Allele of Tc00.1047053510769.120 (EAN95591.1)
[f]Allele of Tc00.1047053510509.9 (EAN84224.1)
[g]Allele of Tc00.1047053503543.10 (EAN81606.1)

T. cruzi apparently lacks some of the proteins that control influx of Ca^{2+} across the plasma membrane in higher eukaryotes. There is no evidence of receptor-operated (Ca^{2+} influx after receptor stimulation) or store-operated Ca^{2+} channels (Ca^{2+} influx initiated by depletion of intracellular stores) (Cahalan 2009) in *T. cruzi*. There are no orthologs in the *T. cruzi* genome to the proteins STIM (the ER Ca^{2+} sensor) and ORAI (the Ca^{2+} channel forming subunit), which are involved in store operated Ca^{2+} entry in higher eukaryotes (Cahalan 2009). A putative transient receptor potential (TRP) calcium channel has been identified in the *T. cruzi* genome (Table 1). Demonstration of this gene product as a functional calcium channel awaits direct analysis by electrophysiology.

Eukaryotic cells typically export Ca^{2+} by the action of an Na^+/Ca^{2+} exchanger and a Ca^{2+}-ATPase (PMCA). There are no reports of the presence of Na^+/Ca^{2+} exchangers in early eukaryotes (Pozos et al. 1996). In contrast, a PMCA-type Ca^{2+}-ATPase (Tca1) has been characterized and localized in the plasma membrane and acidocalcisomes of *T. cruzi* (Lu et al. 1998) (Table 1). Benaim et al. (1991) reported evidence for calmodulin (CaM) stimulation of this pump although Tca1 appears to lack a typical CaM-binding domain (Lu et al. 1998). This suggests that Tca1 contains a noncanonical CaM-binding domain. A gene coding for another putative PMCA is also in the *T. cruzi* genome (Table 1). The predicted amino acid sequence has 32% identity to Tca1 (Lu et al. 1998). It is possible that expression of these genes is stage-specific or that the proteins have different localizations as is the case with *T. brucei* (Luo et al. 2004).

3 Ca^{2+}-Binding Proteins

Inside the cell, Ca^{2+} interacts with soluble Ca^{2+}-binding proteins or is sequestered within intracellular organelles in complexes with storage proteins or polyphosphate. The *T. cruzi* genome project uncovered a wide variety of Ca^{2+}-binding

Table 2 Calcium-binding proteins annotated in *T. cruzi* (CL strain)

Type	GenBank ID	GeneDB ID number	Expression
Calreticulin	EAN90720.1[*]	Tc00.1047053509011.40[a]	Yes[*]
Flagellar calcium-binding protein			
	EAN95149.1	Tc00.1047053507491.162	Yes[#]
	EAN83725.1	Tc00.1047053507891.56	Yes[#]
	EAN95148.1	Tc00.1047053507491.151	Yes[#]
	EAN83722.1	Tc00.1047053507891.29	Yes[#]
	EAN83723.1	Tc00.1047053507891.38	Yes[#]
	EAN83724.1	Tc00.1047053507891.47	Yes[#]
	EAN83057.1	Tc00.1047053506749.20	Yes[#]
Calmodulin (CaM)	EAN86242.1	Tc00.1047053507483.39[*b]	Yes[*]
	EAN83393.1	Tc00.1047053506391.20[*c]	Yes[*]
Proteins with similarities to CaM (annotated as calmodulin)			
	EAN86239.1	Tc00.1047053507483.50 (4 EF)[j,d]	No
	EAN93967.1	Tc00.1047053511233.80 (4 EF)[e]	No
	EAN99779.1	Tc00.1047053508461.380 (4 EF) (H)	No
	EAN81822.1	Tc00.1047053504075.3 (no EF)	No
	EAN93486.1	Tc00.1047053509683.50 (2 EF)[f]	No
	EAN90166.1	Tc00.1047053506933.89 (2 EF) (H)	No
	EAN87143.1	Tc00.1047053508951.50 (2 EF)[g]	No
	EAN84699.1	Tc00.1047053509353.60 (2 EF)	No
	EAN84433.1	Tc00.1047053511729.9 (5 EF)[h]	No
Calcium-binding proteins	EAN86453.1	Tc00.1047053509391.30 (3 EF)	No
	EAN86963.1	Tc00.1047053507925.60[i]	No
	EAN86455.1	Tc00.1047053509391.10	No
	EAN86454.1	Tc00.1047053509391.20	No

H homozygous; *NA* not available
[*]Similar to AAD22175.1 from *T. cruzi* Tulahuen 2 strain (Labriola et al. 1999)
[#]First gene cloned in *T. cruzi* from Y strain (Gonzalez et al. 1985)
[*]Sequences identical to CAA36316.1 from *T. cruzi* CL strain (Chung and Swindle 1990)
[a]Allele of Tc00.1047053510685.10 (EAN82340.1)
[b]Allele of Tc00.1047053506391.10 (EAN83392.1)
[c]Allele of Tc00.1047053507483.30 (EAN86238.1)
[d]Allele of Tc00.1047053506389.79 (EAN86831.1)
[e]Allele of Tc00.1047053506963.90 (EAN89727.1)
[f]Allele of Tc00.1047053508731.30 (EAN84774.1)
[g]Allele of Tc00.1047053510121.50 (EAN94615.1)
[h]Allele of Tc00.1047053506835.60 (EAN91696.1)
[i]Allele of Tc00.1047053509059.30 (EAN89046.1)
[j]Number of EF hand domains is between parentheses

proteins (Table 2), many of which are uncharacterized and share little or no homology with nonkinetoplastid proteins. Among the Ca^{2+}-binding proteins of *T. cruzi* are calmodulin (CaM), a cytosolic Ca^{2+} receptor, and calreticulin, a Ca^{2+} storage protein found within the endoplasmic reticulum. *T. cruzi* CaM (TcCaM) has been purified from epimastigotes (Tellez-Iñón et al. 1985; Benaim et al. 1991) and can stimulate the PMCA Ca^{2+} ATPase (Benaim et al. 1991) and cyclic AMP phosphodiesterase (Tellez-Iñón et al. 1985). TcCaM has four calcium-binding sites (EF-hand domains), is 92% identical to human CaM (Chung and Swindle 1990), and is present in several copies in the genome (Table 2). Rohloff et al. (2004) used antibodies against human CaM to localize TcCaM to the spongiome of the contractile vacuole complex of *T. cruzi* epimastigotes. The CaM inhibitor trifluoperazine inhibits Ca^{2+} release from the endoplasmic reticulum and mitochondria while calmidazolium releases Ca^{2+} from both compartments (Vercesi et al. 1991b). However, these compounds also inhibited respiration and collapsed the mitochondrial membrane potential of *T. cruzi*, indicating that these inhibitors also drive nonspecific effects unrelated to CaM (Vercesi et al. 1991b). A number of genes annotated as calmodulins are present in the *T. cruzi* genome (proteins with similarity to CaM, Table 2). EF-hand domains are lacking in some of these putative calmodulins, and others have 2–5 of these calcium-binding domains. The specific roles of each protein are unclear, but it is likely that they bind calcium with different affinities and modulate regulatory activity. *T. cruzi* calreticulin is involved in quality control of glycoprotein synthesis (Conte et al. 2003), and is localized in the endoplasmic reticulum (Furuya et al. 2001) but no studies have been done in *T. cruzi* concerning its Ca^{2+} storage properties. A number of other hypothetical proteins with calcium-binding domains have also been found (Table 3).

An interesting Ca^{2+}-binding protein in *T. cruzi* is the flagellar Ca^{2+}-binding protein (FCaBP; Engman et al. 1989). Multiple copies of the gene encoding this protein are present in the genome (Table 2). This protein is *N*-myristoylated and palmitoylated and associates with the flagellar membrane in a calcium-dependent manner reminiscent of the recoverin family of calcium-myristoyl switch proteins (Godsel and Engman 1999). The function of this protein remains unknown although its gene was the first cloned from *T. cruzi* (Gonzalez et al. 1985). Genes encoding other calcium-binding proteins have also been found in the genome of *T. cruzi* but have not been studied in detail (Table 2).

4 Ca^{2+} and Cell Signaling

Ca^{2+} can regulate and interact with a number of signaling pathways. Two main Ca^{2+}-sensitive proteins that decode Ca^{2+} signals are protein kinase C (PKC) and Ca^{2+}/calmodulin-dependent kinase (CaMK). A PKC was characterized biochemically in *T. cruzi* epimastigotes (Gomez et al. 1989, 1999). This enzyme requires phosphatidylserine and Ca^{2+} for activity and is stimulated by diacylglycerol. However, although a group of AGC kinases was identified in the *T. cruzi* genome, it was not possible to assign them to the PKC family by sequence alone (Parsons et al. 2005). A Ca^{2+}/CaM kinase activity was also detected in *T. cruzi* (Ogueta et al. 1994), and the soluble enzyme was partially purified and characterized (Ogueta et al. 1996, 1998). Several genes encoding putative Ca^{2+}/CaM regulated kinases have been identified in the genome of *T. cruzi* (Parsons et al. 2005), but no biochemical studies have been reported with the recombinant proteins (Table 4). Ca^{2+} also activates ion channels and genes encoding Ca^{2+}-activated K^+ channel are present in the genome of *T. cruzi* (Table 4).

Table 3 Hypothetical proteins with calcium ion binding sites as determined by predicted or annotated GO function. Predicted and annotated GO functions were extracted from the *Trypanosoma cruzi* database (beta.tritrypdb.org) on July 7, 2009

Gene	Predicted or annotated GO function
Tc00.1047053503455.20	Calcium ion binding
Tc00.1047053504427.120	Calcium ion binding
Tc00.1047053506435.420	Calcium ion binding
Tc00.1047053507165.30	Calcium ion binding
Tc00.1047053509245.20	Calcium ion binding
Tc00.1047053509611.170	Calcium ion binding
Tc00.1047053510525.130	Calcium ion binding
Tc00.1047053510741.140	Calcium ion binding
Tc00.1047053510879.190	Calcium ion binding
Tc00.1047053511391.210	Calcium ion binding
Tc00.1047053510769.70	Calcium ion binding
Tc00.1047053511131.40	Calcium ion binding
Tc00.1047053503967.4	Calcium ion binding
Tc00.1047053506247.130	Calcium ion binding
Tc00.1047053506577.160	Calcium ion binding
Tc00.1047053506607.10	Calcium ion binding
Tc00.1047053506957.120	Calcium ion binding
Tc00.1047053507083.80	Calcium ion binding
Tc00.1047053507625.130	Calcium ion binding
Tc00.1047053508231.180	Calcium ion binding
Tc00.1047053508277.320	Calcium ion binding
Tc00.1047053508815.80	Calcium ion binding
Tc00.1047053508851.90	Calcium ion binding
Tc00.1047053509153.80	Calcium ion binding
Tc00.1047053509453.79	Calcium ion binding
Tc00.1047053509937.190	calcium ion binding
Tc00.1047053509997.40	Calcium ion binding
Tc00.1047053510101.380	Calcium ion binding
Tc00.1047053510329.310	Calcium ion binding
Tc00.1047053510741.30	Calcium ion binding
Tc00.1047053510741.90	Calcium ion binding
Tc00.1047053510797.30	Calcium ion binding
Tc00.1047053511127.20	Calcium ion binding
Tc00.1047053511733.40	Calcium ion binding
Tc00.1047053511809.110	Calcium ion binding
Tc00.1047053511811.20	Calcium ion binding
Tc00.1047053511867.200	Calcium ion binding
Tc00.1047053510323.80	Calcium ion binding, acyltransferase activity
Tc00.1047053506753.70	Calcium ion binding, cAMP-dependent protein kinase regulator
Tc00.1047053508461.210	Calcium ion binding, cAMP-dependent protein kinase regulator
Tc00.1047053506247.250	DNA binding, adenylate kinase activity, calcium ion binding
Tc00.1047053508543.60	Hydrolase activity, calcium ion binding
Tc00.1047053505071.40	Phosphoinositide binding, calcium ion binding, protein binding
Tc00.1047053508479.180	Protein binding, calcium ion binding
Tc00.1047053504035.130	Zinc ion binding, protein binding, calcium ion binding
Tc00.1047053504021.149	Calcium ion binding
Tc00.1047053508799.260	Calcium ion binding
Tc00.1047053509647.190	Calcium ion binding

Table 4 Proteins potentially modulated by Ca^{2+} identified in *T. cruzi* at the molecular level

Type	GenBank Number	GeneDB ID Number	Expression
Ca^{2+}/CaM dependent PK	EAN89956.1	Tc00.1047053508601.90[a]	No
	EAN94435.1	Tc00.1047053506513.50[b]	No
	EAN88177.1	Tc00.1047053506465.40[c]	No
	EAN89603.1	Tc00.1047053503925.30[d]	No
	EAN90816.1	Tc00.1047053506493.50[e]	No
	EAN98183.1	Tc00.1047053506679.80 (H)	No
	EAN86819.1	Tc00.1047053503635.10[f]	No
	EAN91788.1	Tc00.1047053509213.160[g]	No
	EAN88257.1	Tc00.1047053510525.10[h]	No
Ca^{2+} activated K^+ channel	EAN98530.1	Tc00.1047053511585.220[i]	No
	EAO00090.1	Tc00.1047053506529.150[j]	No
	EAN96201.1	Tc00.1047053511245.30[k]	No
PI-PLC	EAN96260.1	Tc00.1047053504149.160	Yes[q]
Calcineurin B subunit	EAN90858	Tc00.1047053510519.60[l]	Yes[r]
Caltractin	EAN90592.1	Tc00.1047053510181.150[m]	No
Centrin	EAN89811.1	Tc00.1047053509161.40 (H)	No
	EAN99948.1	Tc00.1047053506559.380 (H)	No
	EAN91314.1	Tc00.1047053508323.60[n]	No
	EAN91315.1	Tc00.1047053508323.70[o]	No
	EAN84631.1	Tc00.1047053508727.18[p]	No

H homozygous
[a]Allele of Tc00.1047053511801.14 (NA)
[b]Allele of Tc00.1047053508919.70 (EAN93275.1)
[c]Allele of Tc00.1047053507317.60 (EAN86500.1)
[d]Allele of Tc00.1047053510347.60 (EAN85912.1)
[e]Allele of Tc00.1047053510121.130 (EAN94623.1)
[f]Allele of Tc00.1047053511001.60 (EAN97774.1)
[g]Allele of Tc00.1047053510257.130 (EAN89913.1)
[h]Allele of Tc00.1047053511817.80 (EAN97504.1)
[i]Allele of Tc00.1047053510155.210 (EAN98275.1)
[j]Allele of Tc00.1047053510885.60 (EAN94300.1)
[k]Allele of Tc00.1047053506661.130 (EAN92131.1)
[l]Allele of Tc00.1047053506869.50 (1EAN86693.1)
[m]Allele of Tc00.1047053503431.10 (EAN87386.1)
[n]Allele of Tc00.1047053511825.40 (EAN94634.1)
[o]Allele of Tc00.1047053511825.50 (EAN94635.1)
[p]Allele of Tc00.1047053503797.20 (NA)
[q]Similar to AAD12583.1from *T. cruzi* Y strain (Furuya et al. 2000)
[r]Identical to CAI48025

An adenylyl cyclase (D'Angelo et al. 2002) (AAC61849.1) and a cyclic AMP phosphodiesterase (Tellez-Iñón et al. 1985) from *T. cruzi* are also stimulated by Ca^{2+}. Additionally, the phosphoinositide phospholipase C from *T. cruzi* (TcPI-PLC) appears to be active at low Ca^{2+} levels (Furuya et al. 2000).

A number of proteins that are related to Ca^{2+}-dependent, cytosolic, cysteine peptidases (calpains) are present in the genome of *T. cruzi*. However, these calpain-like proteins probably cannot bind Ca^{2+} because they lack EF-hand motifs observed in the domain IV of conventional calpains (Ersfeld et al. 2005).

Complexes of Ca^{2+}/CaM control the activity of calcineurin, a heterotrimeric protein formed by a catalytic subunit (calcineurin A, CnA) and a regulatory subunit (calcineurin B, CnB). *T. cruzi* CnA lacks CaM and autoinhibitory domains, and CnB has only two out of the four EF-hand domains characteristic of other calcineurin B proteins (Moreno et al. 2007). However, *T. cruzi* calcineurin activity requires Ca^{2+}. Regulation of calcineurin activity by Ca^{2+} likely occurs via CnB, which can stimulate CnA by binding Ca^{2+} (Moreno et al. 2007; Araya et al. 2008). This activation seems important for invasion of host cells as treatment of trypomastigotes with Cn inhibitors cyclosporin or cypermethrin or reducing CnB expression with phosphorotioate oligonucleotides strongly inhibited entry of host HeLa cells (Araya et al. 2008).

Finally, centrins are Ca^{2+}-binding proteins involved in a number of cellular processes, such as DNA repair, mRNA export, organelle duplication, and signal transduction (Shi et al. 2008). Several centrins and a related caltractin have been identified in the genome of *T. cruzi* (Table 4) but little is known about their function. Centrins in *T. brucei* are involved in coordination of nuclear and cell division (Shi et al. 2008) and organelle segregation (Selvapandiyan et al. 2007).

5 Calcium Storage Compartments

5.1 Endoplasmic Reticulum

Early evidence of a SERCA-type Ca^{2+}-ATPase in *T. cruzi* was based on low capacity, high affinity, orthovanadate-sensitive Ca^{2+} uptake in permeabilized epimastigotes, and the ability of these cells to buffer $[Ca^{2+}]_i$ in the range of 0.05–0.1 μM (Vercesi et al. 1991b) - features characteristic of SERCA Ca^{2+}-ATPases of animals cells (Carafoli and Brini 2000).

Furuya et al. (2001) provided molecular evidence for the presence of this pump (TcSCA) in *T. cruzi*. The gene encoding this pump complemented yeast deficient in Ca^{2+} pumps. It also restored growth of the same yeast on medium containing Mn^{2+}, suggesting a role in Mn^{2+} uptake. The enzyme localizes to the endoplasmic reticulum (ER) at all stages of *T. cruzi* and forms a 110 kDa phosphoprotein in the presence of [γ-^{32}P]ATP and Ca^{2+}. Phosphorylation of TcSCA is sensitive to cyclopiazonic acid and hydroxylamine but unaffected by thapsigargin, supporting observations that activity of the pump is thapsigargin-insensitive (Furuya et al. 2001). A gene coding for another putative SERCA (Table 1) is also in the *T. cruzi* genome (Table 1). The predicted amino acid sequence has 30% identity to TcSCA1 (Furuya et al. 2001).

Ca^{2+} release from the ER of eukaryotic cells is mediated by ryanodine (RyR) or inositol 1,4,5-trisphosphate ($InsP_3R$) channels. RyR are activated by a rise in $[Ca^{2+}]_i$ (Ca^{2+}-induced Ca^{2+} release, CICR). In addition, there are RyR-like channels activated by cyclic ADP-ribose (cADPR), sphingosine, and nicotinic acid adenine dinucleotide phosphate (NAADP) (Cahalan 2009). *T. cruzi* phosphoinositide-specific phospholipase C (TcPI-PLC, Tc00.1047053504149.160) - the enzyme that generates the second messengers $InsP_3$ and diacylglycerol – was characterized by Furuya et al. (2000). The enzyme was located in the plasma membrane of amastigotes and contains *N*-myristoylation and palmitoylation consensus sequences that have not been described in any other PI-PLC. The enzyme is myristoylated and palmitoylated in vivo (Furuya et al. 2000), and this lipid modification is important for its plasma membrane localization (Okura et al. 2005). The second messenger $InsP_3$ and its precursor (phosphatidylinositol 4,5-bisphosphate, or PIP_2) have been detected in epimastigotes (Docampo and Pignataro 1991), amastigotes (Moreno et al. 1992), and trypomastigotes (Docampo et al. 1993) although experiments examining Ca^{2+} release from intracellular stores using $InsP_3$ have been unsuccessful (Moreno et al. 1992; Docampo et al. 1993).

In recent years the involvement of the intramembrane aspartyl protease presenilin in Ca^{2+} homeostasis has been described (Hass et al. 2009; Green and LaFerla 2008). The presenilins were identified in 1995 as multimembrane-spanning proteins localized predominantly in the ER. They were postulated to be involved in the pathogenesis of Alzheimer's disease when it was found that they form the catalytic core of the γ-secretase complex, which releases amyloid β ($A\beta$) from the amyloid precursor protein (APP) (Hass et al. 2009). Presenilins have also been suggested to carry out a wide range of other functions. For example, they may interact with the SERCA to modulate Ca^{2+} influx into the ER, participate in extrusion of Ca^{2+} from the ER via ryanodine and $InsP_3$ receptors, or affect endogenous leak channels from the ER (Green and LaFerla 2008). Presenilins are present in the genome of *T. cruzi* (Table 1), but their function as Ca^{2+} leak channels in the ER or as modulators of SERCA pumps or calcium channels have not been studied.

We have identified a putative $InsP_3$/ryanodine receptor ($TcInsP_3R$) among the proteins annotated as "hypothetical" in the *T. cruzi* genome. A homolog is also present in *T. brucei*. $TcInsP_3R$ (Table 1) possesses a series of conserved domains including putative $InsP_3$-binding, ATP-binding, ryanodine homology (RIH, Ponting 2000), and transmembrane domains (Fig. 2). While the transmembrane domain does contain a motif for a Ca^{2+}-specific selectivity filter (GGVGD), residues important for $InsP_3$ binding in mouse $InsP_3$ receptors (Yoshikawa et al. 1996) are not well conserved in the predicted $InsP_3$-binding domain of $TcIP_3R$. It is possible that this protein binds another second messenger with greater affinity. Apart from two studies that describe $InsP_3$ and ryanodine receptors in *Paramecium* (Ladenburger et al. 2006, 2009), nothing is known about these channels in lower eukaryotes. Proteomic data from enriched acidocalcisomal fractions of both *T. cruzi* and *T. brucei* included spectra from the putative $TcInsP_3Rs$, suggesting that these proteins are expressed in subcellular fractions (Ulrich et al. unpublished results). These results, however, have not yet been validated by direct observation, and multiple epitope-tagging attempts of *T. brucei* $InsP_3R$ have failed likely due to dominant negative effects.

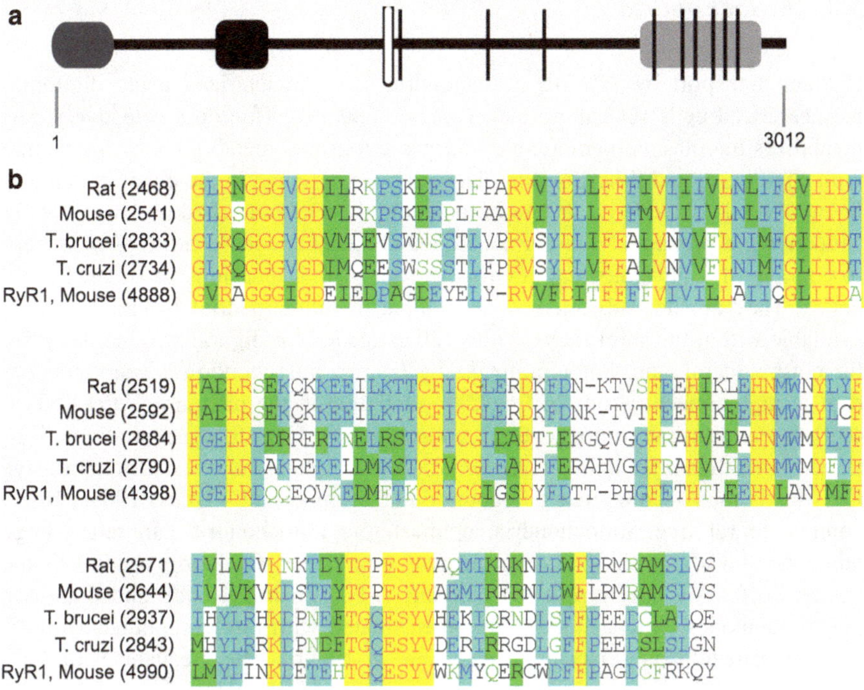

Fig. 2 *Structure of T. cruzi InsP$_3$R showing regions of interest*. (**a**) The *filled dark gray* and *black rectangles* represent a putative potential InsP$_3$ binding site, and an RIH domain, respectively. The *gray rectangle highlights* the region of greatest conservation among known InsP$_3$ and ryanodine receptors. *Vertical black lines* mark predicted transmembrane domains. The *white segment* represents a putative ATP/GTP binding motif. (**b**) Alignment of InsP$_3$ receptors and mouse ryanodine receptor 1 in the conserved region represented by the *gray rectangle* in (**a**). Identical residues are in *yellow*, conserved residues in *cyan*, similar residues in *green*, and different residues in *white*

5.2 Nucleus

Ca^{2+} transport across the nuclear membrane in mammalian cells has been the subject of controversy. It has been reported that the movement of Ca^{2+} into the nucleus may be restricted and require a SERCA-type pump despite other observations that nuclear pores permit movement of large proteins through the nuclear membrane. The nuclear membrane of *T. cruzi* is continuous with the endoplasmic reticulum, and antibodies against markers for the ER (calreticulin, BiP, or TcSCA) also label the nuclear membrane (Furuya et al. 2001). Studies in *T. brucei* using the Ca^{2+}-sensitive protein aequorin (Xiong and Ruben 1998) showed that changes in cytosolic Ca^{2+} levels are closely reflected in the nucleus, ruling out active nuclear accumulation of Ca^{2+}.

5.3 Mitochondria

Calcium transport by *T. cruzi* mitochondria was characterized using digitonin-permeabilized cells (Docampo and Vercesi 1989a, b). Digitonin selectively permeabilizes the plasma membrane to inorganic ions and metabolites by interaction with cholesterol and β-hydroxysterols, which are enriched in the eukaryotic membrane several-fold relative to the mitochondrial membrane (Fiskum et al. 1980). Mitochondria prepared from permeabilized cells experience conditions more representative of a physiological environment than do suspensions of isolated organelles. They are not subjected to the trauma of mitochondrial isolation and are available within the short interval (30–120 s) needed for digitonin to permeabilize the plasma membrane. Calcium uptake by *T. cruzi* mitochondria is energy dependent at high concentrations of free Ca^{2+} (>1 μM) in the medium (Vercesi et al. 1991a). Epimastigote mitochondria can accumulate Ca^{2+} to concentrations 5–10 times higher than mammalian mitochondria and are much more resistant to massive Ca^{2+} loads than mammalian mitochondria (Docampo and Vercesi 1989a, b). In contrast to rat liver mitochondria, epimastigote mitochondria can retain large amounts of Ca^{2+} even in the absence of membrane-stabilizing agents, in the presence of thiols and NAD(P)H oxidants (*t*-butylhydroperoxide and diamide), naphthoquinones (β-lapachone), and when treated with nitrocompounds (nifurtimox or benznidazole) (Docampo and Vercesi 1989b). The mechanism of Ca^{2+} uptake occurs through a uniport system, as evidenced by depolarization of the inner membrane during accumulation of Ca^{2+} (Docampo and Vercesi 1989b; Vercesi et al. 1991b). The results also indicated that mitochondria of *T. cruzi* possess separate pathways for Ca^{2+} influx and efflux as judged by their responses under steady state to additions of Ca^{2+} and EGTA (Docampo and Vercesi 1989b).

5.4 Acidocalcisomes

These acidic calcium-storage organelles are nearly ubiquitous among organisms ranging from bacteria to man (Docampo et al. 2005). The main characteristics of these organelles are their acidity, electron-density, and accumulation of phosphate, pyrophosphate, polyphosphate (poly P), calcium, and magnesium (Docampo et al. 2005) (Fig. 3). Acidocalcisomes are similar to the volutin or metachromatic granules described more than a hundred years ago in trypanosomatids (Swellengrebel 1908). The presence of calcium in these organelles in *T. cruzi* was first detected using X-ray microanalysis (Dvorak et al. 1988).

Acidocalcisomes are the largest calcium reservoir in *T. cruzi*. The number of acidocalcisomes varies in the different stages. Amastigotes contain more acidocalcisomes (~40) than epimastigotes or trypomastigotes (Miranda et al. 2000). However, the volume of the cell occupied by acidocalcisomes is ≤2%. Given their small volume, acidocalcisomes could potentially accumulate calcium to molar levels.

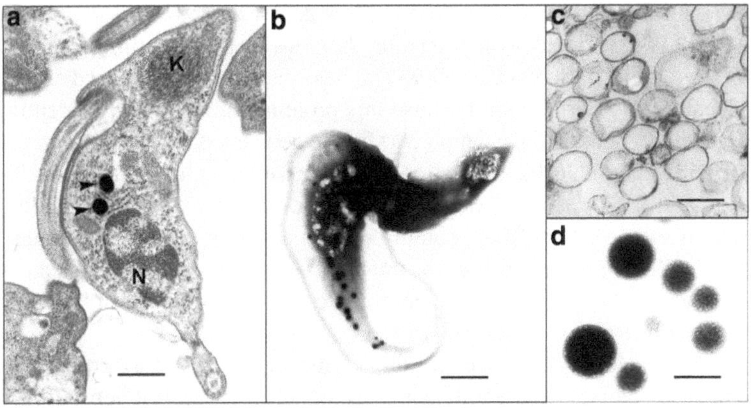

Fig. 3 *Morphology of acidocalcisomes of T. cruzi*. Visualization of acidocalcisomes by different methods. (**a**) trypomastigote as observed by conventional transmission electron microscopy (TEM). (**b**) trypomastigote allowed to adhere to Formvar- and carbon-coated grids and then observed by direct TEM. (**c**) acidocalcisome fraction obtained as described by Scott and Docampo (2000), as observed by conventional TEM. (**d**) acidocalcisome fraction as observed by direct TEM, note the sponge-like structure obtained after submission of the sample to the electron beam. *Scale bars*, (**a, b**) 1 μm; (**c, d**) 0.2 μm. (**a**) and (**b**) are reproduced with permission from Lu et al. (1998) (Copyright © American Society for Microbiology, Lu et al. 1998)

The acidity of acidocalcisomes is easily observed by fluorescence microscopy after incubation of cells with the weak base acridine orange (AO) (Docampo et al. 1995). Acidocalcisomes of *T. cruzi* appear in electron micrographs of thin sections as empty vesicles occasionally bearing electron dense material fixed to the inner face of the membrane (Scott et al. 1997; Miranda et al. 2000). X-ray microanalysis of these organelles revealed considerable amounts of oxygen, sodium, magnesium, phosphorus, potassium, calcium, and zinc (Scott et al. 1997; Miranda et al. 2000). Iron was also detected in acidocalcisomes of bloodstream trypomastigotes (Correa et al. 2002).

T. cruzi acidocalcisomes possess an array of cation and proton transporters. A plasma membrane-type (PMCA) Ca^{2+}-ATPase (Tca1) similar to vacuolar Ca^{2+}-ATPases of other unicellular eukaryotes is involved in Ca^{2+} influx (Docampo et al. 1995) (Lu et al. 1998). Two proton pumps, a vacuolar H^+-ATPase (V-H^+-ATPase) (Docampo et al. 1995; Lu et al. 1998), and a vacuolar H^+-pyrophosphatase (V-H^+-PPase) (Scott et al. 1998) are responsible for acidocalcisome acidification. The Ca^{2+} content of acidocalcisomes is very high, but most of it appears bound to poly P and can be released only upon alkalinization or after poly P hydrolysis (Ruiz et al. 2001). The mechanism for physiological Ca^{2+} release from acidocalcisomes is unknown.

The V-H^+-ATPase activity was first identified in *T. cruzi* by its sensitivity to bafilomycin A_1, an inhibitor that is specific to this pump when used at low concentrations (Bowman et al. 1988). Bafilomycin A_1 causes the release of calcium from an intracellular compartment in intact epimastigotes loaded with the Ca^{2+} indicator fura 2 (Ruiz et al. 2001). The V-H^+-ATPase was also shown to colocalize in acidocalcisomes with the PMCA Ca^{2+}-ATPase (Lu et al. 1998).

A V-H$^+$-PPase activity was also found in acidocalcisomes of *T. cruzi* (Scott et al. 1998). The acidocalcisomal enzyme belongs to the K$^+$-stimulated group of V-H$^+$-PPases (type I) (Scott et al. 1998) and has been successfully used as a marker for acidocalcisome purification because this protein is abundantly concentrated in these organelles (Scott and Docampo 2000). The gene encoding the *T. cruzi* enzyme (*TcPPase* or *TcVP1*) has been functionally expressed in yeast (Hill et al. 2000).

An aquaporin or water channel was also identified in *T. cruzi* acidocalcisomes (Montalvetti et al. 2004). This protein could function as a water channel when expressed in *Xenopus* oocytes but was unable to transport glycerol. This aquaporin (TcAQP1) was also localized to the contractile vacuole complex and has a role in osmoregulation (Rohloff et al. 2004).

A number of genes identified in the genome of *T. cruzi* may code for acidocalcisome transporters. Proteomic analysis of subcellular fractions of *T. cruzi* led to the identification of a putative zinc transporter with no signal peptide and five transmembrane domains (EAN89594.1, Tc00.1047053511439.50) (Ferella et al. 2008). Some of these genes include a putative phosphate transporter (Tc00.1047053508831.60), a putative chloride channel (of eight sequences annotated), and neutral and basic amino acid transporters (of 23 sequences annotated). Polyphosphate synthases (vacuolar transporter chaperones or VTC's) are present in acidocalcisomes of *T. brucei* (Fang et al. 2007) and *T. cruzi* (Ulrich et al. unpublished results). Homologs have also been identified in the genome of *T. cruzi* (TcVTC1, Tc00.1047053511249.44; TcVTC4, Tc00.1047053511127.100). TcVTC4-GFP fusion proteins localize to the acidocalcisomes of *T. cruzi* epimastigotes (Ulrich et al. unpublished). Some or all of these transporters could also be located at the parasite plasma membrane.

Acidocalcisomes of *T. cruzi* are especially rich in pyrophosphate and short chain polyphosphate species (poly P_3, poly P_4, and poly P_5). ^{31}NMR spectra of purified acidocalcisomes indicate that poly P of *T. cruzi* has an average chain length of 3.25 phosphates (Moreno et al. 2000). The concentrations (in terms of Pi monomers) of short-chain poly P (usually less than 50 phosphate units) in epimastigotes, amastigotes, and trypomastigotes are 54.3 ± 0.3, 25.5 ± 5.1, and 3.1 ± 1.4 mM, respectively. Concentrations (in terms of Pi monomers) of long-chain poly P (up to 700–800 phosphate units) are 2.89 ± 0.29, 0.13 ± 0.01, and 0.82 ± 0.005 mM in epimastigotes, amastigotes, and trypomastigotes, respectively. Assuming that the majority of poly P is stored in acidocalcisomes and taking into account the relative acidocalcisomal volume of the *T. cruzi* life cycle (epimastigotes, 0.86%; amastigotes, 2.3%; trypomastigotes, 0.26% of total cell volume) (Miranda et al. 2000) at each stage, concentrations of poly P in the organelle could be as high as 3–8 M (Docampo et al. 2005). These estimates are consistent with detection of solid-state condensed phosphates by magic-angle spinning NMR techniques (Moreno et al. 2002) and the high electron density of these organelles (Scott et al. 1998). Pyrophosphate and short chain poly P are important components of the electron-dense matrix observed in acidocalcisomes, as treatment of fixed epimastigotes with high amounts of yeast pyrophosphatase eliminates the electron dense material (Urbina et al. 1999).

Storage of inorganic phosphate as a polymer is important because it limits the osmotic effects of its accumulation. Short and long chain poly P levels rapidly decrease upon exposure of epimastigotes to agents that mobilize Ca^{2+} such as calcium ionophores (ionomycin) or alkalinizing agents (NH_4Cl, nigericin). Rapid hydrolysis or synthesis of acidocalcisomal poly P occurs when epimastigotes are exposed to hypo-osmotic or hyper-osmotic conditions, respectively, suggesting that poly P is essential for acclimation of parasites to changes in environmental conditions (Ruiz et al. 2001). Synthesis of poly P also increases during the lag phase of growth of epimastigotes and during in vitro differentiation of trypomastigotes into amastigotes (Ruiz et al. 2001).

Acidocalcisomes of *T. cruzi* have low sulfur content (Scott et al. 1997), suggestive of limited protein content within these organelles. Large amounts of arginine and lysine are contained in acidocalcisomes, but these are most likely present as free amino acids (Rohloff et al. 2003). A few enzymatic activities (exopolyphosphatase and polyphosphate kinase) have also been detected (Ruiz et al. 2001), but the molecular nature of various transporters is still unclear. Figure 4 shows a scheme of the known components of acidocalcisomes in *T. cruzi*. Acidocalcisomes

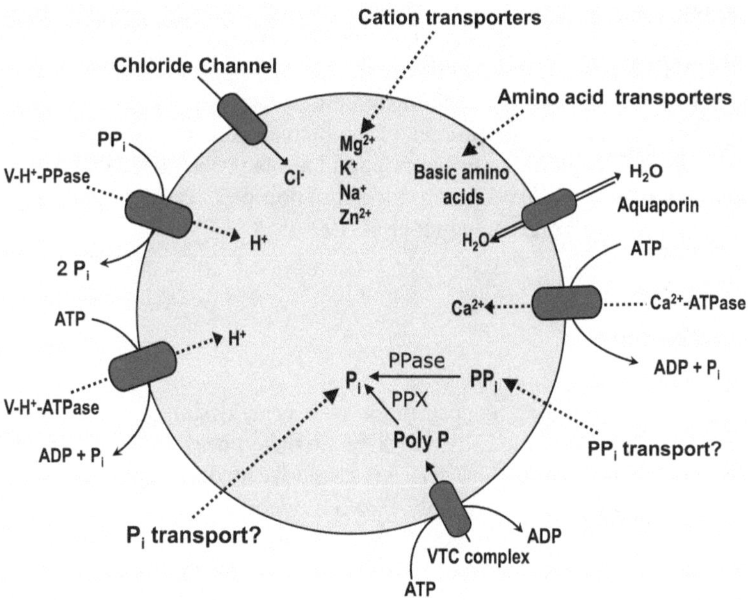

Fig. 4 *Schematic representation of a T. cruzi acidocalcisome.* Ca^{2+} uptake occurs in exchange for H^+ by a reaction catalyzed by a vacuolar Ca^{2+}-ATPase. A H^+ gradient is established by a vacuolar H^+-ATPase and a vacuolar H^+-pyrophosphatase (V-H^+-PPase). An aquaporin allows water transport. Other transporters (i.e., Mg^{2+}, Zn^{2+}, inorganic phosphate (Pi) pyrophosphate (PP_i), and basic amino acids) are probably present. The acidocalcisome is rich in pyrophosphate, short- and long-chain polyphosphate (poly P), magnesium, calcium, sodium, and zinc. An exopolyphosphatase (PPX), a pyrophosphatase (PPase), and a polyphosphate synthase (VTC complex) may also be present. *Question marks* indicate elements for which there is no biochemical evidence yet

have important roles in ion homeostasis and osmoregulation as has been reviewed elsewhere (Docampo et al. 2005; Moreno and Docampo 2009) and are potential targets for chemotherapy (Docampo and Moreno 2008).

6 Ca^{2+} Functions in *T. cruzi*

6.1 Invasion of the Host Cell and Differentiation

The cytosolic Ca^{2+} concentration of *T. cruzi* trypomastigotes increases during interaction with host cells, as demonstrated by digital fluorescence microscopy of tissue culture-derived trypomastigotes (Y strain) loaded with fura-2 (Moreno et al. 1994). When Ca^{2+} transients were prevented by loading the parasites with quin 2-AM or BAPTA-AM at concentrations sufficient to chelate intracellular Ca^{2+}, trypomastigote invasion of host cells was decreased (Moreno et al. 1994). Pretreatment of both tissue culture-derived and bloodstream trypomastigotes (Tulahuén strain) with quin 2-AM or BAPTA-AM decreased their infectivity while treatment with the Ca^{2+} ionophore ionomycin, which elevates $[Ca^{2+}]_i$ in trypomastigotes, significantly enhanced infective capacity of the parasites (Yakubu et al. 1994). These results indicate that the transient Ca^{2+} increase that occurs upon attachment of trypomastigotes to the host cell surface is possibly associated with invasion. The mechanism and sources of the increased $[Ca^{2+}]_i$ are unknown.

A role for Ca^{2+} signaling in differentiation has also been postulated on the basis of changes in $[Ca^{2+}]_i$ observed upon differentiation of *T. cruzi* epimastigotes into metacyclic trypomastigotes (Lammel et al. 1996).

7 Conclusions

Regulation of cytosolic Ca^{2+} concentration in *T. cruzi* is similar to those processes that occur in other eukaryotic cells; yet there are differences that clearly distinguish this parasite. Calcium storage in *T. cruzi* is primarily mediated by acidocalcisomes, and calcium is largely bound to poly P. No evidence is yet available on second messengers involved in Ca^{2+} release from these organelles, and further research is necessary to identify mechanisms of Ca^{2+} and phosphate homeostasis in acidocalcisomes. Although the inositol phosphate/diacylglycerol pathway is present, little is known about the receptors for the second messengers it generates. Among the differences of *T. cruzi* calcium metabolism, *T. cruzi* Ca^{2+}-ATPases vary widely from their mammalian counterparts. The PMCA-type Ca^{2+}-ATPase, an acidocalcisomal protein, does not possess a typical calmodulin-binding domain, and the SERCA-type Ca^{2+}-ATPase is thapsigargin-insensitive. With the information provided by genome sequencing and subcellular proteomics, we hope to discover other functions and exploit them to design effective therapeutic agents for *T. cruzi*.

Acknowledgments This work was supported in part by a postdoctoral fellowship from the American Heart Association (to PU), grant AI-068647 from the National Institutes of Allergy and Infectious Diseases, U.S. National Institutes of Health (NIH) (to RD), and by a NIH Research Supplement to grant AI-068467, to Promote Diversity in Health-Related Research (to RC).

References

Araya JE, Cornejo A, Orrego PR, Cordero EM, Cortez M, Olivares H, Neira I, Sagua H, da Silveira JF, Yoshida N, Gonzalez J (2008) Calcineurin B of the human protozoan parasite *Trypanosoma cruzi* is involved in cell invasion. Microbes Infect 10:892–900

Benaim G, Losada S, Gadelha FR, Docampo R (1991) A calmodulin-activated (Ca^{2+}-Mg^{2+})-ATPase is involved in Ca^{2+} transport by plasma membrane vesicles from *Trypanosoma cruzi*. Biochem J 280:715–720

Bowman EJ, Siebers A, Altendorf K (1988) Bafilomycins: a class of inhibitors of membrane ATPases from microorganisms, animal cells, and plant cells. Proc Natl Acad Sci USA 85: 7972–7976

Cahalan MD (2009) STIMulating store-operated Ca^{2+} entry. Nat Cell Biol 11:669–677

Carafoli E (1987) Intracellular calcium homeostasis. Annu Rev Biochem 56:395–433

Carafoli E, Brini M (2000) Calcium pumps: structural basis for and mechanism of calcium transmembrane transport. Curr Opin Chem Biol 4:152–161

Chung SH, Swindle J (1990) Linkage of the calmodulin and ubiquitin loci in *Trypanosoma cruzi*. Nucleic Acids Res 18:4561–4569

Conte I, Labriola C, Cazzulo JJ, Docampo R, Parodi AJ (2003) The interplay between folding-facilitating mechanisms in *Trypanosoma cruzi* endoplasmic reticulum. Mol Biol Cell 14:3529–3540

Correa AF, Andrade LR, Soares MJ (2002) Elemental composition of acidocalcisomes of *Trypanosoma cruzi* bloodstream trypomastigote forms. Parasitol Res 88:875–880

D'Angelo MA, Montagna AE, Sanguineti S, Torres HN, Flawia MM (2002) A novel calcium-stimulated adenylyl cyclase from *Trypanosoma cruzi*, which interacts with the structural flagellar protein paraflagellar rod. J Biol Chem 277:35025–35034

Docampo R, Moreno SN (2008) The acidocalcisome as a target for chemotherapeutic agents in protozoan parasites. Curr Pharm Des 14:882–888

Docampo R, Pignataro OP (1991) The inositol phosphate/diacylglycerol signalling pathway in *Trypanosoma cruzi*. Biochem J 275:407–411

Docampo R, Vercesi AE (1989a) Ca^{2+} transport by coupled *Trypanosoma cruzi* mitochondria in situ. J Biol Chem 264:108–111

Docampo R, Vercesi AE (1989b) Characteristics of Ca^{2+} transport by *Trypanosoma cruzi* mitochondria in situ. Arch Biochem Biophys 272:122–129

Docampo R, Moreno SN, Vercesi AE (1993) Effect of thapsigargin on calcium homeostasis in *Trypanosoma cruzi* trypomastigotes and epimastigotes. Mol Biochem Parasitol 59:305–313

Docampo R, Scott DA, Vercesi AE, Moreno SN (1995) Intracellular Ca^{2+} storage in acidocalcisomes of *Trypanosoma cruzi*. Biochem J 310:1005–1012

Docampo R, de Souza W, Miranda K, Rohloff P, Moreno SN (2005) Acidocalcisomes – conserved from bacteria to man. Nat Rev Microbiol 3:251–261

Dvorak JA, Engel JC, Leapman RD, Swyt CR, Pella PA (1988) *Trypanosoma cruzi:* elemental composition heterogeneity of cloned stocks. Mol Biochem Parasitol 31:19–26

Engman DM, Krause KH, Blumin JH, Kim KS, Kirchhoff LV, Donelson JE (1989) A novel flagellar Ca^{2+}-binding protein in trypanosomes. J Biol Chem 264:18627–18631

Ersfeld K, Barraclough H, Gull K (2005) Evolutionary relationships and protein domain architecture in an expanded calpain superfamily in kinetoplastid parasites. J Mol Evol 61:742–757

Fang J, Rohloff P, Miranda K, Docampo R (2007) Ablation of a small transmembrane protein of *Trypanosoma brucei* (TbVTC1) involved in the synthesis of polyphosphate alters acidocalcisome biogenesis and function, and leads to a cytokinesis defect. Biochem J 407:161–170

Ferella M, Nilsson D, Darban H, Rodrigues C, Bontempi EJ, Docampo R, Andersson B (2008) Proteomics in *Trypanosoma cruzi*–localization of novel proteins to various organelles. Proteomics 8:2735–2749

Fiskum G, Craig SW, Decker GL, Lehninger AL (1980) The cytoskeleton of digitonin-treated rat hepatocytes. Proc Natl Acad Sci USA 77:3430–3434

Furuya T, Kashuba C, Docampo R, Moreno SN (2000) A novel phosphatidylinositol-phospholipase C of *Trypanosoma cruzi* that is lipid modified and activated during trypomastigote to amastigote differentiation. J Biol Chem 275:6428–6438

Furuya T, Okura M, Ruiz FA, Scott DA, Docampo R (2001) TcSCA complements yeast mutants defective in Ca^{2+} pumps and encodes a Ca^{2+}-ATPase that localizes to the endoplasmic reticulum of *Trypanosoma cruzi*. J Biol Chem 276:32437–32445

Godsel LM, Engman DM (1999) Flagellar protein localization mediated by a calcium-myristoyl/palmitoyl switch mechanism. EMBO J 18:2057–2065

Gomez ML, Erijman L, Arauzo S, Torres HN, Tellez-Iñón MT (1989) Protein kinase C in *Trypanosoma cruzi* epimastigote forms: partial purification and characterization. Mol Biochem Parasitol 36:101–108

Gomez ML, Ochatt CM, Kazanietz MG, Torres HN, Tellez-Iñón MT (1999) Biochemical and immunological studies of protein kinase C from *Trypanosoma cruzi*. Int J Parasitol 29:981–989

Gonzalez A, Lerner TJ, Huecas M, Sosa-Pineda B, Nogueira N, Lizardi PM (1985) Apparent generation of a segmented mRNA from two separate tandem gene families in *Trypanosoma cruzi*. Nucleic Acids Res 13:5789–5804

Green KN, LaFerla FM (2008) Linking calcium to Abeta and Alzheimer's disease. Neuron 59:190–194

Grynkiewicz G, Poenie M, Tsien RY (1985) A new generation of Ca^{2+} indicators with greatly improved fluorescence properties. J Biol Chem 260:3440–3450

Hass MR, Sato C, Kopan R, Zhao G (2009) Presenilin: RIP and beyond. Semin Cell Dev Biol 20:201–210

Hill JE, Scott DA, Luo S, Docampo R (2000) Cloning and functional expression of a gene encoding a vacuolar-type proton-translocating pyrophosphatase from *Trypanosoma cruzi*. Biochem J 351:281–288

Irvine RF (1986) Calcium transients: mobilization of intracellular Ca^{2+}. Br Med Bull 42:369–374

Labriola C, Cazzulo JJ, Parodi A (1999) *Trypanosoma cruzi* calreticulin is a lectin that binds monoglucosylated oligosaccharides but not protein moieties of glycoproteins. Mol Biol Cell 10:1381–1394

Ladenburger EM, Korn I, Kasielke N, Wassmer T, Plattner H (2006) An Ins(1, 4, 5)P_3 receptor in Paramecium is associated with the osmoregulatory system. J Cell Sci 119:3705–3717

Ladenburger EM, Sehring IM, Korn I, Plattner H (2009) Novel types of Ca^{2+} release channels participate in the secretory cycle of *Paramecium* cells. Mol Cell Biol 29:3605–3622

Lammel EM, Barbieri MA, Wilkowsky SE, Bertini F, Isola EL (1996) *Trypanosoma cruzi*: involvement of intracellular calcium in multiplication and differentiation. Exp Parasitol 83:240–249

Lu HG, Zhong L, de Souza W, Benchimol M, Moreno S, Docampo R (1998) Ca^{2+} content and expression of an acidocalcisomal calcium pump are elevated in intracellular forms of *Trypanosoma cruzi*. Mol Cell Biol 18:2309–2323

Luo S, Rohloff P, Cox J, Uyemura SA, Docampo R (2004) *Trypanosoma brucei* plasma membrane-type Ca^{2+}-ATPase 1 (TbPMC1) and 2 (TbPMC2) genes encode functional Ca^{2+}-ATPases localized to the acidocalcisomes and plasma membrane, and essential for Ca^{2+} homeostasis and growth. J Biol Chem 279:14427–14439

Miranda K, Benchimol M, Docampo R, de Souza W (2000) The fine structure of acidocalcisomes in *Trypanosoma cruzi*. Parasitol Res 86:373–384

Montalvetti A, Rohloff P, Docampo R (2004) A functional aquaporin co-localizes with the vacuolar proton pyrophosphatase to acidocalcisomes and the contractile vacuole complex of *Trypanosoma cruzi*. J Biol Chem 279:38673–38682

Moreno SN, Docampo R (2009) The role of acidocalcisomes in parasitic protists. J Eukaryot Microbiol 56:208–213

Moreno SN, Vercesi AE, Pignataro OP, Docampo R (1992) Calcium homeostasis in *Trypanosoma cruzi* amastigotes: presence of inositol phosphates and lack of an inositol 1, 4, 5-trisphosphate-sensitive calcium pool. Mol Biochem Parasitol 52:251–261

Moreno SN, Silva J, Vercesi AE, Docampo R (1994) Cytosolic-free calcium elevation in *Trypanosoma cruzi* is required for cell invasion. J Exp Med 180:1535–1540

Moreno B, Urbina JA, Oldfield E, Bailey BN, Rodrigues CO, Docampo R (2000) ^{31}P NMR spectroscopy of *Trypanosoma brucei*, *Trypanosoma cruzi*, and *Leishmania major* Evidence for high levels of condensed inorganic phosphates. J Biol Chem 275:28356–28362

Moreno B, Rodrigues CO, Bailey BN, Urbina JA, Moreno SN, Docampo R, Oldfield E (2002) Magic-angle spinning ^{31}P NMR spectroscopy of condensed phosphates in parasitic protozoa: visualizing the invisible. FEBS Lett 523:207–212

Moreno VR, Aguero F, Tekiel V, Sanchez DO (2007) The Calcineurin A homologue from *Trypanosoma cruzi* lacks two important regulatory domains. Acta Trop 101:80–89

Nicholls DG, Snelling R, Rial E (1984) Proton and calcium circuits across the mitochondrial inner membrane. Biochem Soc Trans 12:388–390

Ogueta SB, Solari A, Tellez-Iñón MT (1994) *Trypanosoma cruzi* epimastigote forms possess a Ca^{2+}-calmodulin dependent protein kinase. FEBS Lett 337:293–297

Ogueta S, Intosh GM, Tellez-Iñón MT (1996) Regulation of Ca^{2+}/calmodulin-dependent protein kinase from *Trypanosoma cruzi*. Mol Biochem Parasitol 78:171–183

Ogueta SB, Macintosh GC, Tellez-Iñón MT (1998) Stage-specific substrate phosphorylation by a Ca^{2+}/calmodulin-dependent protein kinase in *Trypanosoma cruzi*. J Eukaryot Microbiol 45:392–396

Okura M, Fang J, Salto ML, Singer RS, Docampo R, Moreno SN (2005) A lipid-modified phosphoinositide-specific phospholipase C (TcPI-PLC) is involved in differentiation of trypomastigotes to amastigotes of *Trypanosoma cruzi*. J Biol Chem 280:16235–16243

Parsons M, Worthey EA, Ward PN, Mottram JC (2005) Comparative analysis of the kinomes of three pathogenic trypanosomatids: *Leishmania major Trypanosoma brucei* and *Trypanosoma cruzi*. BMC Genomics 6:127

Ponting CP (2000) Novel repeats in ryanodine and IP3 receptors and protein O-mannosyltransferases. Trends Biochem Sci 25:48–50

Pozos TC, Sekler I, Cyert MS (1996) The product of HUM1, a novel yeast gene, is required for vacuolar Ca^{2+}/H^{+} exchange and is related to mammalian Na^{+}/Ca^{2+} exchangers. Mol Cell Biol 16:3730–3741

Rohloff P, Rodrigues CO, Docampo R (2003) Regulatory volume decrease in *Trypanosoma cruzi* involves amino acid efflux and changes in intracellular calcium. Mol Biochem Parasitol 126:219–230

Rohloff P, Montalvetti A, Docampo R (2004) Acidocalcisomes and the contractile vacuole complex are involved in osmoregulation in *Trypanosoma cruzi*. J Biol Chem 279:52270–52281

Ruiz FA, Rodrigues CO, Docampo R (2001) Rapid changes in polyphosphate content within acidocalcisomes in response to cell growth, differentiation, and environmental stress in *Trypanosoma cruzi*. J Biol Chem 276:26114–26121

Scott DA, Docampo R (2000) Characterization of isolated acidocalcisomes of *Trypanosoma cruzi*. J Biol Chem 275:24215–24221

Scott DA, Docampo R, Dvorak JA, Shi S, Leapman RD (1997) In situ compositional analysis of acidocalcisomes in *Trypanosoma cruzi*. J Biol Chem 272:28020–28029

Scott DA, de Souza W, Benchimol M, Zhong L, Lu HG, Moreno SN, Docampo R (1998) Presence of a plant-like proton-pumping pyrophosphatase in acidocalcisomes of *Trypanosoma cruzi*. J Biol Chem 273:22151–22158

Selvapandiyan A, Kumar P, Morris JC, Salisbury JL, Wang CC, Nakhasi HL (2007) Centrin1 is required for organelle segregation and cytokinesis in *Trypanosoma brucei*. Mol Biol Cell 18:3290–3301

Shi J, Franklin JB, Yelinek JT, Ebersberger I, Warren G, He CY (2008) Centrin4 coordinates cell and nuclear division in *T. brucei*. J Cell Sci 121:3062–3070

Swellengrebel NH (1908) La volutine chez les trypanosomes. C R Soc Biol Paris 64:38–43

Tellez-Iñón MT, Ulloa RM, Torruella M, Torres HN (1985) Calmodulin and Ca^{2+}-dependent cyclic AMP phosphodiesterase activity in *Trypanosoma cruzi*. Mol Biochem Parasitol 17:143–153

Urbina JA, Moreno B, Vierkotter S, Oldfield E, Payares G, Sanoja C, Bailey BN, Yan W, Scott DA, Moreno SN, Docampo R (1999) *Trypanosoma cruzi* contains major pyrophosphate stores, and its growth in vitro and in vivo is blocked by pyrophosphate analogs. J Biol Chem 274:33609–33615

Vercesi AE, Bernardes CF, Hoffmann ME, Gadelha FR, Docampo R (1991a) Digitonin permeabilization does not affect mitochondrial function and allows the determination of the mitochondrial membrane potential of *Trypanosoma cruzi* in situ. J Biol Chem 266:14431–14434

Vercesi AE, Hoffmann ME, Bernardes CF, Docampo R (1991b) Regulation of intracellular calcium homeostasis in *Trypanosoma cruzi* Effects of calmidazolium and trifluoperazine. Cell Calcium 12:361–369

Xiong ZH, Ruben L (1998) *Trypanosoma brucei:* the dynamics of calcium movement between the cytosol, nucleus, and mitochondrion of intact cells. Exp Parasitol 88:231–239

Yakubu MA, Majumder S, Kierszenbaum F (1994) Changes in *Trypanosoma cruzi* infectivity by treatments that affect calcium ion levels. Mol Biochem Parasitol 66:119–125

Yoshikawa F, Morita M, Monkawa T, Michikawa T, Furuichi T, Mikoshiba K (1996) Mutational analysis of the ligand binding site of the inositol 1, 4, 5-trisphosphate receptor. J Biol Chem 1271:18277–18284

Index